Sturge's Statistical and Thermal Physics

Sturge's Statistical and Thermal Physics

Second Edition

Jeffrey Olafsen
Department of Physics
Baylor University

CRC Press
Taylor & Francis Group
Boca Raton London New York

CRC Press is an imprint of the
Taylor & Francis Group, an **informa** business

CRC Press
Taylor & Francis Group
6000 Broken Sound Parkway NW, Suite 300
Boca Raton, FL 33487-2742

Library of Congress Cataloging-in-Publication Data

Names: Olafsen, Jeffrey S., 1970- author. | Sturge, M. D. (Michael Dudley), 1931- Statistical and thermal physics.
Title: Sturge's statistical and thermal physics / by J.S. Olafsen (Department of Physics, Baylor University).
Description: Second edition. | Boca Raton, FL : CRC Press, Taylor & Francis Group, [2019] | Includes bibliographical references and index.
Identifiers: LCCN 2019007064| ISBN 9781482256000 (hardback ; alk. paper) | ISBN 1482256002 (hardback ; alk. paper) | ISBN 9781315156958 (e-Book) | ISBN 1315156954 (e-Book)
Subjects: LCSH: Statistical physics. | Thermodynamics.
Classification: LCC QC174.8 .O44 2019 | DDC 530.15/95--dc23
LC record available at https://lccn.loc.gov/2019007064

Visit the Taylor & Francis Web site at
http://www.taylorandfrancis.com

and the CRC Press Web site at
http://www.crcpress.com

For Susanna, John, and Linda
To the Students of Physics 43, 671, 871, 4340, and 5340

Contents

SECTION I Preamble

SECTION II   The Fundamentals of Modern Thermodynamics

SECTION III The Thermodynamics of Gases

SECTION IV   Modern Thermostatistical Applications

Preface

I came to use the first edition of this textbook in the same manner I think many people do: the search for the best textbook to use in an undergraduate course on thermal and statistical physics. It was after that use that I was approached by the publisher because of some feedback I had given about the textbook. The publisher needed someone to complete the work of a solutions manual for the textbook due to the untimely passing of the author, M. D. Sturge.

Perhaps I should not be surprised, after pursuing that opportunity, that the publisher contacted me about revising the textbook for a second edition. I will be honest that this query made me pause more than when I was asked to author the solutions manual. Certainly, the opportunity to write a second edition of the material would allow for certain typos and mistakes to be fixed, such as the very famous misprint in Chapter 7 of the first edition, which I think is well-known by anyone who has used the text.

Indeed, since teaching the subject for the first time in 2002, it had always been something in the back of my mind to write a textbook for undergraduate statistical and thermal physics. However, this was M. D. Sturge's textbook, and the idea of "leave well enough alone" was certainly forefront in my mind. There was a certain elegant strength to the first edition of the textbook that I did not want to undo. In developing the proposal for the second edition of the textbook, I settled upon retaining those strengths while bringing to the material a few improvements based upon my experiences as an instructor who had used the textbook.

While there are classic textbooks that have been used for decades in electricity and magnetism, classical mechanics, and quantum physics, there is no single agreed upon text for the subject of thermostatistical physics. Certainly Kittel and Kroemer's *Thermal Physics* was a self-described major influence on M. D. Sturge's original text, but there are several other textbooks used for instruction on the subject at the undergraduate level. I have attempted to avoid the conceit of trying to write the ultimate textbook on the subject here.

Rather, I wanted to take M. D. Sturge's original text and make it work better for a broader set of audiences. In his original textbook, M. D. Sturge gave the following description:

This book began as a set of notes for a course for physics majors entitled Statistical Physics which I have taught for a number of years at Dartmouth. The course is designed to provide an introduction to the principles and applications of statistical mechanics and thermodynamics in one term (twenty eight 65 minute lectures). The "Dartmouth plan," under which every student is off campus for at least one quarter in the sophomore or junior year, limits the prerequisites to courses that can be taken in the freshman year. The only prerequisites to the course on which this book is based are: an introductory course in mechanics, in electricity and magnetism, in modern physics, and in multi-variable calculus. In fact, the only essential prerequisite to most of the material in this book is an understanding of what is meant by a "quantum state," of partial derivatives, and of series expansion in a small parameter. Appendices give a brief review of what the student needs to know on these topics.

—**M. D. Sturge**
Preface of First Edition

The situation he described is not universal for the instruction of the subject of statistical and thermal physics. In many programs, the course appears late in the junior year or early in the senior year of the curriculum. However, even in his own description, the material of the book is seen threading together the introductory material on the same topics that a junior or senior level course seeks to unify.

There is a reason thermostatistics draws together subject material from electricity, magnetism, classical physics, and quantum mechanics: in any physical situation, the concepts of entropy and energy hold court. We often obscure this idea in our introductory physics courses where the concept of entropy is confused with that of heat. The reason for this is tied to the essential prerequisite Sturge gives above: the understanding of what is meant by a "quantum state." In most introductory textbooks, thermodynamics is erroneously and historically sandwiched between the material of classical mechanics and that of electricity and magnetism.

Indeed, the reason why modern courses link the title of statistical physics with the more classical title of thermodynamics is not because they are two subjects that mesh well together, but rather they speak to the same universal physics at two different scales: the micro and the macro, which is how M. D. Sturge's notes refer to the original third chapter of his textbook. This idea is one of the motivations to how some of the material in the textbook has been rearranged.

Furthermore, as will be discussed in Chapter 1, there may be an additional reason why there is no one, universally agreed-upon textbook in the subject. In each of the other subjects in physics there is a mathematically based set of equations that can be (overly?) relied upon to sum up the fundamental concepts in the subject. In classical mechanics, this can be the application of Newton's Laws and/or the Hamiltonian; in quantum mechanics, the Schrödinger equation; in electricity and magnetism, Maxwell's equations. In thermodynamics and statistical mechanics, there are instead a set of ideas and principles that must apply over a variety of situations that do not reduce to a simple equation or system of equations.

For these reasons, I have attempted to maintain the spirit of the first edition where possible, particularly in M. D. Sturge's words:

> The mathematics has been kept as simple as possible, and I have made no attempt to achieve a mathematical rigor which, as Max Born has said, is anyway illusory in physics. Nor have I delved into the philosophical implications or logical underpinnings of the basic assumptions of statistical mechanics. As in other branches of physics, for example quantum theory, a student should first achieve facility in handling the theory, and use it to derive concrete results, before inquiring too deeply into its foundations.
>
> Students are often confused by the terminology of thermodynamics and statistical mechanics; all the more since in the literature the same term is used with different meanings, and different terms are used for the same quantity, sometimes even in the same book. I have tried to use a consistent terminology and notation, and to draw careful distinctions where necessary. For example, I distinguish between a "perfect" (often called "noninteracting") gas and an "ideal" gas, which is a perfect gas in the dilute (that is, classical) limit. In order to guide the student through the terminological and notational maze, the book ends with a comprehensive glossary and list of symbols. I have used SI units consistently throughout, even when such consistency is not customary; for example, in this book N_A stands for "Avogadro's kmolar number", that is, the number of molecules in one *kilo*mole (kmole).

—M. D. Sturge
Preface of First Edition

Building upon M. D. Sturge's own notes and description of his original text, I have made a slight rearrangement of the text which I believe will actually improve upon his stated goals (which I have included as an Afterword in this edition of the textbook, so that instructors, if they wish, might use the material in the order originally conceived by M. D. Sturge.)

The material that comprised Chapters 1, 2, and 15 of the original text are now grouped together as the first three chapters in a Preamble in the second edition (Part I). For those who prefer a more classical beginning of the material by starting with macroscopic thermodynamics, these first three chapters of the second edition can be used as a starting point for the textbook.

For those who wish to assume material previously covered in a freshman or sophomore course of introductory physics that includes macroscopic thermal physics, the better starting point of the text is the beginning of Part II: The Fundamentals of Modern Thermodynamics, which contains Chapters 4 through 8 and begins from the microscopic foundations of statistical physics with a more contemporary point of view.

Chapters 9 through 13 cover Perfect (non-interacting), Ideal, Photonic (Blackbody Radiation), Bose, and Fermi gases and are grouped together as Part III: The Thermodynamics of Gases. The grouping of the material has been altered slightly from the first edition in these chapters in order to allow more flexibility for instructors who may wish to emphasize only a subset of the material in this section.

Part IV: Modern Thermostatistical Applications contains Chapters 14 through 18. Here, I have combined material on Transport Processes, Phase Transitions, and Semiconductors that appeared in the first edition, with new material on even more contemporary topics including Biological and Chemical systems and Non-Equilibrium Thermodynamics. The chapters in Part IV are kept separable so that instructors can cover them individually according to their needs and tastes, or as part of a second semester course of advanced topics in thermostatistics.

Finally, I would like to thank not only those people who have been my instructors in thermostatistics classes that I took as a student, but even more so those who have been my students in both the undergraduate and graduate courses I have taught on the material in this textbook. Rarely are we given the opportunity to let our students know how they have helped us to become better teachers. If we do not get to the point where our students and their questions leave us saying, "I don't know" or "I'm not sure" then we have not taught our students everything we know, and that does a disservice to both them and us.

Jeffrey Olafsen

About the Author

Dr. Jeffrey Olafsen grew up in Panama City, Florida and received undergraduate degrees in both physics and mathematics at the University of Southern Mississippi. He later earned a Master's and PhD degrees from Duke University in condensed matter physics working with his research mentors, Dr. Robert Behringer and Dr. Horst Meyer. After a postdoctoral position at Georgetown University and a prior faculty position, Dr. Olafsen has been happily settled at Baylor University in the Department of Physics since 2006, teaching introductory physics, statistical mechanics at both the undergraduate and graduate level, graduate nonlinear dynamics and chaos, and supervising physics research primarily with undergraduates. In his spare time, Dr. Olafsen enjoys all the blessings of living in the great state of Texas with his wife and children, and when possible, sailing with friends on the open water.

I

Preamble

1

Introduction

The grand aim of science is to cover the greatest number of experimental facts by logical deduction from the smallest number of hypotheses or axioms.

Albert Einstein

All things are made of atoms.... In that one sentence...there is an enormous amount of information about the world.

Richard Feynman

Exact answers to approximate models are better than approximate answers to exact models.

Leo Kadanoff

That last quote may take one by surprise. One should pause to reflect upon the difference between exact answers to approximate models and approximate answers to exact models. *What is the difference?* In truth, no matter what your first exposure to physics, it was most definitely a demonstration of an exact answer to an approximate model.

Indeed, that we can model anything simply is a wonder of the universe. Take gravity, for example. The first time gravity is demonstrated, it is often with an object of suitable mass for which air friction is negligible. Like a magician, physicists want to focus your attention on what they want you to see, which, for an instructor, is the phenomenon of interest. Starting a demo about gravity with a feather necessitates discussing air friction, which is not the point of a demonstration of gravity. So, one begins with an approximate model in which air friction is not important, the rock is dropped, an acceleration of motion is observed, and a discussion of gravity begins. A second, different object of a larger or smaller mass (but still negligible air friction) is used to demonstrate that gravity pulls on objects of different sizes (masses) with the same effect. *Then* someone asks about a feather and why it falls slowly. The model of accelerated motion is then modified to include the effect of air friction.

That modified model is yet again an approximate model. To see this more clearly let's understand a detailed (exact) model of what is going on when I drop an object close to the earth: There's a mutual gravitational attraction between the two objects, one of which is spinning on its axis once every twenty-four hours while simultaneously revolving around a star once every year. That star isn't at the center of a circular orbit of the earth, but rather one focus of an ellipse, and the star itself is in a spiral arm of a galaxy[1]

---

[1] A tip of the hat to Douglas Adams.

that is itself rotating about its axis while moving through the universe. Furthermore, the best models tell us that the mass we can see is only five percent of the total mass of the galaxy in order to explain the rate of rotation we measure. So, until we completely understand dark matter . . . the exact model isn't even an exact model, yet.

Oh, and as stated above, it's a mutual attraction, so while the object is being pulled downward by the gravitational attraction of the earth, the earth is simultaneously being pulled up by the object.

Exact models are complicated because they are exact and have to account for everything. Approximate models are beautiful because they are not exact, but they can demonstrate the phenomenon we are trying to focus upon, either as teachers or as students. We can always return to an approximate model and make it better by improving its sophistication to match our ability to measure and thus compare experiment to theory (model).

Like the author of the first edition of this textbook wrote, "thermodynamics and statistical mechanics are to me the most satisfying branches of physics. To set Mark Twain[2] on his head, nowhere else in science can one obtain such wholesale returns of fact out of so trifling an investment of conjecture. However, while both deal with the same subject matter and share many of the same concepts, they differ diametrically in their approach."

That difference in approach has everything to do with perspective, and history. The laws of thermodynamics were developed first, which is why they are often included near the end of the first semester of an introductory sequence in college physics. They are the result of classical physics and a perspective that begins from a macroscopic view of the universe, pushing down to smaller and smaller scales until arriving at situations like the Gibbs paradox that can only be fully understood with the advent of quantum mechanics.

1.1 Two Perspectives

Thermodynamics concerns itself with such questions as, what are the minimum experimental data needed to predict the boiling point of a particular liquid as a function of pressure? What is the maximum possible efficiency of a given design of engine? Will a given gas cool down or heat up when it is expanded through a nozzle? Thermodynamics is beautiful in the way mathematics is beautiful, because it is *true*. Just as no one will ever prove that we've been wrong in believing 1997 to be a prime number, so the second law of thermodynamics will never be proven false. No other branch of physics can make that claim with certainty. Even quantum theory, while it has turned out to be valid at energies and distance scales never even thought of by its progenitors, may ultimately go the way of Newtonian mechanics. Furthermore, quantum theory requires extra theoretical input (models, potentials, Lagrangians) before it can yield results, and such input tends to be ephemeral, rapidly made obsolete both by experiment and theory. For example, few particle theorists expect their "standard model" to be more than a historical curiosity in 50 years' time. Of course, thermodynamics achieves its status at a price; it eschews models and makes statements of the form, "if A is the case, B must follow," which look like propositions of logic rather than of physics, without trying to explain why A is, in fact, true.

Statistical mechanics takes a point of view opposite to that of thermodynamics; it depends on models, and uses them to explain the actual properties of matter, although it does generate some general results which are independent of any model and which provide an atomistic justification for the laws of thermodynamics. A diagrammatic summary of the relationship between thermodynamics and statistical mechanics is given in Table 1.1.

---

[2] "There is something fascinating about science. One gets such wholesale returns of conjecture out of such a trifling investment of fact." M. Twain, *Life on the Mississippi* (Chatto and Windus, 1874).

TABLE 1.1 Two Complementary Ways of Dealing with Systems with Many Degrees of Freedom

Thermodynamics	Statistical Mechanics
Four basic "laws" (axioms based on experiment).	These laws are deduced from the general laws of physics (with some plausible assumptions).
Zero'th law: If A is in equilibrium with B and also with C, then B is in equilibrium with C. This allows us to define a temperature; whatever it is, it's the same for A, B, and C.	Zero'th law follows from the fact that in equilibrium, probability is maximized. Temperature has a direct physical meaning.
First law: Heat and work are equivalent.	First law: Energy is conserved.
Second law: The entropy of a closed system can only increase. (The definition of entropy seems rather artificial in thermodynamics.)	Second law: A closed system tends to the most probable distribution. (Entropy has a simple relation to probability.)
Third "law": All systems in equilibrium at absolute zero have the same entropy. This doesn't have the status of the other laws; there are exceptions.	Everything is in its ground state at absolute zero entropy is zero. Not always true; e.g., a glass has a highly degenerate ground state.
Establishes general relations between physical parameters which must always hold.	Attempts to calculate these parameters from the properties of the microscopic constituents (e.g., atoms).
Deals only with average values.	Allows for fluctuations from the average (important in small systems, or in large systems probed on a small scale).
Within its range of validity, exact.	Almost always approximate.
With sufficient input of data, can cope with any physical system, however complex. Favored by chemists and engineers, who have to deal with complex systems whether they like it or not.	Needs little input of data, but can be accurately applied only to a small class of particularly simple systems, although it can give qualitative understanding of complex ones. Favored by physicists, most of whom are allowed to choose problems they can solve.

Most college-level introductory physics course sequences only introduce classical, macroscopic thermodynamics in the first semester and spend the second semester almost entirely on electricity and magnetism. Students are fortunate if that second semester reaches quantum mechanics, which is often left to a third semester of introductory physics or a course in modern physics. The issue is that one may enter an upper level course on statistical and thermal physics with only an appreciation of the classical macroscopic view never having attached quantum mechanics to anything thermodynamic.

Statistical mechanics makes accurate predictions about systems which may contain 10^{23} or more particles. When one considers how challenging the three body problem is in Newtonian mechanics, and the difficulty that quantum theorists have in accurately calculating the properties of even so simple a system as the hydrogen molecule, this is quite astounding. A sense of disbelief may be partly responsible for the widely held, but, incorrect view that statistical mechanics has to be *difficult*. What has potentially made the subject seem challenging in the past is the belief that because historically (and usually pedagogically) classical mechanics precedes quantum mechanics, statistical mechanics should be taught first from a classical point of view, and quantum ideas introduced later. This approach suffers from two issues. One issue is that, from the beginning, classical statistical mechanics requires concepts such as phase space, which are indeed quite abstract and difficult to grasp in the generality needed. One can go a long way without such concepts if one starts from quantized states. The other, more fundamental, issue is that classical statistical mechanics is internally inconsistent, and can only be made to work by the importation from quantum theory of non-classical ideas such as indistinguishability. This leads to such absurd situations as occur, for example, in one of the most widely used texts on the subject, where a rigorous derivation from classical theory of the entropy of an ideal gas is immediately followed by the (true) statement that "this expression... is *not* correct" (emphasis in the original). Gibbs' paradox, over which the classical books agonize at length, vanishes into thin air in the quantum approach.

As we will see later in this text, a background in thermodynamics exclusively from the classical macroscopic point of view leaves a student with the erroneous belief that entropy is synonymous with heat. A true understanding of entropy only comes from the microscopic treatment of the quantum approach made by statistical mechanics. One could therefore ask why the classical macroscopic approach to thermodynamics continues to dominate instruction at the introductory level. At some point in our future, when a Grand Unification Theory is successful, physics will likely undergo a wholesale change in pedagogy.

Certainly there is the historical aspect of classical physics, but many pedagogical approaches to introductory physics rely on appealing to the common everyday experiences of life. Those experiences are macroscopic, but they are powerful. Since everyone has an experience similar to putting, say a hot hamburger on top of a cold drink, and understanding that the two come to a final temperature in between those two extremes, the second law has an automatic appeal to truth. One does not have the counter-experience of observing the hamburger getting hotter and the drink getting colder, even though that would not be violated by the first law (energy conservation).

What we may fail to impress upon our students as they learn quantum mechanics is that everything that happens at the quantum mechanical level eventually manifests itself in the classical limit at the macroscopic level. This means that all those classical thermodynamic quantities of temperature, heat, and most importantly *entropy* have a correspondence to the physics at the quantum mechanical level. Indeed, statistical mechanics and thermodynamics, as a subject in physics, is where students should come to appreciate the Correspondence Principle in physics most keenly.

When we fail to impress this upon our students, they end up treating statistical mechanics and thermodynamics as two subjects forced together into a single class rather than to corresponding viewpoints of the same physics. Statistical mechanics begins at the quantum mechanical level and drills up to the classical limit to see how it manifests itself in the world we see around us, while classical thermodynamics begins at the macroscopic level drilling down for a better understanding of exactly what entropy is.

Consider the simple process of melting a solid. From the Correspondence Principle we understand that the macroscopic view of watching a material go from a solid phase to a liquid phase is the same physics that results from beginning with inter-atomic or intermolecular bonds at the quantum mechanical level and seeing how it manifests itself in the classical limit. Indeed, a great deal of work must be done in solid state physics to describe the different types of bonds in solids (ionic, covalent, dipole, etc.) to understand the relative strength of each, while thermodynamics in the classical limit offers us a simplistic point of view to measure the relative strengths of the bonds: the higher the melting point, the stronger the bond.

In fact, there is no reason to regard statistical mechanics as any more difficult than any other branch of physics taught at the undergraduate level. In some ways it is easier than electromagnetism or quantum theory, since it can be taught without any mathematics beyond partial differentiation, and the concepts are more concrete than, say, the concepts of vector potential or wave function. On the other hand, it does require a new way of thinking, since it depends on probabilistic arguments from the beginning.

The student is strongly advised to read the appendices carefully, and to work through all the derivations there and in the text. Try to visualize the meaning of the mathematical results obtained in terms of simple models. When terms are said to be small and are neglected, substitute reasonable numbers and convince yourself that this is a valid approximation. In equations involving differentials, think of these as very small increments. Consider, for example, Equation (2.4) for the energy supplied to a thermally isolated gas in a reversible change of volume, $dU = -p\,dV$. What this means is that when the volume V changes by dV, which is so small an amount that the pressure p does not change appreciably, the energy supplied (which in this case is equal to the work done on the gas) is dU. Make sure that you understand the sign; for example, the negative sign in Equation (2.4) is there because if V increases, that is, if $dV > 0$, the gas *does* work on the outside world, so that the energy supplied to the gas is negative (*the gas loses energy*).

1.2 Problems

1.1 Object A is not in thermal equilibrium with object B, which itself is not in thermal equilibrium with object C. What, if anything, can you say (thermally) about objects A and C?

1.2 Thermodynamically, why are the bottoms of expensive cooking pans made of copper, but the handles are made of stainless steel?

1.3 From the Correspondence Principle, thermodynamics must work on both the macroscopic and microscopic length scales. Argue for the length that serves as a transition between these two scales.

1.4 What is the physical significance that your body temperature has a value of 98.6 degrees Fahrenheit?

1.5 Give three characteristics that make for a good thermometer.

1.6 Why do you feel uncomfortable in a room that is at the same temperature as your body temperature, even if you sit still?

1.7 Estimate the change in energy for removing one atom of hydrogen from the sun.

1.8 Make a simple argument for how you know all of the air in a closed room is *not* moving at the same speed.

1.9 Given your argument in problem 8 above, provide a simple argument that the average velocity $< v >$ of the air in a closed room is zero, but the mean square velocity $< v^2 >$ is non-zero.

2

Temperature, Work, and Heat

We define thermodynamics…as the investigation of the dynamical and thermal properties of bodies, deduced entirely from the first and second laws of thermodynamics, without speculation as to the molecular constitution.

J. Clerk Maxwell

Classical thermodynamics…is the only physical theory of universal content which I am convinced…will never be overthrown.

Albert Einstein

2.1 Thermal Equilibrium and the Zero'th Law of Thermodynamics

Thermal physics (Thermodynamics) can be defined as the study of all physical processes involving temperature or heat. From Latin, we know thermodynamics concerns itself with the flow of heat resulting in a change in temperature. However, the precise meaning of these two words needs to be examined carefully. We start with temperature.

We all have an intuitive sense of temperature ("Phew, it's hot in here"), but this sense is of very little use in physics because it is so subjective ("You think so? I find it rather chilly.")[1] To get a more objective view we use a thermometer, but what does it actually measure? What do we mean when we interpret the length of a column of mercury as a measure of temperature? Before we can answer these questions, we need to introduce the concepts of thermal equilibrium and of insulating (or isolating) and diathermal walls.

If we dip a mercury thermometer into coffee straight from the coffee machine, we see that the length of the mercury column changes initially, but soon settles down at a new steady value, the "reading." The thermometer is now said to be in *thermal equilibrium* (which we will often shorten to "equilibrium") with the coffee. Two things may then happen, according to the nature of the container. If the coffee is in a vacuum flask, the length of the column remains (almost) constant for hours, and we can imagine an ideal[2] container in which it would remain constant forever. Such an idealized container is said

---

[1] If you are in any doubt about the subjectivity of one's sense of temperature, try the following experiment. Take three bowls of water, one as hot as you can stand, one ice-cold, and one tepid. Put your right-hand in the hot water and your left in the cold for about a minute. Then put both hands in the tepid water. Your right-hand will tell you that this water is cool, your left that it is warm.

[2] Ideal should immediately raise the concept from Chapter 1 that we are making an approximate model. Indeed the thermometer in this example should also thought to be ideal, in that we do not consider the small amount of heat that makes the column of mercury rise in the thermometer as cooling the coffee.

to have *insulating*[3] walls and its contents are said to be *thermally isolated*. On the other hand, if the coffee is in a metal cup, the column shrinks rapidly to the length it had before it was put into the coffee. One can imagine an ideal container in which this process occurs instantaneously. This idealized container is said to have *diathermal* walls and its contents are said to be in *thermal contact* with the surroundings. With these definitions, we can refine our definition of thermal equilibrium as follows. If a system is thermally isolated from its surroundings it will, if left alone, eventually reach a state in which nothing observable changes. This state is called thermal equilibrium. If two isolated systems are brought into thermal contact with each other (e.g., if hot coffee in a metal cup is placed in a pan of cold water), the properties of both systems will change, but eventually, the change stops. The two systems are then in thermal equilibrium with each other.

Thermal equilibrium also means we can *now* move the thermometer from the coffee to the water and back again, observing no change in the reading of the thermometer. Notice as well that once the two objects (coffee and water) have come into thermal equilibrium, we have lost knowledge of what their initial temperatures were when they were not in thermal equilibrium. The coffee and water, being in thermal equilibrium, can now be thought of as being in the same thermodynamic state that is defined by being at the same temperature. The idea of a thermodynamic state is a very powerful concept that will be extremely useful to us as we take up the ideas of statistical mechanics. However, the concept of the thermodynamic (equilibrium) state comes at a cost: the two objects now in the same state because they are in thermodynamic equilibrium and we lose information about what their prior thermodynamic conditions were. This is why for much (but all) of thermodynamics, we will set aside the quantity of *time*. That is not to say that thermodynamics will not ultimately have some very profound things to say about time, but the idea of the thermodynamic state under the condition of equilibrium lacking information about its past thermodynamic condition (its prior state) will let us set aside the question of time and concern ourselves only with the *initial* and *final* states of the system.

2.2 Gedanken: Empirical Temperature

We now imagine the following experiment[4]. We take our hot coffee in its metal cup, and a vacuum flask of cold water. We dip the thermometer into the coffee and get one reading, and into the water and get another. In each case the thermometer has come into equilibrium with its surroundings: cold water or hot coffee. We now immerse the cup in the cold water and we find that both readings begin to change. After a time, however, they stop changing, indicating that the coffee and the water are in thermal equilibrium. We now find, and this is the vital point, that the two thermometer readings are the same. This result can be generalized as follows: *If two objects (in our example, the water and the coffee) are simultaneously in equilibrium with a third object (here, the thermometer with a particular reading), they are in equilibrium with each other.* Thus thermal equilibrium is a transitive relation.[5] The transitivity of thermal equilibrium, often called the zero'th law of thermodynamics, is an empirical fact, not a logical necessity; after all, two people both in love with a third don't necessarily love each other. If this seems too unscientific an example, think of geographical relationships. For example, New York is within 200 miles of Baltimore, and Boston is within 200 miles of New York, but Boston is not within 200 miles of Baltimore. The zero'th law makes it possible to define *empirical temperature* θ as the property that two objects have in common when they are in thermal equilibrium, and because of the zero'th law this property can be defined in terms only

---

[3] The word "adiabatic" is more commonly used, but in this book we confine adiabatic to a more specialized meaning, which will be discussed in Chapter 5.

[4] We will often begin each chapter with a thought experiment. The thought experiment is an idealization of an experiment we might carry out but can reason through instead. Gedanken is the German word for thought.

[5] Here "transitive" is used in its sense in logic: If a relation R holds between A and B, and R also holds between B and C, then R is transitive if and only if it holds between A and C. A familiar example of such a relation is equality of mass, whose transitivity is assumed whenever one uses a chemical balance.

of the properties of a single object.[6] For example, when the mercury column has a certain length, the temperature of the mercury has a certain value, regardless of what the thermometer is in contact with. A mathematical proof that the zero'th law implies the existence of temperature is given in Appendix C. We shall see later that the zero'th law, and hence the existence of temperature, follows from the more basic assumptions of statistical mechanics in Chapter 5.

While temperature can in principle be defined in terms of many different properties, depending on the substance under consideration, it is customary in introductory thermodynamics to concentrate on compressible fluids such as gases. The state of a given mass of a non-magnetic fluid can be defined by its pressure p and its volume V, and the empirical temperature is a function of these two variables. For other substances, different variables are appropriate, but the analysis is analogous. For a gas, it is convenient to write the equation for θ as $\theta = \theta(p, V)$, using the same symbol for the function and for its value.[7] The equation $\theta = \theta(p, V)$ simply means that the three quantities p, V, and θ are not independent of each other, and in principle we can solve the equation for any one of them. More complex materials, or the presence of external fields, can be treated by simply adding more arguments to the function θ. However, one should also observe that in that growing list of arguments (variables), any time two systems can be described as having the same value for the same variable, there is one less degree of freedom between the two objects. Hence, our discussion above that the coffee and water, clearly two different substances with different densities and volumes, share a common state because their temperatures are the same when they have come into thermal equilibrium. Note that there is nothing that makes one form of $\theta(p, V)$ (i.e., one "scale" of θ) logically preferable over any other, although it is often convenient to use the length of a mercury column as a measure of θ. In general, $\theta(p, V)$ will be a different function for different substances.

In the case of a fluid (gas or liquid), it is customary, and often useful, to express the function $\theta = \theta(p, V)$ in terms of kmolar volume; that is, the volume V_m occupied by N_A molecules. Here, N_A is the number of molecules in 1 kmole, the kmole being defined as M_w kg of the substance, where M_w is the molecular weight. M_w is the mass of the molecule expressed in atomic mass units (amu),[8] where the amu is defined as the mass of the hydrogen atom M_H. In this book we shall generally use the particle density (i.e., the number of particles per unit volume) $n \equiv \frac{N}{V} = \frac{N_A}{V_m}$ as the independent variable rather than V_m. Note that the functions $\theta(p, V_m)$ and $\theta(p, n)$ are *different* (although related) functions, and both, in general, depend on the particular substance. We can solve the equation $\theta = \theta(p, n)$ for n or for p, writing $n = n(p, \theta)$ or $p = p(n, \theta)$. Two samples of the same fluid with different values of p and n will be in thermal equilibrium with each other if $\theta(p, n)$ is the same for both. A line in the (p, n) plane defined by $\theta(p, n)$ = constant, or the corresponding line $\theta(p, V)$ = constant in the (p, V) plane, is called an *isotherm*. In relation to our discussions of a thermodynamic state, the isotherm represents the lack of the degree of freedom in the θ variable.

---

[6] The empirical basis of the zero'th law seems first to have been recognized in the 1750s by the Scottish chemist Joseph Black (1728–1799), who wrote, "We must therefore adopt, as one of the most general laws of heat, that 'all bodies communicating freely with each other, and exposed to no inequality of external action, acquire the same temperature as indicated by a thermometer'." [*Lectures on the Elements of Chemistry*, published from Black's manuscripts by John Robison, 1st US edition (Mathew Carey, 1807) p. 74; reprinted in W. F. Magie, *A Source Book in Physics* (Harvard, 1963) p. 134]. The law was formulated and recognized as an essential basis for thermodynamics in 1884 by Hermann von Helmholtz (1821–1894), the great German physicist and polymath who was one of the first to recognize the equivalence of work and heat, but is better known for his contributions to the physics of perception. See Bailyn, p. 22 (this book is a gold mine of information on the history of thermodynamics).

[7] This is, of course, an offence against mathematical rigor, but is convenient in thermodynamics (see Footnote 3 in Appendix E).

[8] The definition of molecular weight in terms of M_H is inconvenient for precise measurements, and the official unit is 1/12 of the mass of the ^{12}C isotope of carbon. The difference is less than 1% and is irrelevant to the purposes of this book. As pointed out in the preface, we use International System (SI) units in which Avogadro's kmolar number, $N_A = 6.022 \times 10^{26}$, is a factor 10^3 greater than Avogadro's number as usually defined.

Example 2.1: Thermometers and Gases

(a) We have two samples of gas. One is nitrogen with mass density 1.26 kg/m^3 and pressure 1 bar, and the other is helium with mass density 0.36 kg/m^3 and pressure 2 bar. Both are "ideal" gases[9] obeying Boyle's law, which states that $p \propto n$ at constant temperature (or, equivalently, that for a given mass of gas pV = constant). They also obey Avogadro's law, that at given temperature and pressure, equal volumes of different ideal gases contain an equal number of molecules.[10] The two laws together imply that at a given temperature, the ratio $\frac{p}{n}$ is the same for all ideal gases. Are the two gases in thermal equilibrium with each other? The molecular weight of nitrogen is 28 and of helium 4.

The kmolar volume of the nitrogen at the given mass density is $\frac{28}{1.26}$ = 22.2 m^3 while that of the helium is $\frac{4}{0.36}$ = 11.1 m^3. The product pV_m = 22.2 bar m^3 is thus the same for both. Since $V_m \equiv \frac{N_A}{n}$, the two gases have the same ratio $\frac{p}{n}$ and hence, the same temperature. It follows that they are in thermal equilibrium with each other. That is, temperature is not a degree of freedom that differs between them. Their pressures and volumes are different, so they are not in the identical thermodynamic state, but they share a common value for the thermodynamic variable of temperature, and hence would not transfer heat to one another (or to a thermometer at the same temperature used to measure their temperatures). Only when all of the thermodynamic variables have the same values would we claim the two samples are in the same (identical) thermodynamic state.

(b) Now consider a mercury thermometer in which the length ℓ of the column is a measure of the change in volume of mercury from an arbitrary reference value. It is found that $\ell = 0$ when the mercury is in equilibrium with freezing water, and that $\ell = 100$ when in equilibrium with boiling water at atmospheric pressure, where ℓ is measured in some arbitrary units engraved on the thermometer. We define the mercury scale of temperature by $\theta_m \equiv \ell$ where the unit of θ_m is called the degree Celsius. Indeed, there are temperatures colder than zero Celsius, and one should be careful not to attach undue physical meaning to such arbitrarily defined negative temperatures (or even the value of the average human body temperature of 98.6 degree Fahrenheit.

We use the thermometer to measure the temperature of 1 gram (g) of air. When the thermometer comes into equilibrium with the air, it is found experimentally that

$$pV = 0.286(\theta_m + 273). \tag{2.1}$$

where V is volume of the air at pressure p. Note that, because of the definition $\theta_m \equiv \ell$, Equation (2.1) is an empirical relation between quantities that are measurable by purely mechanical means.

The average molecular weight of air is 29, so that 1 g is $\frac{1}{29000}$ kmole, and the kmolar volume is $V_m = 2.9 \times 10^4 V$. We define the *ideal gas scale* of temperature by $\theta \equiv \frac{pV_m}{R_0}$, where the "gas constant" $R_0 \equiv 0.286 \times 2.9 \times 10^4 = 8300$ J kmole^{-1} degree^{-1}.

Find the relationship between the ideal gas scale and the mercury scale θ_m.

---

[9] The term ideal gas will be defined more precisely later. All that matters here is that an ideal gas obeys Boyle's law, which was established empirically for air in the 17th century by Sir Robert Boyle (1627–1691) who, according to Victorian textbooks, was "the father of Chemistry and the son of the Earl of Cork."

[10] The Italian chemist Amedeo Avogadro (1776–1856) put forward this hypothesis in 1811. However, it was not generally accepted until 1860, when his compatriot Stanislau Cannizzaro (1826–1910) showed that it could bring order into the current chaos of chemical notation. This was *after* Maxwell had shown that the Avogadro's law follows directly from the kinetic theory of ideal gases. A German chemistry professor (whom I have been unable to identify) is reported to have said that when he heard Cannizzaro's exposition of Avogadro's theory at the Karlsruhe Congress in 1860, "it was as though scales fell from my eyes, doubt vanished, and was replaced by a feeling of peaceful certainty." Soon after this, the Austrian chemist Josef Loschmidt (1821–1895) made a good order of magnitude estimate of the actual number of molecules in unit volume of an ideal gas, but it would be nearly 50 years before a reasonably accurate value was obtained.

From Equation (2.1), this relationship is $\theta = \theta_m + 273$. Note that this θ is not an arbitrary temperature scale:

$$pV = 0.286(\theta).$$

First, there are no negative temperatures (or pressures or volumes) defined by the above relationship. Second, a temperature of zero on the ideal gas scale is only possible when one of the other two thermodynamic variables is zero. In relation to the gases it describes, one should imagine a case where the gas is allowed to expand infinitely large such that its pressure approaches zero, wherein the temperature of the gas would approach zero as well.[11] This is a good situation to keep in mind in terms of the variables p and V, one tends to zero as the other approaches infinity, but they do so in such a way that the temperature, the product of p, and V remain finite.

(c) We now use the thermometer to measure the temperature of 1 g of ethyl alcohol. When the thermometer comes into equilibrium with the alcohol, it is found experimentally that the volume V_a of the alcohol is given, over a certain limited range of temperature and pressure, by[12]:

$$V_a = (1 + 10^{-3}\theta_m + 2 \times 10^{-6}\theta_m^2)V_0, \qquad (2.2)$$

where V_0 is the volume of the alcohol when in equilibrium with freezing water. Find the relationship between θ_m and the "alcohol" scale θ_a, where

$$\theta_a \equiv 10^3 \left(\frac{V_a}{V_0} - 1 \right).$$

Substituting for V_a in Equation (2.2), we find $\theta_a = \theta_m + 2 \times 10^{-3}\theta_m^2$. Thus we can expect significant deviations between the readings of a mercury and an alcohol thermometer unless we calibrate carefully.

Note that because these are empirical relations they are only valid over the range actually studied experimentally; for example, they cannot be expected to apply down to $\theta = 0$ for a real gas.

At present we have no reason other than convenience for preferring one scale over another. This flexibility in one respect can be seen in the same manner as our freedom to set the zero of potential energy wherever we wish. One knows that in certain situations, where one sets the zero of potential energy can have rewards of elegance in a solution. This is also true in our selection of a temperature scale. The ideal gas scale θ depends on the properties of a particular group of substances (ideal gases). However, we shall see in Chapter 10 that the ideal gas scale is identical with the absolute or Kelvin scale of temperature T, whose definition does not depend on the properties of any particular substance. Absolute temperature will be defined in Chapter 5.

Yet, readers who have been exposed to the elementary kinetic theory of ideal gases may already understand the connection at this point. They have probably learnt that in kinetic theory the ideal gas scale is privileged, θ being proportional to the mean (average) kinetic energy of a molecule in the gas. Kinetic theory [see Equation (H.6)] shows that for an ideal gas:

$$p = \frac{1}{3}nm\langle v^2 \rangle, \qquad (H.6)$$

---

[11] While one could make the same argument with p and V reversed, the example used of a gas expanding to the extreme that p and T go to zero is far more physically intuitive than the case of trying to compress any real gas to a zero volume with an infinite pressure and expecting the temperature to remain finite.

[12] The empirical constants in this equation are taken from the *American Institute of Physics Handbook*, 3rd edition, ed. D. E. Gray (McGraw-Hill, 1972), pp. 4–141, and are rounded off to one significant figure.

where n is the number of molecules per unit volume, m is the mass of a molecule, and $\langle v^2 \rangle$ is the mean square molecular velocity. In view of the empirical fact, embodied in Equation (2.1), that the product pV is a linear function of temperature as measured by a mercury thermometer,[13] it is natural and convenient to define our temperature scale by $\theta \equiv \frac{pV_m}{R_0}$, as in Example 2.1. If we substitute for p and V_m ($V_m = \frac{N_A}{n}$), we find that $\theta = \frac{m\langle v^2 \rangle}{3k_B}$, where Boltzmann's constant $k_B \equiv R_0/N_A$. Thus θ is directly proportional to the mean kinetic energy of a molecule, $\frac{1}{2}m\langle v^2 \rangle$.

While we shall see in later chapters that θ in the ideal gas scale does indeed have more fundamental significance, the implication of this fact for the meaning of the concept of temperature is misleading. *It is entirely wrong to identify temperature with an energy* such as the mean kinetic energy of a molecule. This energy happens to provide a temperature scale, but only in the special case of an ideal gas. It is essential to realize that temperature is merely what bodies in thermal equilibrium with each other have in common, and it does *not*, in general, have a simple relationship to kinetic energy or to any other quantity that might loosely be called the amount of energy contained in a body. To illustrate the distinction between temperature and energy, consider a solid such as ice initially below its melting point. We supply energy at a constant rate to the material measuring its temperature by a mercury thermometer as we do so. Initially the temperature increases, but at the melting point it ceases to rise until a substantial amount of energy (the latent heat, to be discussed in Chapter 15) has been injected. Once the solid is entirely melted, the temperature starts to increase again. Such a heating curve is illustrated in Figure 2.1.

This example illustrates the fact that except in special cases, such as the ideal gas, there is no one-to-one relationship between energy content and temperature; they are quite distinct concepts.[14]

FIGURE 2.1 A typical heating curve (temperature θ versus time t) for a substance undergoing a phase transition. Since energy is being supplied at a constant rate, the t axis is a measure of the amount of energy taken up by the substance.

---

[13] Charles' law states that for an ideal gas at constant density, p is a linear function of θ (as measured with a mercury thermometer). It was first hypothesized on the basis of very few data by G. Amontons (1663–1705) at the end of the seventeenth century and was firmly established by J.-L. Gay-Lussac (1778–1850) a century later. The eponymous attribution to J. A. C. Charles is incorrect [see D. S. L. Caldwell, *From Watt to Clausius* (Iowa State University Press, 1989), p. 129].

[14] The distinction between heat and temperature is generally credited to Joseph Black, who was the first to make quantitative calorimetric measurements and to formulate the concept of latent heat [see Footnote 6, and J. S. Dugdale, *Entropy and Its Physical Meaning* (Taylor & Francis, 1996)]. In fact, the distinction seems to have been recognized centuries earlier by the monk Walter of Evesham (*circa* 1280–1330) [*Dictionary of Scientific Biography*, ed. C. C. Gillispie (Scribner, 1970–1980)]. However, confusion of the two persists in some places to this day (for some examples, see the article by Zemansky referred to in Footnote 16).

2.3 Work and Heat: The First Law of Thermodynamics

Now let's go back to our cup of coffee. We've let it grow cold and want to warm it up again. One way of warming it up is to put it into a vacuum flask (an insulating container) and insert an electric heater. This takes electrical energy, which was generated from mechanical work in a power station and somehow transfers it to the coffee. For our purposes, we need not distinguish between mechanical and electrical energy, and we call them both *work W*. As a result of the transfer of energy, we see that θ (as measured by our mercury thermometer) increases to a new value. It is found experimentally that *so long as an object is thermally isolated from its surroundings, the same net amount of work produces the same rise in temperature regardless of how it is supplied*: by stirring, by friction, electrically, by compression (as in a bicycle pump), or whatever. It follows that there exists a property of an object, its *internal energy U*, defined (within an additive constant) by $\Delta U = W$, where ΔU is the change in U produced by the input of work W. Because the work done to get from one state to another does not depend on the route taken, so long as the object is isolated, U is a property of the object, like its temperature, and can be defined in terms of the state variables (e.g., the pressure and density of a gas). In thermodynamics we do not inquire further into what U might consist of. In statistical mechanics we identify U as the mechanical energy (i.e., the kinetic and potential energy) of the atoms of which our object is composed, excluding any contributions due to motion of the object as a whole and to external forces on it.

So long as the object is in an insulating container, its internal energy change ΔU for a given change in state is simply W, the work we have had to put in from outside to produce the change in state. On the other hand, we might have put the coffee into a metal cup (a diathermal container) and immersed it in a bowl of hot water, with exactly the same effect on the coffee. Because ΔU is the same, but $W = 0$, we have to suppose that the hot water supplied this energy. We call energy transferred from one body to another through a diathermal wall or by mixing, without any mechanical (or electrical) energy being involved, *heat Q*. That this is a valid procedure is shown by the empirical fact that if the hot water is isolated from the surroundings, with no external source of energy, it is found to lose the same amount of internal energy that the coffee gains, so that in the special case of a thermally isolated system in which no work is done, $\Sigma Q = 0$, the sum being over all the different bodies in the isolated system. *Heat is the energy which is transferred from one object to another without mechanical or electrical means.* Note that heat is defined only in the context of such a process of energy transfer; for this reason, some authors insist that the word "heat" should be used only as a verb, never as a noun.[15] This insistence leads to some awkward circumlocutions[16] and will not be adhered to in this book, but the reader must remember that heat is not a substance and that it makes no sense, for example, to talk of the "quantity of heat" in a body.

For a general process in which work is done and energy is transferred as heat, the change in internal energy is equal to the sum of both contributions:

$$\Delta U = W + Q. \qquad (2.3)$$

Equation (2.3) is the most elementary form of the first law of thermodynamics. It says that *energy is conserved if heat is taken into account*. Note that our sign convention is such that W is positive when work

---

[15] See, for example, R. H. Romer, "Heat is not a noun," *Am. J. Phys.* **69**, 107 (2001).

[16] As can indeed be seen from the article referred to in Footnote 15, which calls for a "good noun for 'energy transferred by virtue of a temperature difference," but does not suggest one. Even this lengthy phrase is not strictly accurate; for example, in a thermocouple *electrical* energy is transferred "by virtue of a temperature difference." Nor would simply deciding not to use "heat" as a noun obviate the difficulties, which arise from confused thinking rather than from terminology. For example, Mark Zemansky, in his excellent article "The use and misuse of the word 'heat' in physics teaching," *Phy. Teach.* **8**, 295–300 (1970), finds fault with the use of "heat" as a *verb*, and quotes many examples of the confusion endemic in elementary teaching about heat.

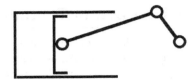

FIGURE 2.2   Model system which extracts work from the expansion of a gas.

is done *on* the object of interest, and negative when work is done *by* it; many writers use the opposite convention. In thermodynamics, Equation (2.3) is to be regarded as an empirical generalization, but in statistical mechanics we shall find that it is a direct and obvious consequence of the fundamental principle of conservation of energy.

While the laws of thermodynamics, such as Equation (2.3), are quite general, it is often convenient to express them in terms appropriate to the compressible fluid mentioned in the previous section, whose state can be specified by two variables, p and n or V, where $V = \frac{N}{n}$ is the volume occupied by a given number N of particles. In particular, we shall find it useful to have an expression for the work W. We imagine the fluid to be contained in a cylinder, one wall of which is a piston that can move without friction against an external force as illustrated schematically in Figure 2.2. This model, which is used in all but the most abstract expositions of thermodynamics, betrays the origin of the science of thermodynamics in 19th century engineering. Nevertheless, it remains the easiest way to visualize the results, which are independent of the model.

From the definition of pressure as force per unit area, the force of the gas on the piston is $f = pA$, where A is the area of the piston, and in equilibrium, this force must be balanced by an equal and opposite external force. We now suppose that the piston moves an infinitesimal distance dx to the right. If there is no friction, the work done by the gas against the external force is:

$$f \, dx = pA \, dx = p \, dV,$$

where dV is the change in volume of the gas. Because the work $f \, dx$ represents energy lost by the gas, it is negative by our convention (see Equation 2.3). Thus we have for the infinitesimal change in internal energy, if the system is thermally isolated and there is no friction:

$$dU = đW = -p \, dV. \tag{2.4}$$

We use the notation $đW$ for the infinitesimal amount of work in Equation (2.4) to indicate that, unlike p and V, work W is not a unique function of the state of the system as defined by p and V; that is, it is not a "state function." What this means is the following. We saw that one can make an object go from one state to another state with different p and V in different ways: work may be done by or on the object, energy may flow in or out as heat, or any combination of these satisfying Equation (2.3). Hence, the amount of work done on or by the object depends on the path taken; there is no quantity that we can call "work" that is property of the object and a function only of p and V. Another perspective that may be helpful is to understand that U, p, and V are quantities the system possesses at a point in phase space determined uniquely by the values p and V, so:

$$U = U(P, V).$$

One would not ask how much work the system possesses at that point in phase space, rather that work could be done by or done on the system to move it from one point in phase space to another. Similarly, one would say that the system could have some amount of heat added or withdrawn to move it to another

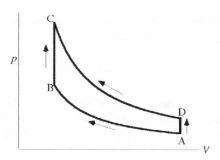

FIGURE 2.3 Example of a (p, V) diagram for a gas, showing different paths between two given states A and C: (1) ABC and (2) ADC.

point in phase space. Heat and work are not variables possessed by the state of the system, but are quantities of change [17] that are involved in moving from one state to another.

This is illustrated in Figure 2.3, in which pressure p is plotted against volume V of a fixed mass of a gas. The curves AB and CD are typical isotherms for two different temperatures, while the vertical lines BC and AD represent heating or cooling at constant volume, during which no work is done since the piston does not move. Two possible paths from state A to state C are shown, path 1 via B and path 2 via D. Although the initial and final states are the same, the work done on the gas on path 1 via B is less than that on path 2 via D, since the pressure is always lower. (This is made quantitative for the case of an ideal gas in Example 2.2 below, where it is shown that the work done in a given isothermal change of volume is proportional to the initial pressure.) The difference in work is balanced by an equal and opposite difference in the heat transferred to or from the surroundings, so that Equation (2.3) is satisfied.

In mathematical language, $đW$ is an *inexact differential*, which means that it is not the differential of a function of the defining variables (p and V in the present case) and hence cannot in general be integrated (see Appendix E). On the other hand, we found earlier that the internal energy U *is* a state function; an object in a given state has a given internal energy, so that U is a property of the object. Hence, dU in Equation (2.4) is an exact differential which can be integrated with respect to V to give the change ΔU in $U(p, V)$, so long as no heat flows in or out of the object. If we reverse the direction of motion of the piston we simply change the sign of dV, and hence of $đW$. If there is no friction, the internal energy is converted back into work; the gas is "springy."

Example 2.2: Work and Energy

The piston of a car engine has an area 50 cm^2. The pressure after the fuel has all burned is 50 bar (1 bar = 10^5 Pa). Assuming that the change in volume is so small that the pressure can be taken to be constant, how much work is done when the piston moves down by 1 mm? If the walls are effectively insulating, what is the change ΔU in the internal energy of the air in the cylinder? Neglect friction.

The work done by the gas $= -W = p\, dV$.
$A = 5 \times 10^{-3}$ m^2, $\Delta x = 10^{-3}$ m, so that $\Delta V = A\,\Delta x = 5 \times 10^{-6}$ m^3;
$p = 5 \times 10^6$ Pa, so that the work done by the gas, which is $-W$, is 25 J.
If no heat is transferred, $\Delta U = W = -25$ J.

---

[17] Hence we speak of state variables having fixed values such as isotherm or isobar, but when there is no heat flow the term is not *isoheat but adiabatic*.

If an amount of heat Q is supplied, integrating Equation (2.4) and substituting in Equation (2.3) gives for the change in internal energy if there is no friction:

$$\Delta U = Q - \int_{V_1}^{V_2} p \, dV, \tag{2.5}$$

where V_1 and V_2 are respectively the initial and final volume. If friction is neglected, the process described by the second term in Equation (2.5) can go in either direction; also Q can have either sign. Note the negative signs in Equations (2.4) and (2.5); when a gas is compressed, the integral is negative, so that the internal energy increases.

Equation (2.5) is more commonly and usefully written in differential form:

$$\boxed{dU = \dalembertian Q - p \, dV.} \tag{2.6}$$

Note that Equation (2.6) only holds if we neglect friction, since only then is $p \, dV$ the work done. Furthermore, $\dalembertian Q$, like $\dalembertian W$, is an inexact differential; heat, like work, is not a state function, but depends on the path taken. And yet we should not be dismissive of what we have been able to accomplish here. We have found that it is possible for mechanical work on a gas to replace the inexact differential of work with a term that is made up of a combination of state variables, pdV. Moreover, let us spend another moment examining the above relationship. If we rearrange the above equation like this:

$$dU + p \, dV = \dalembertian Q. \tag{2.7}$$

Note that everything on the left-hand side of the equation is now expressed in terms of only state variables. If everything on the left-hand side of the equation can be written in terms of state variables, it means it can be replaced by a state variable. That is, everything on the left-hand side can be integrated along a line of constant pressure from V_1 to V_2 and energy U_1 to energy U_2 between points 1 and 2 in phase space as shown in Figure 2.4. Or another manner in which to say this is that the isobar of value p connects states 1 and 2 in phase space. The points 1 and 2 are then distinguishable from one another by the state variables V_1, U_1 and V_2, U_2 with a common pressure p (the pressure is not a degree of freedom relative to the two states). This would imply that while the right-hand side of the equation is an inexact differential that it can, at least in this instance for mechanical work, be replaced by an *exact* differential. Just as we replaced $\dalembertian W$ with an exact differential, perhaps the above equation gives us inspiration to be able to replace $\dalembertian Q$ with an exact differential as well.

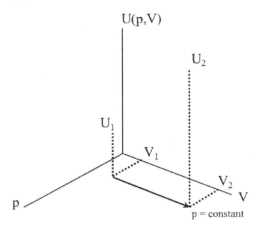

FIGURE 2.4   The energy is changed in a thermodynamic system from a value of U_1 to a value of U_2 along an isobar p by changing the volume from V_1 to V_2. In this case, the heat that flowed into the system is expressible entirely in terms of state variables.

Work and heat both have units of energy, but that alone does not make them state variables. However, the product of pressure and volume also has units of energy and are state variables. This suggests that perhaps there is another product of state variables that could, at least under certain circumstances, replace $đQ$. Indeed, if we integrate the left-hand side of the above equation one would obtain $U + pV$ as a general form of a state variable relationship. At the end of this chapter we'll see this state variable is the *enthalpy*. Here, however, we wish to begin to think of other variables that as a product could replace the inexact differential of heat with an exact differential relationship. If we imagine fixing the volume of an ideal gas in a container, so that $p\,dV = 0$ above, we obtain once more the situation we took up in the beginning of the section, namely, increasing the internal energy of the gas by adding heat and not allowing the gas to do any work. Clearly, we would expect the result of such a process to raise the temperature θ of the gas. Since the melting point and boiling point of water (at standard pressure) is a reproducible value despite what temperature scale one uses, it is easy to see that temperature is a state variable. Near the end of this chapter we will get another hint in this direction when we discuss *heat capacity*. We will see (properly, from the point of view of statistical mechanics) in Chapter 5 that there is a product of temperature and another state variable that allows us to replace the inexact differential of heat in the above relationship.

Before that, we want to take up the ideas of reversible and irreversible processes and have a closer look at that state variable of enthalpy that we've alluded to above:

Example 2.3: Work and Heat

(a) If the cylinder walls in Example 2.2 were diathermal, and the expansion were to occur so slowly that the temperature remained constant (this is not the case in a real engine, of course) how much heat would be transferred to the air in the cylinder from outside? Assume that air is an ideal gas, for which U is a function of temperature only.

Because the temperature is assumed to be constant, U is also constant so that $\Delta U = 0$, and energy equal to the work done by the gas must be transferred to it in the form of heat:

$$Q = p\,\Delta V = 25\text{ J}.$$

(b) With the same assumptions as in (a), how much work is done by the gas in the entire expansion if the initial pressure is $p_1 = 50$ bar, and the volume starts at $V_1 = 50$ cm^3 and expands to $V_2 = 500$ cm^3?

An ideal gas obeys Boyle's law: pV = constant at constant temperature. Hence $p = p_1 V_1/V$.

Work done by the gas $= -W = \int_{V_1}^{V_2} p\,dV = p_1 V_1 \int_{V_1}^{V_2} \frac{dV}{V} = p_1 V_1 \ln\left(\frac{V_2}{V_1}\right)$, where the minus sign is present because W is defined as work done *on* the gas. Substituting $p_1 = 5 \times 10^6$ Pa, $V_1 = 5 \times 10^{-5}$ m^3, $\frac{V_2}{V_1} = 10$, the work is:

$$-W = 250\ln 10 = 576\text{ J}.$$

In Chapter 3, we shall find the work done in the more realistic situation where no heat is transferred and the temperature falls during the expansion.

2.4 Reversible and Irreversible Processes

The mathematical property of $đW$, that it cannot in general be integrated, has a physical correlate. The process by which we derived Equation (2.4) was *reversible*; this means that reversing the direction of motion of the piston simply changes the sign of dU. However, if there is friction, the work we obtain in an expansion is less than $p\,dV$, and the work that we must do in a compression is more than $|p\,dV|$. Thus the process is *irreversible*; if, after compressing the gas, you reverse the direction of motion and let it expand, you don't get as much work back as you put in. Thus, Equation (2.4) applies *only to reversible processes*.

In general, one can get from one state of the system to another by either type of process, and because $đW$ is different for the two types, it cannot be a state function.

Equation (2.6) is a very important equation which we will use often, and it is essential to clearly understand that, like Equations (2.4) and (2.5), it only holds for reversible changes. We can illustrate this limitation as follows, using Figure 2.4. There is another way we could have changed the volume of the gas. Suppose that instead of the piston, we had a fixed insulated cylinder of volume V_2, and a thin diaphragm dividing it into volumes V_1 and V_0, where $V_0 = V_2 - V_1$. The gas is initially confined to the left-hand side of the diaphragm, so that its initial volume is V_1; there is a vacuum in V_0. We now remove the diaphragm. The gas rushes into the empty space, and its volume increases to V_2. Even though $\int p \, dV \neq 0$, no work is extracted from the gas, so that Equations (2.4) through (2.6) do not hold. Because no heat is transferred, and no work is done, the internal energy of the gas cannot change; the expansion is a constant U process. The essential difference between the expansion into a vacuum and expansion against a frictionless piston is that the process is *irreversible*; we cannot restore the initial state of the gas just by putting the diaphragm back. Other irreversible processes, such as friction, also invalidate these equations by reducing the work done. Thus, we see that Equations (2.4) through (2.6) only apply to reversible processes.

Example 2.4: An Isoenergetic Ideal Gas

Suppose that the cylinder walls in Figure 2.5 are diathermal, surrounded by a constant temperature bath, and that the gas is ideal. One property of an ideal gas is that U is a function only of temperature. How much heat is transferred through the walls when the process described in the previous paragraph occurs?

Since U does not change in this irreversible process, the temperature is unchanged and *no* heat is transferred in or out.

Equation (2.4) is the expression for work done on or by a compressible fluid. However, thermodynamics can in principle be applied to any system; we merely have to find the expression for work appropriate to it. For example, if a rubber band or a spring of length ℓ is stretched an amount $d\ell$ by a force f, the expression for the work done on the material is:

$$đW = f \, d\ell. \tag{2.8}$$

Note the sign difference from Equation (2.4). When a rubber band shrinks under tension, its internal energy decreases and it cools down, because it has to do work against the stretching force. Try it and see.[18]

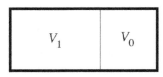

FIGURE 2.5 A gas of volume V_1 isolated from its surroundings by an insulating wall (thick line) and separated by a diaphragm (thin line) from a vacuum of volume V_0, where $V_1 + V_0 = V_2$.

---

[18] Take a flat rubber band at least 5 mm wide, stretch it quickly, and touch it to your lips. You should feel a slight warming. Hold it fully stretched for a few seconds to allow it to equilibrate with the air, then allow it to shrink quickly, but *smoothly* (not snap back) to its original size. Touch it gently to your lips; you should feel a noticeable cooling. The effect is relatively large in rubber because the change in length is large. This experiment was first described in 1806 by John Gough, who interpreted the rise in temperature on stretching as the squeezing of "caloric" out of the pores in the rubber. The correct interpretation in terms of the laws of thermodynamics (see Example 3 of Chapter 5) was given 50 years later by Lord Kelvin [see P. J. Flory, *Principles of Polymer Chemistry* (Cornell, 1953) p. 434].

2.5 Magnetic Work

Another form of work is magnetic work: the work done by or on an electromagnetic field when the magnetization of a body changes. Magnetic work is important not only because magnetic systems are interesting in themselves, but also because a simple magnetic system can provide a useful model for visualizing many of the basic concepts of statistical mechanics, as we shall see in the next chapter. We will now derive an expression for the work done when a magnetic body develops a magnetic moment M in a field H (not to be confused with the enthalpy H, defined in the next section). The derivation of this expression requires some care, because we have to distinguish the work that goes into the internal energy of the body itself from that which goes into the energy of the magnetic field, which is altered by the presence of the magnetized body. Only the former should appear in the expression analogous to Equation (2.4).

Our derivation follows Mandl's.[19] Figure 2.6 shows schematically a cylindrical magnetic body of cross sectional area A and length a. It completely fills a solenoid which has N turns and which is supplied by a constant current source. The current I in the solenoid generates a uniform field:

$$H = \frac{NI}{a},\tag{2.9}$$

if one ignores end corrections. The extension to non-uniform field is simple in principle.[20] The resulting magnetization per unit volume of the body is $\frac{M}{Aa}$, so that the flux density inside the body is:

$$B = \mu_0 \left(\frac{NI}{a} + \frac{M}{Aa} \right).\tag{2.10}$$

If M changes, while I is kept constant, the flux through the solenoid changes, generating a reverse voltage:

$$V = NA\frac{dB}{dt} = \frac{\mu_0 N}{a}\frac{dM}{dt}.\tag{2.11}$$

The power drawn from the electrical source is $IV = \frac{\mu_0 NI}{a}\frac{dM}{dt}$. Hence, the work done by the source when M changes by dM is:

$$đW_s = IV\,dt = \frac{\mu_0 NI}{a}\,dM,$$

$$= \mu_0 H\,dM\tag{2.12}$$

by Equation (2.9).

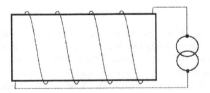

FIGURE 2.6   A cylindrical body being magnetized by current from a constant current generator.

---

[19] Section 1.4 and Appendix C of Mandl.

[20] V. Heine, "The thermodynamics of bodies in static electromagnetic fields," *Proc. Cam. Philo. Soc.* **52**, 546 (1956) (see also Pippard, pp. 23–28).

The work in Equation (2.12) includes the field energy U_{MH} due to the field's interaction with the magnetization. This is given in SI units by:[21]

$$U_{MH} = \mu_0 MH. \tag{2.13}$$

Hence, the work done on the body itself is:

$$\begin{aligned} \dbar W &= \dbar W_s - dU_{MH} \\ &= \mu_0 H \, dM - \mu_0 M \, dH - \mu_0 H \, dM \\ &= -\mu_0 M \, dH. \end{aligned} \tag{2.14}$$

Unless we are considering a ferromagnet or a superconductor, the magnetization (moment per unit volume) is small, so that $\mu_0 H = B$ and we can write:

$$\dbar W = -M \, dB. \tag{2.15}$$

The work given by Equation (2.14) is what we shall actually calculate from statistical mechanics in the following chapters.[22] It follows from Equation (2.14), that for a magnetic system the expression corresponding to Equation (2.6) is:

$$dU = \dbar Q - p \, dV - \mu_0 M dH, \tag{2.16}$$

where the third term is the magnetic contribution to the change in internal energy. For a careful discussion of the difference between Equations (2.15) and (2.12), and of the various terms in the field energy, see Mandl.[15]

Example 2.5: An Example in Magnetism

In most non-ferromagnetic materials, the magnetization $M \equiv \frac{M}{V}$ (the magnetic moment per unit volume), is proportional to B. The constant of proportionality is called the magnetic susceptibility χ, so that $M = \chi V B$. The kmolar susceptibility[23] of cupric potassium sulfate (a material used to achieve very low temperatures by adiabatic demagnetization; see Problem 6.4) at a temperature of 1 K is $\chi V_m = 4500 \, \text{A m}^{-1} \, \text{T}^{-1} \, \text{kmole}^{-1}$, where V_m is the kmolar volume. The molecular weight is 442.

(a) What is the change in the internal magnetic energy of 0.1 kg of cupric sulf1., isolated from its surroundings, when a field of 2 T is applied?

---

[21] See any textbook of electromagnetism. *M, H*, and *B* are really vectors, and we assume here that they are parallel to each other (as they will be if the magnetic susceptibility and demagnetizing factor are isotropic). More generally, the field energy U_{MH} is given by the scalar product $\mu_0 \mathbf{M} \cdot \mathbf{H}$. In addition there is the vacuum self-energy of the electromagnetic field, $\frac{1}{2}\mu_0 \int H^2 \, dV$, but since this is independent of *M*, it need not concern us.

[22] Purely thermodynamic treatments (e.g., Pippard, pp. 23–28, 63–68) usually define the internal energy *U* to include the interaction energy U_{MH}; this is convenient since $\mu_0 H \, dM$ in Equation (2.12) is analogous to $-p \, dV$ in Equation (2.4); *M* being an extensive quantity (i.e., one whose value is proportional to the size of the system) like *V*, and *H* is an intensive quantity (whose value is independent of the size of the system) like *p*. However, the choice is arbitrary, merely interchanging the definitions of internal energy *U* and enthalpy *H* (see Equation 2.20). The thermodynamic definition is inconvenient in statistical mechanics; for example, Equations (6.11) and (6.29), which give the internal energy and free energy in terms of the partition function, would not apply to a magnetic system if we were to use this definition.

[23] The susceptibilities are taken from the *American Institute of Physics Handbook*, 3rd edition, ed. D. E. Gray (McGraw-Hill, 1972), pp. 5–238.

Integrating Equation (2.16) with $dQ = 0$, $dV = 0$, gives the change in U when the field is increased from zero to B:

$$\Delta U = -\int_0^B M\, dB = -\chi V \int_0^B B\, dB = -\frac{1}{2}\chi V B^2.$$

For 0.1 kg, $\chi V = \frac{0.1 \times 4500}{442} = 1.0\ \mathrm{A\,m^{-1}\,T^{-1}}$, $B = 2$ T, so that $\Delta U = -2.0$ J.

ΔU is negative because lining up the magnetic ions in the field lowers their energy.

(b) A material with positive χ is called a paramagnet. Diamagnets, which have negative χ, also exist; for example, bismuth. If a bismuth sample is placed in a uniform magnetic field, how are the magnetic lines of force distorted?

Because the internal energy of the bismuth is increased by the field, it can be reduced by excluding it. Thus, total energy is minimized when the magnetic lines of force bend slightly to reduce the field inside the material (bending stretches the field lines and requires field energy, and a balance is achieved between the decrease in internal energy of the bismuth and the increase in field energy). An extreme case is a Type I superconductor, which has a very high diamagnetic susceptibility and excludes magnetic fields altogether, so long as the field is not too large (see Chapter 15).

The above discussion implicitly assumes that at a given temperature and magnetic field, the magnetization of any particular object is always the same and does not depend on the history of the material. However, the magnetization of many ferromagnetic materials does depend on history. These materials often have a permanent magnetic moment even in the absence of a magnetic field, and exhibit hysteresis; the magnetization at a given field depends on how the field was reached. Figure 2.7 illustrates the difference between the magnetization curve $M(H)$ of such a hysteretic material (full curve) and a material that does not show hysteresis (dashed curve). The hysteretic material at any given field has a magnetization which depends on whether the field is increasing or decreasing. Since the magnetic state of a hysteretic material is not defined uniquely by the field, but depends on its history, the magnetization is not a state function and Equation (2.16) does not apply.

2.6 Enthalpy

It is not always easy in practice to carry out processes at constant volume, and it may be dangerous even to try; for instance, if you heat a liquid past its boiling point. This is particularly true in chemistry, which is a hazardous enough pursuit as it is. On the other hand, a condition of constant pressure is easy and usually safe to maintain, especially if that pressure is atmospheric. In order to deal with constant pressure processes, it has been found convenient to define a quantity *enthalpy*[24] H by:

$$H \equiv U + pV. \tag{2.17}$$

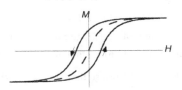

FIGURE 2.7 Full line: typical magnetization vs. field curve of a hysteretic material. Dashed line: magnetization vs. field curve of a non-hysteretic material.

---

[24] Enthalpy is sometimes called "total heat," a misleading name which we will not use.

We came across this relationship as a state variable back in Equation (2.7). Here, we will take a closer look at it physically. Do not confuse enthalpy with entropy (which will be introduced in Chapter 5), or its symbol H with the magnetic field H.

From the product rule:

$$dH = dU + p\,dV + V\,dp.$$

Substituting for dU from Equation (2.6), we obtain:

$$dH = đQ + V\,dp. \tag{2.18}$$

The reason for defining H by Equation (2.17) can be seen by considering a simple example: the boiling of a liquid at atmospheric pressure inside a cylinder like that in Figure 2.2. Suppose that no force other than atmospheric pressure is applied to the outside of the piston. When a given mass of liquid is completely converted into gas, its internal energy increases by a certain amount ΔU, and the volume increases by ΔV. To achieve this increase, we have to supply not only the energy ΔU, but also the work needed to push the piston outward against the atmosphere. Because p is constant, from Equation (2.4) this work is $p\,\Delta V$, so that the energy needed to convert the liquid to gas (the latent heat L) is $\Delta U + p\,\Delta V = \Delta H$, where H is defined by Equation (2.17). Even if the piston were not there, this work would still have to be done against the atmosphere. The same is true of any process, such as a chemical reaction, which takes place at constant pressure with a change in volume. For this reason, tabulated latent heats and heats of reaction are always expressed as enthalpy changes. In statistical mechanics, on the other hand, we usually calculate internal energies, and in comparing with experiment, one must remember to add the term $p\,\Delta V$ to the calculated change in U.

2.7 Heat Capacity

The quantity of energy that must be supplied to raise the temperature of an object of unit size by one unit is called its *heat capacity*.[25] This is a misnomer; it should really be energy capacity since, as pointed out above, heat is not "contained" in a body, but we are stuck with it for historical reasons. The heat capacity will, in general, depend on the conditions under which the energy is supplied. For example, in a compressible fluid the heat capacity depends on whether the measurement is made at constant volume or constant pressure; in the latter case, work is done against outside pressure and this extra energy must be supplied, usually, but not necessarily as heat.

Example 2.6: Heat Capacities at Constant Volume and Constant Pressure

The kmolar heat capacity (the heat capacity of N_A molecules) measured at constant volume is called C_V, and that measured at constant pressure is called C_p.

(a) Show that $C_V = \left(\frac{\partial U}{\partial \theta}\right)_V$ and that $C_p = \left(\frac{\partial H}{\partial \theta}\right)_p$.

We supply an infinitesimal amount of heat $đQ$ to raise the temperature by $d\theta$.

If V is constant, $dV = 0$ in Equation (2.6), so that $đQ = dU$ and $C_V \equiv \frac{đQ}{d\theta} = \left(\frac{\partial U}{\partial \theta}\right)_V$.

If p is constant, $dp = 0$ in Equation (2.18), so that now $đQ = dH$ and $C_p \equiv \frac{đQ}{d\theta} = \left(\frac{\partial H}{\partial \theta}\right)_p$.

---

[25] The heat capacity per mole, or "molar heat capacity," is often called the "specific heat," but this term is ambiguous because it is often used for heat capacity per unit mass or volume, and we do not use it in this book.

(b) A gas is defined as "ideal" if its internal energy at a given temperature is independent of the volume, and if it obeys Boyle's law $p \propto n$ at constant temperature. Air approximates to an ideal gas.

The ideal gas scale of temperature is defined by $\theta \equiv \frac{p}{k_B n}$, where p is the pressure, n the particle density of an ideal gas, and k_B is Boltzmann's constant. Note that by this relationship it is clear that since p is a state variable, so is θ, and it may therefore be treated with an exact differential above in defining C_p and C_V. Find $C_p - C_V$ (kmolar) for an ideal gas.

$$dU = \left(\frac{\partial U}{\partial \theta}\right)_V d\theta + \left(\frac{\partial U}{\partial V}\right)_\theta dV.$$

Differentiating both sides with respect to θ at constant p then gives

$$\left(\frac{\partial U}{\partial \theta}\right)_p = \left(\frac{\partial U}{\partial \theta}\right)_V + \left(\frac{\partial U}{\partial V}\right)_\theta \left(\frac{\partial V}{\partial \theta}\right)_p.$$

In an ideal gas U is independent of V at given θ, so that $\left(\frac{\partial U}{\partial V}\right)_\theta = 0$.

Hence, for an ideal gas,

$$\left(\frac{\partial U}{\partial \theta}\right)_p = \left(\frac{\partial U}{\partial \theta}\right)_V = C_V.$$

From Equation (2.17), the enthalpy is $H = U + pV$, so that, from (a):

$$C_p = \left(\frac{\partial H}{\partial \theta}\right)_p = \left(\frac{\partial U}{\partial \theta}\right)_p + p\left(\frac{\partial V}{\partial \theta}\right)_p = C_V + p\left(\frac{\partial V}{\partial \theta}\right)_p.$$

Since $p = nk_B\theta$ from the definition of θ, for one kmole $V = V_m = \frac{N_A}{n} = R_0\theta/p$, where $R_0 = N_A k_B$, so that:

$$\left(\frac{\partial V}{\partial \theta}\right)_p = \frac{R_0}{p}.$$

Hence, for kmolar heat capacities,

$$C_p - C_V = R_0. \tag{2.19}$$

Besides its value in dealing with processes occurring at constant pressure, enthalpy is also a useful concept in continuous flow processes. An example will be given in Chapter 5.

In a magnetic system, the electromagnetic field plays a role analogous to that of the atmosphere in the case of a gas, and the corresponding definition of enthalpy for a magnetic system is:

$$H = U + pV + MB, \tag{2.20}$$

whence:

$$dH = đQ + V\,dP + B\,dM. \tag{2.21}$$

The work done by the supply, $đW_s$ in Equation (2.12), is thus the increase in enthalpy.

2.8 Envoi

In this chapter, we have introduced the concepts of temperature, heat, and work. In order to do this, we formulated the zero'th and first laws of thermodynamics as generalizations of experience. The zero'th law states if two objects are simultaneously in thermal equilibrium with a third object, they are in equilibrium with each other. This permits the definition of empirical temperature as the property that two objects in thermal equilibrium have in common. The first law states that energy is conserved if heat is taken into account, heat being that form of energy which is transferred when two objects come into thermal equilibrium.

Equation (2.6), and its magnetic analog Equation (2.16), which follow from the first law in the form Equation (2.3), are important steps forward, but are unsatisfactory for two connected reasons. The first is that they only apply to reversible processes. The more fundamental reason is that they contain in $đQ$ an inexact differential, which cannot be integrated because Q is not a state function.

However, we have seen that in Equation (2.7) that under specific conditions there were expression(s) of the inexact differential $đQ$ that could be expressed as a linear combination of exact differentials. Once a form of reversible work was defined to replace the inexact differential $đW$, a relationship was found that implied the same could be accomplished for $đQ$ under those same conditions. However, since such a relationship can be obtained for different forms of work, $-pdV$, $-M\,dB$, and $f\,dℓ$, it implies there is a more general form for an exact differential to replace $đQ$ because it has to relate to heat regardless of the manifestation of the reversible work.

Moreover, we could note that each time we replaced the inexact differential $đW$ with an exact form of reversible work, it was always with a pair of thermodynamic variables: pressure and volume, force and length, magnetic moment, and magnetic field. In addition to being a pair of thermodynamic variables whose product had units of energy, the combination follows a pattern of pairing an *extensive* variable that is dependent upon system size (volume, length, magnetic field) with an *intensive* variable that is not dependent upon size (pressure, force, the magnetic moment of a material). This pairing of such variables to produce an energy that allows the inexact differential of work to be replaced with an exact differential is more profound than it may initially appear if one is only looking at the terms mathematically instead of physically.

In order to obtain an exact differential to replace $đQ$ we would follow a historical treatment wherein we would argue counter to our discussion in regard to Figure 2.1 that in a specific system, such as an ideal gas, the temperature could be directly proportional to energy of the system (i.e., a system without phase changes that would release a latent heat). We would then use the idea of heat capacity to argue that in an ideal gas if the gas does no work but we continue to extract heat from the gas and thus lower its temperature, we would obtain a one-to-one relationship between the energy of the ideal gas and its temperature based upon its heat capacity at constant volume.

If this book were only a textbook of thermodynamics, we would then introduce a new quantity, called entropy and denoted S, which, unlike Q, is a state function (see Appendix E, particularly Footnote 5). We would then determine the absolute scale of temperature defined by $T \equiv \left(\frac{\partial U}{\partial S}\right)_V$, and that Equation (2.6) can be written in terms of T and S [see Equation (5.13)]. We denote absolute temperature by T rather than θ in order to distinguish it from the arbitrary scales, dependent as they are on the properties of particular substances. This is the approach we will take in the next chapter in order to discuss more classical applications of thermodynamics such as engines and efficiency.

However, the concept of entropy is most easily (and best) understood if we start from the atomic level and use statistical arguments, which form the substance of Chapter 4. This is the shortcoming of most introductory physics textbooks used at the freshman/sophomore level: Entropy is an *ad hoc* concept if defined macroscopically and misleads students into believing that entropy is synonymous with heat. A true understanding of entropy is better left to the microscopic picture of statistical physics. The laws of thermodynamics can then be expressed in terms of entropy as a manifestation of microscopic (quantum) effects at the macroscopic (classical) level, and a beautiful example of the Correspondence Principle.

2.9 Problems

2.1 (a) Does the refrigerator in your kitchen cool the room (on average), heat it, or have no effect? Assume that the refrigerator door has been closed long enough to ensure that the contents are at a constant temperature

   (b) Suppose now that the door is left open so that the thermostat keeps the compressor running continuously in an attempt to keep the contents cold. Does the average temperature of the room rise, fall, or stay the same as in (a)?

2.2 Which of the following processes is reversible and which not? Note that "reversibility" is an idealization; the question really means can the actual process be approximated by a reversible process?

   (a) The gas in a cylinder (Figure 2.2) with insulating walls is compressed by pushing in the piston. Assume that the piston moves without friction, and that the seal between it and the cylinder wall is perfectly leak-proof

   (b) The piston seal develops a leak, so that the high pressure gas leaks out

   (c) The same as (a), but the cylinder now has perfectly diathermal walls and is surrounded by a *heat bath* (a very large system whose temperature does not change appreciably when energy flows in or out). Assume that the compression is so slow that heat has time to transfer freely between the cylinder and the heat bath without any appreciable temperature rise, so that the gas in the cylinder is always in equilibrium with the heat bath

   (d) Heat is transferred by conduction from a hot to a cold body

   (e) In Figure 2.4, the volumes on either side of the diaphragm are at the same pressure, but contain different gases. The diaphragm is removed and the two gases interdiffuse

   (f) The same as (e), but the two volumes are now filled with the *same* gas

   (g) A magnetic material is magnetized by slowly increasing the applied magnetic field. Assume that the material does not show hysteresis

2.3 When 1 kmole of a given liquid evaporates at atmospheric pressure (1 bar, or 10^5 Pa), 20 m³ of gas is produced. What is the difference (in J/kmole) between the enthalpy of the gas and its internal energy, assuming that the volume of the liquid is negligible relative to that of the gas?

2.4 A gas thermometer consists of a bulb of fixed volume containing a certain quantity of an ideal gas. A pressure gauge is attached

   (a) If we use the pressure in the bulb to define a temperature scale, what scale is it?

   (b) The bulb of the gas thermometer is immersed in 1 kg of ice which is in an insulated container. The pressure in the bulb is initially 1 bar. An electric heater immersed in the ice is switched on and delivers a power of 1 kW. The ice immediately begins to melt, and after 330 seconds it has just completely turned to water. After another 330 seconds, the pressure is 1.3 bar. What is the pressure at the end of the *first* 330 seconds? What is the latent heat of melting per kg of ice? If the temperature of melting ice is defined as 273 degrees on the ideal gas scale, what is the heat capacity of liquid water in $\mathrm{J\,kg^{-1}\,degree^{-1}}$? Assume that thermal equilibrium is always maintained throughout the container, and that the heat capacity of water is independent of temperature over the range of this experiment, while that of the container is negligible

2.5 Any free surface has a certain energy, proportional to the area of the surface. The energy per unit area is called the surface energy (or surface tension) σ. Its temperature dependence is usually small and can often be neglected

   (a) A soap film, which has two such surfaces, is held in a frame as illustrated in Figure 2.8a. The width of the film is increased by an amount dx by pulling on one side of the frame with a force f. The length of this side is ℓ. How much work is done, and what is f in terms of σ and ℓ?

   (b) Calculate the work done when the radius of a spherical drop of an incompressible fluid is increased from r to $r + dr$ by injecting fluid slowly with a hypodermic needle (see Figure 2.8b).

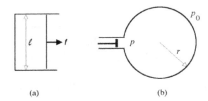

(a)   (b)

FIGURE 2.8 (a) Soap film on a frame, one side of which can be moved by exerting a force f, so that the film is stretched in the x-direction. (b) Spherical liquid drop of radius r, into which liquid is being pumped. The pressure p inside differs from that outside (p_0) because of the surface energy.

The pressure outside the drop is p_0. Hence, show that the pressure inside the drop is $p = p_0 + \frac{2\sigma}{r}$. (Note that in this case there is only one surface)

2.6 (a) A rubber band obeying Hooke's law (extension proportional to force) with force constant 100 N/m, initially unstretched, is stretched slowly by an amount 0.1 m. The final temperature is the same as the initial. How much energy in the form of heat is transferred to the surrounding air?

    (b) The same rubber band, stretched as in (a), is isolated from its surroundings and allowed to return slowly to its original length. If the mass of the band is 10^{-3} kg and the heat capacity per unit mass of rubber is 2 kJ/kg/degree Celsius, what is the change in the temperature of the band?

2.7 It is shown in Appendix H that an ideal non-relativistic monatomic gas ("non-relativistic" means that the atomic velocities are much less than the velocity of light) exerts a pressure $p = \frac{2u}{3}$, where $u \equiv \frac{U}{V}$ is the internal energy density

    (a) Show that for such a gas, $\frac{C_p}{C_V} = \frac{5}{3}$

    (b) Suppose that this gas is in an insulating container and expands reversibly from volume V to $V + dV$. Because no heat is transferred, the first law of thermodynamics requires that the work done by the gas be equal to $-dU$, the decrease in U [see Equation (2.4)]. Differentiate $U = uV$, using the product rule, and substitute for u to find a differential equation for $p(V)$. Integrate this equation to show that for such an expansion, $p \propto V^{-5/3}$

        This equation is a special case of the general result (see Section 9.7) for the reversible expansion of an ideal gas in an insulating container: $p \propto n^\gamma$, where $\gamma \equiv \frac{C_p}{C_V}$. For reasons which will become clear in Chapter 4, such an expansion is called isentropic

2.8 Air approximates an ideal gas obeying Boyle's law $p \propto n$ at constant temperature, and its kmolar heat capacity at constant volume is $C_V = \frac{5}{2}R_0$, independent of temperature (see Section 9.5). 1 kmole of air, initially at a pressure p_0 and a temperature θ_0 on the ideal gas scale ($\theta \equiv \frac{p}{k_B n}$), is subjected to the following cycle. First, it is heated at constant volume until its pressure has increased by a factor r. It is then heated at constant pressure until its volume has increased by a factor of r. It is then cooled at constant volume to its initial pressure, and finally cooled at constant pressure to its initial volume. Assume that each process is carried out reversibly

    (a) Draw the p, V diagram analogous to Figure 2.3 for the cycle

    (b) Calculate the work done by the gas in one cycle (Caution: watch signs!)

    (c) Calculate the heat input to the gas in the first two stages, and the heat output in the second two. Show that the first law of thermodynamics is obeyed, and show that the efficiency η, defined as the ratio of work out to heat in, is $\frac{2(r-1)}{7r+5}$.

2.9 1 kmole of air is subjected to the cycle shown in Figure 2.3, in the order ABCD, where the compression (AB) and the expansion (CD) are both isothermal

    Calculate the work done by the gas, the heat input, and the efficiency η in terms of the compression ratio r (defined as the ratio of the volume at A to that at B)

    Show that if $(r - 1) \ll 1$, $\eta \approx 2(r - 1)/5$

2.10　A monatomic "van der Waals" gas (see Chapter 15) has internal energy:

$$U = N \left(\frac{3}{2} k_B \theta - an \right), \tag{2.22}$$

where θ is the temperature on the ideal gas scale, $n = \frac{N}{V}$ is the particle density, and a is a constant. The first term represents the kinetic energy of the molecules and the second the potential energy due to the attractive forces between them. The gas is initially enclosed in the left-hand part of the insulating container shown in Figure 2.4, with density n_1, and suddenly expands into the vacuum so that its density decreases to n_2, no heat being transferred in or out. What is the change in θ? Explain in physical terms why the temperature of this imperfect gas changes, whereas the temperature of an ideal gas (in which there are no interactions between molecules) would not

2.11　An incompressible magnetic material has heat capacity per unit volume C (measured using a temperature scale θ) and magnetic susceptibility χ. If the material is thermally isolated and the field is increased from zero to B, what is the change in θ? Assume that C and χ are independent of field and of temperature

2.12　A low density approximation to $p(\theta, n)$ for an imperfect gas is:

$$p = nk_B\theta[1 + B_2(\theta)n],$$

where B_2 is a virial coefficient (see Section 4.11), and θ is the temperature on the ideal gas scale. This expression only holds when $B_2(\theta)n << 1$. The bulk expansion coefficient α_b is defined by

$$\alpha_b = \frac{1}{V} \left(\frac{\partial V}{\partial \theta} \right)_p = -\frac{1}{n} \left(\frac{\partial n}{\partial \theta} \right)_p.$$

Show that for this gas, to first order in n,

$$\alpha_b = \frac{1}{\theta} \left[1 + n \left(\theta \frac{dB_2}{d\theta} - B_2 \right) \right]. \tag{2.23}$$

2.13　In the experiment on a rubber band described in Footnote 18, what would you find if you simply released the stretched rubber band, allowing it to snap back to its unstretched length, rather than allowing it to shrink smoothly?

2.14　Calculate the change in enthalpy for adding 1kJ of heat to a $1m^3$ volume of 1 mole of a gas if it increases the pressure by 10% from an initial value of 1 atm. Compare this to the change in energy for adding the same amount of heat to the initial volume increasing it by 10% at a constant pressure of 1 atm

2.15　A certain gas has a heat capacity as a function of temperature given by $C_v = A\theta + B(\theta)^2$. Determine the energy of the gas as a function of temperature up to an additive constant. If the gas were ideal, explain why the additive constant is superfluous information

2.16　A piston head is placed above a large block of ice that maintains the coexisting volume of air at 0 degree Celsius. Assuming the air as an ideal gas, what is the result of pressing down on the piston head compressing the gas column that sits above the block of ice? Is this process reversible if the piston head is then withdrawn to its original position?

2.17　What is the ratio of the root mean square (RMS) speed of nitrogen atoms in 1 mole of the gas on the hottest day recorded in North America 56.7 degrees Celsius to that of the coldest day recorded in North America (−62 degrees Celsius)?

2.18　How long does a 5kW alternating current (AC) voltage water heater take to warm a 200 kg hot water tank of water from 25 degrees Celsius to 50 degrees Celsius?

2.19　Justify the approximation for solids that $C_p \approx C_V$

2.20　Starting at Equation (2.20) show the validity of Equation (2.21)

3

Heat Engines

We are quite ignorant of the condition of energy in bodies generally. We know how much goes in, and how much comes out, and know whether at entrance or exit it is in the form of heat or of work. That is all.

P. G. Tait, *Sketch of Thermodynamics* (1877)

3.1 Gedanken

We saw in the last chapter that heat and work, two different forms of energy, can give us two paths in phase space between two different states. Along one of these paths, there was less work to do than along the other because the pressure was lower. As mechanical work is reversible it would then seem plausible to build a cycle wherein the path of lesser work was the input and the greater work the output. These two paths would be synonymous without the ability to separate them (moving between them) via the non-work change of energy for the system: adding and then removing heat at the end of the lesser and greater work paths, respectively. Therefore, before any gear, any wheel, any process by which work is implemented, the maximum of what we can possibly hope for from any engine is determined by the spacing between these two paths determined by the heat we add and the heat we exhaust in each cycle of the engine (the energy difference that does not involve work).

3.2 Calorimetry and Entropy

As we said at the end of the previous chapter, there is a historical derivation of entropy as a macroscopic quantity that we will discuss here in terms of the heat capacity of an ideal gas. The ideal gas is used because only where there are no changes in phase (no interaction potentials to consider in the microscopic details of the substance) can a temperature scale be defined synonymously with energy. It is this calorimetric derivation of entropy that historically leads students to think of entropy as heat. However, once we begin a microscopic treatment of systems in Chapters 4 and 5, one will see that this is an *ad hoc* treatment of entropy. We'll use one of our *gedanken* situations to realize that entropy is a much broader concept than heat, best defined by working up from the microscopic details using statistical mechanics.

We recall our general version of the first law from Equation (2.7):

$$dU + p\,dV = đQ. \tag{3.1}$$

As everything on the left-hand side of the equation is a state variable we could integrate each term independently and obtain:

$$U + p\,V + c = \delta Q, \tag{3.2}$$

where the value of c is a constant of the indefinite integration. Since the left-hand side is a set of state variables, the quantity must be the result of the integration of some state variable on the right-hand side, even though $đQ$ is an inexact differential. We propose, therefore to examine the system in the absence of any work being done (so, for instance, an ideal gas at constant volume). Here, it is easier to see that the inexact differential $đQ$ is synonymous with the exact differential of the energy change in the system:

$$dU = đQ. \tag{3.3}$$

It is here that we wish to appeal to one's experience to understand that a material that loses heat (without undergoing a change in phase) is observed to lower its temperature as the heat is removed. By this, we propose that there is an exact function S, which like U is extensive because we observe that temperature is intensive, and our experiences in Chapter 2 demonstrate that our thermodynamically intensive variables appear in the first law with an extensive companion. So we propose that there is an extensive quantity S such that:

$$\boxed{TdS = đQ,} \tag{3.4}$$

so that we can now define the absolute temperature scale T by the behavior that $T = 0$ when there is no longer any heat that can be withdrawn from the material. Hence, the zero on this temperature scale is an absolute zero, and the scale is an absolute temperature scale.

Moreover, upon rearrangement:

$$dS = đQ/T, \tag{3.5}$$

we have now defined an exact differential dS by the amount of heat extracted from the material at temperature T. One will immediately scratch their head upon this extremely *ad hoc* idea of a change in the value of some quantity S because of moving an amount of heat $đQ$ into or out of a material at a constant temperature T. However, we will invoke here the even more specific case that:

$$đQ = mC_v dT, \tag{3.6}$$

from our first look in Chapter 2 to understand that in this case, just as there was the special case of volumetric work in a gas that allowed $đW$ to be replaced by $-pdV$, here we replace the amount of heat that moves into or out of the material with the more understandable constant volume heat capacity of the material that results in a temperature change for any real material. Later, we'll use the idea of a thermal reservoir to make this idea of the term TdS more universal than this special case.

Now one sees that under the constant volume heat capacity assumption both the right and left sides are well defined exact differentials as:

$$dS = T^{-1}mC_v dT. \tag{3.7}$$

Students with experience in high school physics and chemistry will recognize this expression from their work in calorimetry. Namely, from the above equation we see assuming that C_v is not a function of temperature that:

$$S = mC_v ln(T_f/T_i), \tag{3.8}$$

when integrating from the initial temperature T_i to the final temperature T_f. A process for which $dS = 0$ for which by the above relationships, $đQ$ would also be zero is said to be isentropic or adiabatic. While one may think of these quantities as somehow synonymous, they are only in the *ad hoc* manner in which we have dealt with entropy here. Entropy is a much broader concept than simply heat, as we will see in Chapter 5.

3.3 Isentropic Expansion of an Ideal Gas

The equation of state of an ideal gas is $pV = Nk_BT$ so that in an isothermal expansion Boyle's law holds:

$$p \propto n = N/V. \tag{3.9}$$

In an isentropic expansion, the temperature changes, but it is still possible to find a relation between pressure and density, n. To do this, we start from the first law of thermodynamics in the form given in Equation (3.1) modified by Equation (3.4):

$$dU = T\,dS - p\,dV,$$

and for definiteness, we consider ν kmoles of gas, so that $V = \nu V_m$.

In an isentropic change (for which $dS = 0$), Equation (4.13) becomes:

$$dU = \nu C_V\,dT = -p\,dV. \tag{3.10}$$

The equation of state $pV = \nu R_0 T$ can be differentiated to give:

$$p\,dV + V\,dp = \nu R_0\,dT. \tag{3.11}$$

Eliminating dT from Equations (3.10) and (3.11) gives:

$$p\,dV + V\,dp = -\frac{R_0}{C_V}p\,dV,$$

or

$$(C_V + R_0)p\,dV + C_V V\,dp = 0. \tag{3.12}$$

Note that ν has cancelled out in Equation (3.12).

From Equation (2.19), $C_V + R_0 = C_p$, so that if we divide Equation (3.12) by $C_V pV$, we obtain a differential equation connecting p and V:

$$\gamma\frac{dV}{V} + \frac{dp}{p} = 0, \tag{3.13}$$

where $\gamma \equiv \frac{C_p}{C_V}$.

Since $n \propto V^{-1}$, $\frac{dV}{V} = -\frac{dn}{n}$, so that Equation (3.13) can be written:

$$\frac{dp}{p} = \gamma\frac{dn}{n}, \tag{3.14}$$

which we integrate to give:

$$\ln p = \gamma \ln n + \text{constant}, \tag{3.15}$$

or

$$\boxed{p \propto n^\gamma.} \tag{3.16}$$

This is the isentropic (commonly called the adiabatic) law of expansion of an ideal gas, analogous to the isothermal law [Equation (3.9)]. Equation (3.16) is usually written as $pV^\gamma = $ constant.

If we substitute $p = nk_BT$ [Equation (9.11)] in Equation (3.16), we obtain the isentropic relation between density and temperature:

$$\boxed{T \propto n^{\gamma-1}.}$$

(3.17)

Thus, when an ideal gas is compressed isentropically to a factor r times its initial density, the absolute temperature rises by a factor $r^{\gamma-1} = r^{0.4}$ for air. This rise in temperature is familiar to anyone who has used a bicycle pump. The rise can be substantial; for example, in a Diesel engine for which a typical compression ratio $r = 15$, the temperature of the air in the cylinder rises during the compression stroke from 300 K to about 900 K.

Example 3.1: A Gun Example

In a gun, explosion of the charge generates a hot gas at high pressure. The gas expands, forcing the projectile out of the barrel.

A medium sized naval gun has a projectile weighing 50 kg. On firing, 0.25 kmole of charge is completely converted into a compressed gas. While the burning continues, the gas expands at approximately constant pressure, driving the projectile along the barrel. The burning ceases when the projectile is half way along the barrel: the temperature is then 2500 K. Assuming that the subsequent expansion is isentropic, that the combustion products are all polyatomic ideal gases with $\gamma = 1.33$, that the volume of the charge before firing is negligible, and that 25% of the energy delivered to the projectile is lost (in friction and to work done against the atmosphere), find the velocity V at which the projectile leaves the gun.

If the volume of the charge when burning ends is V_0, the temperature T_0, and the number of molecules is N, the pressure during burning is $p_0 = \frac{Nk_BT_0}{V_0}$. The work done by the gas on the projectile before losses is:

$$W_{\text{out}} = p_0V_0 + \int_{V_0}^{rV_0} p\,dV,$$

where $pV^\gamma = p_0V_0^\gamma$ and rV_0 is the final volume.

Hence,

$$W_{\text{out}} = p_0V_0 + \frac{p_0V_0^\gamma}{1-\gamma}\left[(rV_0)^{1-\gamma} - V_0^{1-\gamma}\right] = \frac{p_0V_0}{\gamma-1}\left(\gamma - r^{1-\gamma}\right),$$

where $p_0V_0 = Nk_BT_0$.

In the present case, $r = 2$, $N = 0.25N_A$, $T_0 = 2500$ K, so that the kinetic energy of the projectile when it leaves the gun barrel is:

$$\frac{1}{2}Mv^2 = 0.75 \times \frac{0.25 \times 6 \times 10^{26} \times 1.38 \times 10^{-23} \times 2500}{0.33}(1.33 - 20^{-0.33})$$

$$= 6.3 \text{ MJ}.$$

$M = 50$ kg, so that $v = \left(\frac{2 \times 6.3 \times 10^6}{50}\right)^{1/2} = 500$ m/s.

3.4 The Carnot Engine

The invention of the steam engine in the eighteenth century ushered in the Industrial Revolution. This revolution was made by practical engineers, most of whom, like Thomas Edison a century later, "measured everything by the size of a silver dollar" (or, since many of them were British, the size of a gold sovereign), and had little use for what Edison called "the bulge-headed fraternity of the savanic (*sic*) world." However, Carnot[1] was an exception: He founded the science of thermodynamics in an attempt to find out if there is a fundamental limit to the efficiency with which heat can be converted into work. This search led to the discovery of the second law of thermodynamics, which does indeed impose such a limit, and he was able to enunciate the second law even before the first law was recognized.

A *heat engine* is any device that converts heat (usually generated by burning some fuel) into work. An idealized model of a heat engine is shown in Figure 3.1a. While the figure is based on the steam engine, we shall see that the analysis can be applied to any heat engine; for example, the internal combustion engine. With certain modifications it can also be applied to chemical, physiological, and meteorological processes, although we do not do this in this book.[2]

The engine contains a fixed amount of *working substance* (e.g., water in a steam engine), which undergoes various processes in a continuously repeated cycle. The working substance is recycled, finishing each cycle of processes in the same state that it started. In the *Carnot cycle*, heat is taken in at a constant temperature T_1 and given out at a lower constant temperature T_2. The processes are assumed to occur reversibly and are as follows:

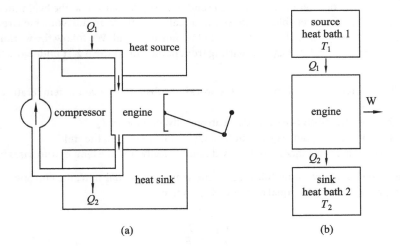

(a)   (b)

FIGURE 3.1 (a) Schematic heat engine. Energy flow is represented by labelled arrows, substance flow by unlabelled ones. The work needed to drive the compressor comes from the engine. (b) Idealization of the essential processes of the Carnot cycle: in one cycle the engine takes in heat Q_1 at T_1, rejects heat Q_2 at T_2, and delivers net work $W = Q_1 - Q_2$.

---

[1] Sadi Carnot (1796–1832), French military engineer, son of the brilliant general and polymath Lazare Carnot ("L'organisateur de la victoire"). He was led to his study of heat engines by his conviction that the recent defeat of France by Britain and her allies was due at least in part to France's inadequate utilization of steam in her industries. His seminal essay, *Reflections on the motive power of fire and on the machines fitted to develop this power*, [transl. R. H. Thurston (Peter Smith, 1990)] was published in 1824.

[2] See the books by McGlashen, by Edsall and Gutfreund, and by Bohren and Albrecht in the bibliography.

1. The working substance is initially at a high pressure and at temperature T_1. Heat Q_1 is supplied to it by heat bath 1 (the "source"), whose temperature is T_1. During this process, the working substance expands isothermally
2. The working substance is disconnected from the heat bath and is allowed to expand isentropically, doing work and cooling to temperature T_2
3. The working substance, now at a low pressure, is transferred to heat bath 2 (the "sink"), which is at temperature T_2, where it gives up heat Q_2 and contracts isothermally
4. The working substance is then recompressed isentropically, its temperature rising to T_1, and is returned to heat bath 1 in its initial condition

The overall process of conversion of heat into work can be schematized still further, as in Figure 3.1b. The net work output (i.e., the work obtained in the expansion less the work needed to do the recompression), is W. Note that here, W is defined as the work *out* and Q_2 the heat *out*, whereas in Chapter 2, W and Q represented the work and heat put *into* the system of interest. These sign conventions are chosen so that in a heat engine, all quantities are positive (when we come to consider heat pumps and refrigerators, the sign conventions will be reversed, for the same reason). Since Q_2, the heat given up to the sink by the engine, is lost (or at least cannot be used to generate work), the efficiency is defined as the work output W divided by the heat input Q_1, and we are concerned to calculate this efficiency. A reversible cycle, such as this one, in which all heat transfer is isothermal is called a Carnot cycle.

The expansion and recompression of the working substance are both isentropic, since (ideally) no heat flows during these processes. We can plot the cycle on a graph of temperature versus entropy (T-S diagram), as shown in Figure 3.2. Since S is a state function, the T-S diagram for the working substance is necessarily closed, its entropy being the same at the end of the cycle as it was at the beginning, whether or not the actual processes are reversible. Of course, in a real cycle, there are unavoidable irreversible processes, but the entropy so generated is discharged to the outside world. We follow the working substance round the Carnot cycle in the T-S diagram, starting from point a in Figure 3.2. It undergoes the following processes:

1. a to b: The source supplies heat Q_1 to the working substance. Since the temperature is constant, $Q_1 = T_1(S_1 - S_2)$
2. b to c: The working substance expands isentropically, cooling to T_2
3. c to d: The working substance gives out heat $Q_2 = T_2(S_1 - S_2)$ to the sink
4. d to a: The working substance is compressed isentropically, and the temperature rises from T_2 to T_1

Since the final state of the working substance is the same as its initial state, the entropy changes in the two isothermal processes must be equal and opposite, so that:

$$\frac{Q_1}{Q_2} = \frac{T_1}{T_2}. \tag{3.18}$$

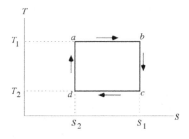

FIGURE 3.2 T-S diagram for the Carnot cycle.

Furthermore, the internal energy is unchanged, so that the first law of thermodynamics requires that:

$$W = Q_1 - Q_2 \tag{3.19}$$
$$= (T_1 - T_2)(S_1 - S_2).$$

This is just the area enclosed within the loop in the *T-S* diagram.

From Equations (3.18) and (3.19), the efficiency η_c of the engine, defined as the work out divided by the heat in, is:

$$\boxed{\eta_c \equiv \frac{W}{Q_1} = \frac{T_1 - T_2}{T_1}.} \tag{3.20}$$

This is called the Carnot efficiency. We now show that the second law of thermodynamics requires that *no* engine whose source of heat is at T_1 and whose sink is at T_2 can have an efficiency which exceeds η_c, and that *all* reversible engines working between the same two temperatures have this efficiency. To do this, we introduce the concept of a heat pump.

A heat pump is a heat engine run backwards, so that all the arrows in Figures 3.1 and 3.2 are reversed. Examples are refrigerators, whose purpose is to extract heat from a bath at a low temperature, and the heat pumps used to heat large buildings, whose purpose is to deliver heat at a temperature above ambient more efficiently than a simple furnace can. In either case, work W is supplied from outside in order to remove heat Q_2 from heat bath 2 and discharge heat Q_1 to heat bath 1. Since the processes are assumed to be reversible, Equation (3.18) holds, while the sign of each term in Equation (3.19) is reversed. The coefficient of performance h_c, defined as the heat delivered divided by the work in,[3] for this Carnot heat pump is:

$$h_c \equiv \frac{Q_1}{W} = \frac{1}{\eta_c}. \tag{3.21}$$

Suppose that we were to devise some engine, using the same source and sink, that has an efficiency η. We will now show that the second law forbids η from exceeding η_c, the Carnot efficiency. We suppose that this hypothetical engine is used to drive a Carnot heat pump as in Figure 3.20, in which the heat source for the engine is the sink for the heat pump, and vice versa. The entire system of engine, heat pump, and heat baths is assumed to be completely isolated from its surroundings. The hypothetical engine takes heat Q_1 from heat bath 1 and generates work $W = \eta Q_1$, discharging heat $Q_2 = Q_1 - W$ into heat bath 2. The work W is used to drive the Carnot heat pump, which discharges $Q_1' = h_c W$ into heat bath 1, extracting $Q_2' = Q_1' - W$ from heat bath 2. Energy conservation requires that $Q_2 - Q_2' = Q_1 - Q_1'$, which is the net transfer of heat from bath 1 to bath 2. The change in the combined entropy of the two heat baths in one cycle is:

$$\Delta S = \frac{Q_1' - Q_1}{T_1} - \frac{Q_2' - Q_2}{T_2} = (Q_1' - Q_1)\left(\frac{1}{T_1} - \frac{1}{T_2}\right).$$

Since $T_1 > T_2$, the factor $\left(\frac{1}{T_1} - \frac{1}{T_2}\right)$ in the expression for ΔS is negative. However, since the system is isolated, and the entropy of the working substance is unchanged after a complete cycle, the second law forbids ΔS to be negative. Hence, the first factor must also be negative, or at most, zero. It follows

---

[3] This definition is appropriate for a heat pump, where we are concerned with the heat delivered to the outside world at the high temperature. In a refrigerator, we are interested in how much heat can be extracted at the low temperature, and the coefficient of performance is defined as $\frac{Q_2}{W}$. This complicates the algebra slightly.

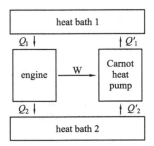

FIGURE 3.3   Heat engine driving a Carnot heat pump.

that $Q_1' \leq Q_1$. Substituting $Q_1 = \frac{W}{\eta}$ and $Q_1' = h_c W$, we find that $h_c \eta \leq 1$; that is, $\eta \leq \eta_c$ since $h_c = \frac{1}{\eta_c}$. Hence, the efficiency of our hypothetical engine, whether reversible or not, cannot exceed the Carnot efficiency.

A similar argument can be used to show that all *reversible* engines working between the same two temperatures have the Carnot efficiency. To prove this, we take our hypothetical engine to be reversible. The system shown in Figure 3.3 can then be run in reverse, the hypothetical engine operating as a heat pump driven by a Carnot engine. The system now transfers $Q_1 - Q_1' = (1 - h_c \eta)Q_1$ from the cooler heat bath to the hotter. The combined entropy of the heat baths changes by:

$$\Delta S = (Q_1 - Q_1') \left(\frac{1}{T_1} - \frac{1}{T_2} \right).$$

Since ΔS cannot be negative, $Q_1' \geq Q_1$ and $h_c \eta \geq 1$. Since we have seen that $h_c \eta \leq 1$, it follows that $\eta = h_c^{-1} = \eta_c$. Of course, no real engine is perfectly reversible, so its efficiency is always less than the Carnot efficiency; similarly, the coefficient of performance of a real heat pump is always less than the Carnot value.

The above arguments enable us to formulate the second law of thermodynamics as an impossibility statement.[4] One such formulation is due to Clausius: "It is impossible to devise an engine which, working in a closed cycle, shall produce no other effect than the transfer of heat from a colder to a hotter body;" another is Kelvin's: "It is impossible to devise an engine which, working in a closed cycle, shall produce no other effect than the extraction of heat from a heat bath and the performance of an equivalent amount of mechanical work." Both these formulations, in Park's phrase, "smell of engine oil," and the formulation which we use, that the entropy of an isolated system cannot decrease (see Section 5.2), is preferable to a physicist.

Since the properties of water are rather complicated, Carnot simplified his cycle conceptually by considering an ideal gas as his working substance. The *T-S* diagram is still Figure 3.2, but it is helpful also to plot the processes on a *p-V* diagram (often called the *indicator diagram*) as shown schematically in Figure 3.4. The processes shown in Figure 3.4 are the same as those in Figure 3.2, and are itemized in the discussion of that figure. Note that since the isentropic bulk modulus of an ideal gas is always larger than the isothermal one (see Section 10.8), the isentrope (line of constant entropy) through any point on the *p-V* diagram is necessarily steeper than the isotherm.

The heat input during the isothermal expansion, Q_1, is equal to the work done *by* the gas during this expansion, $\int_a^b p \, dV$, since for an ideal gas U does not change. Similarly, the heat output during the

[4] For a broad discussion of the use of impossibility statements in physics, see David Park, "When Nature Says No," in *No Way, on the Nature of the Impossible*, ed. Davis and Park (Freeman, 1986), p. 139.

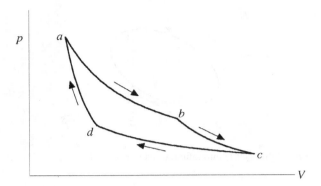

FIGURE 3.4 *p-V* diagram for a Carnot cycle using an ideal gas as the working substance. The slope of the isentropes ($d \to a$ and $b \to c$) is exaggerated for clarity.

isothermal compression, Q_2, is equal to the work done *on* the gas, $-\int_a^b p\, dV$. The equation of state for N atoms of an ideal gas is $pV = Nk_B T$, where N is the number of molecules, so that the heat input at T_1 is:

$$Q_1 = \int_a^b p\, dV = Nk_B T_1 \int_a^b \frac{dV}{V} = Nk_B T_1 \ln\left(\frac{V_b}{V_a}\right). \tag{3.22}$$

The heat output at T_2 is:

$$Q_2 = -\int_c^d p\, dV = Nk_B T_2 \int_c^d \frac{dV}{V} = Nk_B T_2 \ln\left(\frac{V_c}{V_d}\right). \tag{3.23}$$

From Equation (3.17), $TV^{\gamma-1}$ is a constant in the isentropic expansion, so that:

$$\frac{V_b}{V_c} = \left(\frac{T_1}{T_2}\right)^{\frac{1}{(1-\gamma)}} = \frac{V_a}{V_d}, \tag{3.24}$$

whence:

$$\frac{V_b}{V_a} = \frac{V_c}{V_d}.$$

The work done is:

$$W = Q_1 - Q_2 = Nk_B(T_1 - T_2) \ln\left(\frac{V_b}{V_a}\right), \tag{3.25}$$

and the efficiency is:

$$\eta \equiv \frac{W}{Q_1} = \frac{T_1 - T_2}{T_1}$$
$$= \eta_c,$$

as it must be, since this is a Carnot cycle. It can easily be shown (see Problem 3.7) that $W = \oint p\, dV$ (i.e., the area enclosed by the loop in the *p-V* diagram), and that this is equal to the area enclosed in the curve in the *T-S* diagram.

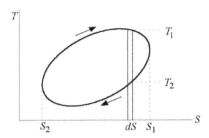

FIGURE 3.5 Slicing up an arbitrary cycle into infinitesimal Carnot cycles.

The Carnot ideal gas cycle is not realizable in practice,[5] since it is difficult to transfer heat isothermally to a gas. However, as we have seen, *any* reversible cycle in which heat is supplied and discharged isothermally has the Carnot efficiency, and can be called a Carnot cycle. Furthermore, any closed curve in the T-S plane can be split up into Carnot cycles, as illustrated in Figure 3.5. This figure shows the T-S diagram for an arbitrary cycle, in which the maximum entropy of the working substance is S_1 and the minimum S_2. Consider a vertical slice in the T-S diagram, of width dS, cutting the actual cycle at T_1 and T_2, and imagine a Carnot cycle in which infinitesimal amounts of heat are supplied at T_1 and discharged at T_2. The other two (vertical) sides of the cycle are isentropic and no heat is transferred. The heat input is $đQ_1 = T_1\,dS$ and the heat output is $đQ_2 = T_2\,dS$. By energy conservation, the work done in this infinitesimal cycle is $đW = đQ_1 - đQ_2 = (T_1 - T_2)\,dS = \eta_c\,đQ_1$. The complete cycle is just the integral over these infinitesimal cycles, so that the total work done is:

$$W = \int_{S_2}^{S_1} (T_1 - T_2)\,dS = \oint T\,dS. \qquad (3.26)$$

This is the area enclosed by the loop in the T-S diagram and is equal to $\oint S\,dT$.

Example 3.2: Cooling a Debye Solid

Suppose that you have a heat bath at a temperature T_1 which is sufficiently low that the heat capacity of a certain insulating solid is given by Debye's T^3 law (Equation 7.44), and that you have designed an ideal Carnot refrigerator to cool the solid from T_1 to T_2, where $T_2 \ll T_1$. The refrigerator discharges heat to the heat bath at T_1. Calculate the work needed to cool 1 kmole of the solid from T_1 to T_2.

When the solid is at temperature T, the coefficient of performance of the refrigerator (defined as heat extracted for unit work in: see Footnote 3) is $h_c = \frac{T}{T_1 - T}$. Hence, the work needed to remove heat $đQ$ is $đW = \frac{T_1 - T}{T}\,đQ$. If the heat capacity is $C(T)$, the change in temperature is $dT = -\frac{đQ}{C(T)}$. Hence, total work done is:

$$W = -\int_{T_1}^{T_2} C(T)\frac{T_1 - T}{T}\,dT.$$

From Equation (7.44) $C(T) = AT^3$, where $A = \frac{12\pi^4}{5}R_0\theta_D^{-3}$, which is constant.

---

[5] The nearest practical realization to the Carnot ideal gas cycle is the Stirling cycle; see, for example, P. C. Riedi, *Thermal Physics* (Oxford, UK, 1988), p. 57.

Hence:

$$W = -A \int_{T_1}^{T_2} T^2(T_1 - T_2)\,dT \approx \frac{AT_1^4}{12},$$

where we have neglected terms in T_2^3 and higher.

3.5 The Steam Engine

Most of the electrical generation in industrialized countries is powered by steam turbines, although the fuel (coal, oil, gas, or nuclear) may vary. The flow diagram for a steam power station is shown schematically in Figure 3.6. Water at high pressure in the boiler is heated to its boiling point at that pressure and vaporized. The high pressure (HP) steam thus produced is sent to a turbine (in practice, a succession of turbines designed to operate at successively lower pressures) in which it expands, doing work that can be transmitted to an alternator to generate electricity. The "spent" low pressure (LP) steam, usually at a pressure well below atmospheric, then goes to a condenser where it is condensed to water, the latent heat produced being removed by a coolant which may be river water, or water that is allowed to cool by evaporation in a cooling tower. The low pressure water is then pumped back into the boiler, the work needed to do this being obtained from the turbine. This cycle differs from that used in modern power stations in that for simplicity it omits superheating, which in the engineering sense means heating the steam above the boiling point at the high pressure (do not confuse this usage with that of Chapter 15, where the same word is used for the metastability of a phase above its transition temperature). Superheating not only raises the efficiency (see Example 3.2), but also prevents troublesome condensation of water in the turbines. The cycle is called the *Rankine cycle*.[6] Its *p-V* and *T-S* diagrams are shown in Figure 3.7. The processes in the Rankine cycle are as follows:

1. *e* to *a*: Liquid water at high pressure p_1 and initially at a low temperature (point *e*) is heated in the boiler at constant pressure to its boiling point T_1 (point *a*); there is little change in volume, so that these two points almost coincide on the *p-V* diagram. However, there is an increase in entropy since liquid water has a substantial heat capacity.

2. *a* to *b*: The water is vaporized in the boiler, with a large increase in volume and entropy, but no change in temperature or pressure.

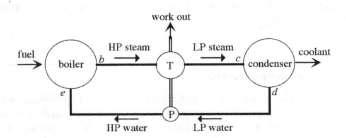

FIGURE 3.6 Schematic flow diagram for a steam power generator. Full lines represent fluid flow and double lines represent power transmission. T = turbine, P = pump, LP = low pressure, HP = high pressure. The letters *b* through *e* refer to points on the cycle (see Figure 3.7).

---

[6] W. J. Macquorn Rankine (1820–1872) was a Scottish railway engineer turned professor of engineering. He was the first to apply Carnot's ideas to the steam engine, and his many contributions to physics include the first use of the term "energy" in its modern scientific sense.

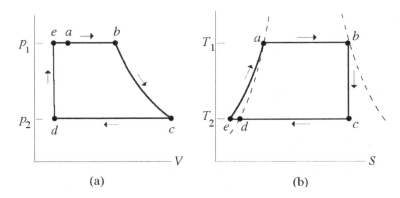

FIGURE 3.7 (a) p-V and (b) T-S diagrams for the Rankine (steam) cycle. The dashed lines in the T-S diagram indicate the limits of the coexistence region.

3. b to c: The high pressure steam expands isentropically through the turbines, its pressure falling to p_2 and its temperature to T_2.
4. c to d: The steam is then condensed to liquid, giving up its latent heat to the outside world.
5. d to e: The resulting liquid water is recompressed to pressure p_1 by the pump. Since water is virtually incompressible, there is little change in either temperature or entropy on compression, so that points d and e are almost coincident in the diagram.

If the heat capacity of liquid water were zero, all the input heat would be supplied at T_1 and all the output heat would be discharged at T_2. In this case, the Rankine cycle would have the same T-S diagram as the Carnot cycle, and the ideal efficiency would be the Carnot efficiency $\frac{T_1 - T_2}{T_1}$. However, the non-zero heat capacity of the liquid means that some heat has to be supplied at a lower temperature, lowering the efficiency. The efficiency is, of course, also lowered by irreversible processes.

The Rankine cycle suffers from two disadvantages. One is that some steam condenses during the expansion; as mentioned above, this can be avoided by superheating, in the engineering sense of heating above the boiling point. A much more serious problem is that T_1 cannot be raised indefinitely: The critical point (see Chapter 15) of water is 647 K, at which the latent heat of vaporization vanishes. This limits the ideal efficiency to ~45% and the actual efficiency to ~35%. Modern power stations use pressures above critical and achieve actual efficiencies of up to 45%.

Example 3.3: The Superheated Rankine Cycle

A steam engine operates on the following *superheated Rankine cycle*: Liquid water on the gas-liquid coexistence curve at temperature T_2, and pressure p_2 is pumped isothermally to pressure p_1. It is then heated to its boiling point T_b, vaporized completely, and heated (superheated) to a temperature T_1, all at constant pressure p_1. The steam then expands isentropically to pressure p_2 and temperature T_2 so that liquid just begins to form; it is then condensed at constant temperature and pressure and returned to the pump.

(a) Draw a T-S diagram for the cycle, showing the limits of the coexistence region

(b) Assuming that steam is an ideal gas, and that liquid water is incompressible, calculate the ideal thermal efficiency η in terms of T_1, T_2, the heat capacity of steam at constant pressure C_p, and the latent heat L. Assume for simplicity that L is independent of temperature over the relevant range and that the specific heat of liquid water C_ℓ is equal to C_p. These assumptions are connected

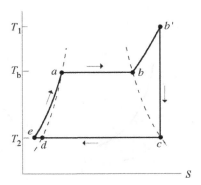

FIGURE 3.8 The superheated Rankine cycle. The heavy lines show the cycle and the dashed lines show the coexistence region. From c to d the pressure is p_2, while from e to b' it is p_1.

(see Problem 15.1), and neither is in fact correct. Since C_p is assumed to be the same for liquid and gas:

heat in $\quad Q_1 = C_p(T_1 - T_2) + L$;

heat out $\quad Q_2 = L$;

work out $\quad W = Q_1 - Q_2$.

Since the liquid is assumed to be incompressible, no work is done by the compressor, and there is no entropy change (points d and e in Figure 3.8 should be coincident on this assumption). Hence, the efficiency is:

$$\eta = \frac{Q_1 - Q_2}{Q_1} = \frac{C_p(T_1 - T_2)}{L + C_p(T_1 - T_2)}.$$

3.6 Internal Combustion Engines: The Otto and Diesel Cycles

In an internal combustion engine, the working substance is air, which is essentially an ideal gas. There are two main types of internal combustion engine: the spark-ignition gasoline engine which operates (ideally) on the Otto cycle,[7] and the Diesel engine, whose operation is idealized in the Diesel cycle.[8] Since in both cases the working substance is drawn in from the environment at the beginning of the cycle and discharged to it at the end, neither cycle is in fact closed, a necessary condition for application of the thermodynamic principles discussed here. However, they can be conveniently replaced by an equivalent closed cycle.[9]

Figure 3.9 illustrates the operation of an engine running on the Otto cycle. The cylinder, with its inlet and exhaust valves (labelled I and E), piston, and crankshaft, is shown schematically in (a), and the p-V and the T-S diagrams in (b) and (c). There are four "strokes" of the piston in one cycle, corresponding to two rotations of the crankshaft, but only two of these, the compression and power strokes, are relevant

---

[7] This cycle was conceived by the French engineer Alphonse Beau de Rochas (1815–1893) in 1862, who was apparently the first to recognize that the air-fuel mixture in an internal combustion engine must be compressed before ignition. It is eponymously credited to the German engineer Nikolaus Otto (1832–1891), who made the first working model based on this cycle in 1876.

[8] Invented by Rudolf Diesel (1858–1913), German engineer, in 1892. The modern fuel-injection gasoline engine is a hybrid.

[9] The justification for, and pitfalls of, this substitution are carefully discussed in Chapter 4 of Haywood.

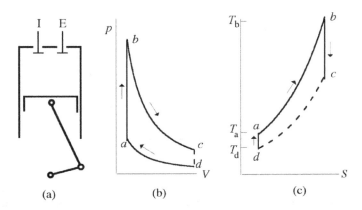

FIGURE 3.9 (a) Schematic of a cylinder in an internal combustion engine. I and E are the input and exhaust valves. (b) Ideal p-V diagram for the Otto cycle, omitting the inlet and exhaust strokes. (c) Equivalent T-S diagram. The dashed ($c \rightarrow d$) part of the cycle in figures (b) and (c) corresponds to the exhaust of the spent fuel-air mixture and the induction of a fresh charge (see text).

to the simple picture of a closed thermodynamic working cycle given here. In the inlet stroke, the inlet valve I is opened when the piston is at the top of its stroke, and as the piston moves down a combustible mixture of air and fuel is drawn in through the inlet valve. When the piston reaches the bottom of its stroke (point d on the p-V and T-S diagrams, which omit the inlet and exhaust strokes), the inlet valve closes. The mixture is now at pressure p_d and temperature T_d, both near atmospheric, and the working cycle begins. It consists of the following processes:

1. d to a: The compression stroke. The mixture is compressed isentropically,[10] reducing the volume by a factor r, where r is the compression ratio. The temperature rises to T_a
2. a to b: The mixture is now ignited by a spark and burns, the temperature rising to T_b. Most of the burning occurs at the top of the stroke where the piston is reversing direction, so that the heating is approximately at constant volume, as indicated by the line $a \rightarrow b$ in Figure 3.9b
3. b to c: The power stroke. The piston now moves downwards, the air and combustion products expanding isentropically back to the original volume, cooling to T_c
4. c to d: At c, the exhaust valve E opens and the spent air-fuel mixture is discharged to the atmosphere by the upwards exhaust stroke. Although in practice a fresh charge of air is drawn after each cycle, it is convenient to treat the exhaust process as the cooling of the working substance at constant volume to its original state at point d, thus replacing the real open cycle by an imaginary closed one. Since no work is done in the step $c \rightarrow d$ of the closed cycle, the work done $\oint p \, dV$ is the same as in the open cycle. The heat input, which is entirely in step $a \rightarrow b$, is also the same. Since energy must be conserved, the heat output in step $c \rightarrow d$ of the closed cycle must equal the energy given up to the atmosphere by the exhaust, so that the two cycles are thermodynamically equivalent

The ideal thermal efficiency is $\eta = \frac{W}{Q_1}$, where W is the work obtained and Q_1 is the heat input, both per cycle. By energy conservation, $W = Q_1 - Q_2$, where Q_2 is the heat discharged in step $c \rightarrow d$. Since both heating and cooling are assumed to take place at constant volume,

$$Q_1 = C_V(T_b - T_a); \quad Q_2 = C_V(T_c - T_d),$$

per kmole of air if C_V is independent of temperature.

---

[10] The assumption that the compression, and the expansion in the power stroke, is isentropic is, of course, only a rough approximation (see Haywood). A large amount of heat is conducted through the cylinder walls to the coolant, as any car owner knows.

Hence,

$$\eta = 1 - \frac{Q_2}{Q_1} = 1 - \frac{T_c - T_d}{T_b - T_a}.$$

Compression and expansion are assumed to be isentropic, and the compression ratio is:

$$r = \frac{V_d}{V_a} = \frac{V_c}{V_b},$$

so that, from Equation (3.17),

$$\frac{T_d}{T_a} = \frac{T_c}{T_b} = r^{(1-\gamma)}.$$

For any numbers s, t, u, v for which $\frac{u}{v} = \frac{s}{t}$, it is elementary that $\frac{s-u}{t-v} = \frac{s}{t}$. Hence,

$$\frac{T_c - T_d}{T_b - T_a} = \frac{T_d}{T_a} = r^{(1-\gamma)}$$

and the ideal efficiency is:

$$\eta = 1 - r^{(1-\gamma)}. \tag{3.27}$$

Substituting $r = 8$, a typical value for a gasoline engine, and taking $\gamma = 1.4$ for air, we find $\eta \approx 0.56$. The actual efficiency is substantially lower than this for several reasons. A significant fraction of the work is converted back into heat as air is drawn into the cylinder; this loss is maximized when the throttle is closed, and provides a braking force when a vehicle is slowed by gearing down. The expansion and compression are not truly isentropic (see Footnote 10). There is heat loss and mechanical friction; furthermore, γ tends to decrease at high temperature and is also reduced by the presence of polyatomic substances such as gasoline and its reaction products (see Section 3.3).[11] Note that this is not a Carnot cycle, since the heat is not transferred at constant temperature, and even the ideal efficiency is much less than that of a Carnot engine working between the same maximum and minimum temperatures.

While the efficiency increases with compression ratio, it cannot be increased indefinitely, since T_a, the temperature at the end of the compression stroke, increases with r. If T_a is too high, the fuel-air mixture ignites too early, a phenomenon called *pre-ignition* or *pinking*.[12] This difficulty is avoided in the Diesel engine, where the fuel is injected at the end of the compression stroke, rather than being sucked in during the inlet stroke. The compression ratio is then limited by the maximum permissible pressure rather than by temperature, and high values of r (15–20) are common. Because of this limitation by the maximum permissible pressure, the Diesel engine is designed so that burning is ideally at constant pressure rather than constant volume (in practice, this ideal is only achieved in slow-speed marine engines). The cycle is shown schematically in Figure 3.10. Point d corresponds to the beginning of the

---

[11] It is interesting to note that if the electronic and vibrational internal degrees of freedom of the molecules contributed to the heat capacity, as required by the classical equipartition theorem (see Chapter 7 and Appendix G), $\gamma \to 1$ and $\eta \to 0$. The internal combustion engine could not exist in a classical world!

[12] The addition of tetra-ethyl lead to gasoline raises its ignition temperature by suppressing the formation of free radicals, and thus permits a higher compression ratio. Out of the hydrocarbons normally present in gasoline, octane has the highest ignition temperature, and the resistance of a particular fuel to pre-ignition is often expressed as an *octane rating*, which is the percentage of octane in an octane/heptane mixture having the same ignition temperature as the fuel being rated. Pinking is not to be confused with knocking, which is due to explosive detonation rather than controlled burning of the fuel-air mixture.

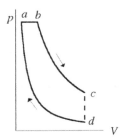

FIGURE 3.10   *p-V* diagram for the ideal Diesel cycle. The *T-S* diagram is qualitatively the same as for the Otto cycle (see Figure 3.9c).

compression stroke, with the cylinder filled with air at approximately atmospheric pressure ($p_d \approx 1$ bar) and temperature T_d. The air is compressed isentropically to point a, the volume being reduced by a factor $\frac{1}{r}$. The pressure increases to its maximum value p_a bar, and the temperature increases to T_a. From Equations (3.16) and (3.17), $p_a = r^\gamma p_d$, $T_a = r^{(\gamma-1)}T_d$. At this point, fuel is injected into the cylinder. Because r is large, T_a is high enough (typically ~900 K) to ignite the fuel-air mixture. As the piston starts down on the power stroke, the mixture burns, the rate of fuel injection being controlled to maintain an approximately constant pressure while the temperature rises to T_b. At point b the burning is over, and the air continues to expand isentropically until the end of the power stroke is reached at point c, when the exhaust valve opens. As in the case of the Otto cycle, the real open cycle can be modelled by a hypothetical closed one. The efficiency is more complicated to calculate than for the Otto cycle (see Example 3.3) and is in general higher because of the larger compression ratio. Another advantage of the Diesel engine is the fact that the power of a given engine can be increased when needed (at the cost of lower efficiency), without raising the maximum pressure, by prolonging the burning stage. In the Otto cycle, increasing the heat input raises the pressure in the cylinder. The main disadvantage of the Diesel is its dependence on achieving a sufficiently high temperature at the end of the compression stroke to ignite the fuel. If the air is initially too cold, the ignition temperature may not be reached. The modern fuel-injection gasoline engine combines the advantages of both the Diesel and Otto cycles.

Example 3.4: Power and Efficiency of a Diesel Engine

In an engine operating on the Diesel cycle, with compression ratio r, burning stops when the volume of the cylinder has increased to s times its minimum value. The ratio s/r is called the cut-off. Find the ideal efficiency, the temperature at the end of the burning stage, and the work done per cycle, per liter of cylinder volume.

We use the notation of Figure 3.10:

The heat input per kmole is $Q_1 = C_p (T_b - T_a) = \gamma C_V (T_b - T_a)$.

The heat rejected is $Q_2 = C_V (T_c - T_d)$.

At constant pressure, $T \propto V$, so that:

$$\frac{T_b}{T_a} = \frac{V_b}{V_a} = s.$$

In isentropic compression, $T \propto V^{(1-\gamma)}$, so that:

$$\frac{T_a}{T_d} = \left(\frac{V_a}{V_d}\right)^{1-\gamma} = r^{\gamma-1}.$$

In isentropic expansion,

$$\frac{T_c}{T_b} = \left(\frac{V_c}{V_b}\right)^{1-\gamma} = \left(\frac{s}{r}\right)^{\gamma-1}.$$

Hence,

$$\frac{T_b}{T_d} = sr^{\gamma-1}, \quad \frac{T_c}{T_d} = s^{\gamma}.$$

$$\eta = 1 - \frac{Q_2}{Q_1} = 1 - \gamma^{-1}\frac{T_c - T_d}{T_b - T_a} = 1 - \frac{s^{\gamma} - 1}{\gamma r^{(\gamma-1)}(s-1)}.$$

The temperature at the end of burning is $T_b = sr^{(\gamma-1)}T_d$.

The cylinder contains $\frac{V_d}{V_m}$ kmoles of air, where $V_m = \frac{R_0 T_d}{p_d}$ is the kmolar volume at pressure p_d and temperature T_d. Hence, the work done per cycle is:

$$W = \frac{V_d}{V_m}(Q_1 - Q_2) = \frac{V_d}{V_m}C_V T_d\left[\gamma r^{(\gamma-1)}(s-1) - (s^{\gamma} - 1)\right].$$

From Equation (2.18), $C_V = \frac{R_0}{\gamma-1}$. Hence, $\frac{V_d}{V_m}C_V T_d = \frac{p_d V_d}{\gamma-1}$, and

$$W = \frac{p_d V_d}{\gamma - 1}\left[\gamma r^{(\gamma-1)}(s-1) - (s^{\gamma} - 1)\right] \quad \text{J·liter}^{-1}\text{·cycle}^{-1}.$$

$V_d = 1$ liter $= 10^{-3}$ m^3, $p_d \approx 1$ bar $= 10^5$ Pa, and $\gamma = 1.4$ for air. A typical Diesel engine for a passenger car has $r = 20$ (corresponding to a pressure at the end of the compression stroke of about 65 bar) and $s \approx 2$. These give an ideal efficiency $\eta \approx 0.65$, $T_b \approx 2000$ K and $W \approx 750$ J/cycle. Hence, a 1 liter engine running at 3600 rpm (30 cycles per second in a four-stroke engine) would ideally generate 22.5 kW. In practice, the output before losses (the indicated horsepower) is higher than this because the pressure rises considerably during the burning of the fuel, so that the true cycle is intermediate between the Otto and Diesel cycles; on the other hand, the output is reduced by thermal and mechanical losses. For reasons dating back to eighteenth century Scotland, where steam was first replacing muscle as the primary source of power, the power of an engine is usually expressed in horsepower (1 HP = 750 W), so that this engine can be expected to generate about 30 HP per liter.

3.7 Refrigerators and Heat Pumps: The Vapor-Compression Cycle

As remarked earlier, a refrigerator or heat pump is a heat engine run backwards. However, this does not mean that any of the heat engines described previously could make useful refrigerators. For one thing, it is obviously desirable that the heat being extracted from the cold heat bath be taken in at constant temperature. This immediately rules out the Otto and Diesel cycles, and limits the working substance to a

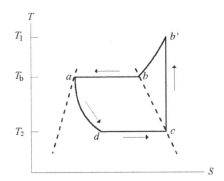

FIGURE 3.11 The quasi-ideal vapor-compression cycle (full lines). The dashed lines indicate the limits of the vapor-liquid coexistence region.

vapor condensible in the relevant temperature range (e.g., ammonia or freon[13] for refrigerators generating temperatures in the region of 0°C). The cycle, known as the vapor-compression cycle, resembles the superheated Rankine cycle (see Figure 3.8).[14] However, it differs from this cycle in one important respect. It is not practical to extract work by the isentropic expansion of a liquid, which is almost incompressible, so that one cannot simply reverse the compression phase of the Rankine cycle. Instead, the expansion phase of the cycle is achieved by allowing the high pressure liquid to expand irreversibly but adiathermally through a valve or porous plug, as in Figure 5.6. This process is called throttling. It will be shown in Chapter 5 that such a process is isenthalpic; that is, H is conserved. Because the process is irreversible, entropy is necessarily generated. For this reason the cycle is called *quasi-ideal*, to distinguish it from the ideal cycles in which the only entropy generation comes from neglected processes such as friction and heat loss. The quasi-ideal vapor-compression cycle is shown in Figure 3.11. Liquid at its boiling point T_b at high pressure (point a) is throttled isenthalpically to point d. Some liquid evaporates in this process, cooling the resulting vapor-liquid mixture to T_2, the desired low temperature. The mixture is then passed through the cooling coils of the refrigerator, in which the liquid entirely evaporates, extracting heat from the refrigerator compartment and reaching point c. The saturated vapor is then compressed isentropically to point b', its temperature rising to T_1. The hot high pressure vapor is then passed through a condenser, in which it cools to T_b and then condenses, giving up heat to the outside world and ending as saturated liquid at point a.

---

[13] Freon is the generic name for a group of organic compounds based on carbon, chlorine, and fluorine. Their low boiling points, toxicity, surface tension, and viscosity, combined with the fact that they are odorless and non-flammable, make them especially useful refrigerants. However, it has been found that the presence of these and related compounds (collectively known as CFC's) in the atmosphere is leading to the destruction of the ozone layer, which protects us from the sun's ultraviolet radiation, and their use is now being phased out by international treaty. Fortunately, less harmful substitutes, principally hydrofluorocarbons (HFCs), have been discovered; see C. Baird, *Environmental Chemistry*, 2nd ed. (Freeman, 1999), pp. 60–66, and A. Sekiya, "A continuing search for new refrigerants," *Chemtech* **26** (12), 44–48 (1996). See also F. Pearce, "A lucky escape," *New Scientist* **171** (2300), p. 5 (July 21, 2001), who points out that early signs of other environmental disasters could be going unnoticed.

[14] The Stirling cycle (see Footnote 5 of this chapter) has been used to generate temperatures in the range 4 K–190 K; see J. W. L. Kohler, "The Stirling refrigerator cycle," *Scientific American* **212** (4), pp. 119–127 (April 1965), and Haywood, Chapter 10.

Example 3.5: The Coefficient of Performance of a Quasi-Ideal Refrigerator

Using the same assumptions as those in Example 3.2, find the coefficient of performance (heat extracted per unit work in) for the quasi-ideal vapor-compression cycle shown in Figure 3.11.

Let the fraction of vapor at point d be f. Then the heat extracted is $Q_2 = (1 - f)L$. The heat discharged to the condenser is $Q_1 = C_p(T_1 - T_b) + L$. By the first law of thermodynamics, the work in is $W = Q_1 - Q_2$.

We calculate f as follows. Since L is assumed to be independent of T, the energy required to evaporate the fraction f is fL. Since the process is isenthalpic, this energy has been provided by cooling the working substance from T_b to T_2. Hence, $fL = C_p(T_b - T_2)$, where we have assumed for simplicity that the heat capacity of the liquid is the same as that of the gas. Hence $Q_2 = L - C_p(T_b - T_2)$, and the coefficient of performance is:

$$h = \frac{Q_2}{Q_1 - Q_2} = \frac{L - C_p(T_b - T_2)}{C_p(T_1 - T_2)}.$$

3.8 Envoi

The ideal and quasi-ideal cycles discussed in this chapter are only rough approximations to reality, and their chief usefulness to an engineer is to provide a criterion of success in designing a practical piece of equipment. In his book *Analysis of Engineering Cycles*, listed in the bibliography, the Cambridge University engineer R. W. Haywood talks of "Thermotopia, that idyllic land of the thermodynamicist in which all processes are reversible," and points out that "striving towards Thermotopian perfection ... without regard to social context can be as much the mark of a well-informed barbarian as that of an educated engineer." The reader is referred to Haywood's book for a thorough discussion of real as well as ideal cycles in a wide variety of heat engines and refrigerators.

3.9 Problems

3.1 (a) In the Otto cycle, what is the work W done per cycle per unit volume of the cylinder, if the maximum permissible pressure in the cylinder is p_m bar and the compression ratio is r? Assume that the fuel-air mixture is ideal, with fixed γ, and that it is initially at atmospheric pressure (1 bar) and at temperature T_0. Neglect all losses. Calculate the power and the peak temperature in the cylinder of a high performance 4-stroke, 1 liter engine running on the Otto cycle at 9000 rpm, given that $p_m = 100$ bar, the air is initially at 300 K, $\gamma = 1.4$, $r = 10$

(b) In some applications (e.g., in hand-held equipment such as a chainsaw), a large ratio of power to weight is more important than fuel efficiency. Show that for the Otto cycle, if the maximum pressure p_m (measured in bar) is large relative to atmospheric, the work W per cycle, for a given cylinder volume and given p_m, has a maximum when $r^{\gamma-1} \approx \gamma$, and find this output if the other parameters are the same as in (a). Note that in the transcendental equation for r, obtained by maximizing W, one term is much smaller than the others (if p_m is large), so that the equation can be solved by successive approximation. Explain qualitatively in physical terms why $W(r)$ has a maximum

3.2 (a) In Example 3.3, show that as $s \to 1$, the ideal efficiency of a Diesel engine approaches $1 - r^{(1-\gamma)}$, while if $s = r$, corresponding to 100% cut-off, and $r \gg 1$, the efficiency is $1 - \gamma^{-1}$. Compare these numbers for $r = 20$, $\gamma = 1.4$

(b) If the maximum permissible gas temperature in the Diesel engine of part (a) is 3000 K, find the maximum s, the maximum work done per cycle, and the efficiency.

3.3 In a gas turbine or jet engine the working substance is air, which is taken in at atmospheric pressure and compressed before heating. Heat is supplied at approximately constant pressure, as in the Diesel engine. However, the volume to which the air can expand is not limited by cylinder size, so that it can expand to atmospheric pressure while still doing useful work. In a jet engine, this expansion occurs through a nozzle so that a high exhaust velocity is achieved. The open side of the cycle can thus be replaced by cooling at constant pressure rather than constant volume as in the Diesel cycle. This cycle is called the Brayton cycle. Draw the p-V diagram and show that the ideal efficiency is $1 - s^{(1-\gamma)/\gamma}$, where s is the ratio of final to initial pressure in the compressor

3.4 In some high performance gasoline engines, the air exhausted from the cylinder is passed through a turbine to generate extra work. Draw the effective p-V diagram for the cylinder plus turbine, assuming that the cylinder operates on the Otto cycle, that the input pressure to the turbine is equal to the pressure in the cylinder at the end of the expansion stroke, and that the pressure after expansion through the turbine is atmospheric. Show that the overall ideal efficiency is increased from $1 - r^{(1-\gamma)}$ to $1 - \gamma \frac{s^{1/\gamma} - r}{s - r^{\gamma}}$, where r is the compression ratio and s is the ratio of the maximum pressure in the cylinder to atmospheric

3.5 In an air compressor, air (an ideal gas) is compressed isentropically. The air then is cooled to its original temperature at constant pressure by passing through tubes surrounded by flowing cold water
   (a) Draw the p-V and T-S diagrams for the air in this compressor (note that this is *not* a closed cycle). Draw the corresponding diagrams for a hypothetical reversible compressor generating the same pressure, but in which the compression is isothermal and reversible. Show that, for a given final pressure, the work input to the actual compressor per unit mass of air compressed is greater than that needed by the reversible compressor
   (b) What is the sign of ΔS_{air}, the overall change in entropy of the air being compressed? Is ΔS_{air} the same or different (per unit mass of air compressed) in the two compressors? What is the sign of the change in the entropy of the universe, in each case? Explain the difference

3.6 (a) A Carnot heat pump which takes in heat at temperature T_0 and delivers it at temperature T_2 is driven by Carnot engine operating between temperatures T_1 and T_0, where $T_1 > T_2 > T_0$. Find the coefficient of performance, defined as the ratio of heat delivered by the pump at T_2 to the heat supplied to the engine at T_1, if $T_1 = 600$ K, $T_2 = 300$ K, $T_0 = 273$ K
   (b) Now be more realistic. Suppose that the heat pump has a coefficient of performance $f_1 h_c$ and the engine has an efficiency $f_2 h_c$, where h_c and η_c are the Carnot values. What is the minimum value the product $f_1 f_2$ can have if this arrangement is to save energy (i.e., if it is to have a coefficient of performance greater than unity), for the temperatures given in (a)?
   (c) If $f_1 f_2 = 0.5$, what is the increase in entropy of the universe per kilowatt-hour of heat delivered (1 kW h = 3600 kJ)?

3.7 (a) Prove that the work done in any reversible closed cycle is given by $W = \oint p \, dV$; that is, the area enclosed by the loop in the p-V diagram
   (b) Show by direct calculation that the work done in the ideal Otto cycle is equal to the area enclosed by the loop in the T-S diagram.

3.8 (Thermal pollution.) A major problem for steam power stations is disposal of the heat discharged in the condenser. One common solution is to divert water from a river through the cooling coils of the condenser, which is why so many power stations are located on rivers. However, the resulting heating of the river water is often ecologically unacceptable. If the output power of the station is 1500 MW and the efficiency is 35%, find the rise in water temperature when the river flow is 50 m³/sec. What would the temperature rise be if the efficiency were raised to 45%, the generated power remaining the same? The heat capacity per unit volume of liquid water is 4 MJ K⁻¹ m⁻³

3.9 (Recycling heat.) Instead of being discharged to the atmosphere or to a river, the waste heat from a power station can be used to heat buildings if the power station is in a populated area. However, since this heat must be delivered at a temperature well above ambient, the efficiency of the power station is reduced by the higher discharge temperature

Consider two schemes for simultaneously generating power and supplying heat. In one, heat is supplied at temperature T_1 and discharged at T_2, the waste heat being used to heat buildings. In the other scheme, heat is supplied at T_1, but is discharged at T_0 ($T_0 < T_2$). The extra power generated is used to drive heat pumps that deliver heat to the buildings, taking in heat at T_0 and discharging it at T_2. Show that if Carnot engines and heat pumps are used throughout, the heat input required to generate a given amount of power (over and above the power needed to drive the heat pumps), and to supply a given amount of heat to the buildings, is the same for both schemes. Do this in two ways: (a) by considering each process in turn, and (b) by using the fact that in a reversible process, the net production of entropy must be zero. Thus, the choice between the two schemes cannot be made on the basis of ideal thermodynamic efficiency, but on the actual performance of heat pumps, on the losses involved in distributing heat from a central facility, capital cost, and other practical engineering considerations

3.10 Carnot cycles, the most efficient manner in which to perform thermodynamic work, are comprised of two adiabatic and two isothermal processes Describe which portion of a Carnot cycle could be linked to an absolute zero temperature bath. Discuss the implications for reaching absolute zero in the most efficient manner possible

3.11 Given Figure 15.2, show that for the Carnot cycle that $\delta S = S_2 - S_1 = Nk_B ln(V_a/V_b)$

3.12 Beer bubbles rise in a glass that is 15 *cm* tall. Would an adiabatic or isothermal argument make a better prediction for the change in the bubble size from the bottom of the glass to the top?

3.13 Calculate the work done in an isothermal expansion at room temperature that expands the gas to 10 times its original volume

3.14 Calculate the work done in an adiabatic expansion of the same gas as in Problem 13

II

The Fundamentals of Modern Thermodynamics

4

Macrostates and Microstates

The true logic of this world is the calculus of probabilities.

J. Clerk Maxwell

Nothing is proverbially more uncertain than the span of an individual human life. Little is more certain than the financial stability of life insurance companies.

Attributed to Sir Arthur Eddington

4.1 Gedanken

Imagine watching your child making a tower or wall of wooden blocks. The energy of the system grows as the work done in stacking each block adds to the potential energy of the higher and higher structure. The blocks are in thermal equilibrium with the air in the room, so with no heat flow, the change in energy of the system is solely due to the work done. Now the child knocks over the tower or wall of blocks. Some of the potential energy goes into the sound of the collapsing structure, but eventually the blocks lay in a scattered mess on the floor. So, where did the potential energy go? Into the kinetic energy of the falling blocks? Where did the kinetic energy go? Students are tempted to say that the rest of the energy is lost as *heat*. Hence, the process is irreversible. So, let's look at that. Let's imagine instead that I ask my child to very carefully unstack the structure of blocks and lay them on the floor (however the child likes). Do I get exactly the same final state? Probably not, as there's a myriad of ways to lay the blocks gently on the floor, even to do so in a very neat and orderly (and very un-childlike) way. So, the blocks are still not in *exactly* the same state they were if the child knocked them over (irreversibly). But in both cases the blocks are all at rest. So, the work I would have to do, for instance, to take the random distribution of knocked over blocks and reproduce the ordered state of unstacking the blocks is an amount of energy that has nothing to do with *heat*. It has everything to do with a sense of order. Entropy is not heat, it is something much more subtle and can only be truly understood by looking at the microscopic (quantum) scale.

4.2 Macrosystems, Microsystems, and Multiplicity

In thermodynamics we are concerned with the macroscopic world. The dictionary definitions of "macroscopic" and "microscopic" are, respectively, "visible" and "invisible" (to the naked eye), but in statistical mechanics the usage is somewhat different, though analogous. Here, *macroscopic* means having a very large number of components, so that it would require an impossibly large amount of information to specify its state precisely. An object can be quite small on a human scale and still be macroscopic by our

55

definition; for instance, a bubble of an ideal gas 1 μm in diameter (the size of a small bacterium), at room temperature and pressure, contains $\sim 10^7$ molecules. The state of a macroscopic system, or *macrosystem*, is specified by a small number of parameters which can be measured (at least in principle) by ordinary mechanical or electromagnetic means; for example, the properties of our bubble of gas may be specified by its pressure and volume. A state defined by such properties is called a *macrostate*. The instantaneous positions and velocities of the individual molecules of which the macrosystem consists are beyond our ability to specify.

We use *microscopic* to mean on an atomic (or molecular) scale; a microscopic system or *microsystem* has so few degrees of freedom that it is possible (at least in principle) to specify the state of all its components exactly: in classical physics by its position, velocity, and perhaps other variables such as angular momentum; in quantum physics by its quantum numbers. The microsystems in which we are interested may be particles, such as electrons, atoms, or molecules; we will also discuss quantum oscillators, such as the normal modes of vibration of the atoms in a solid, and spins (elementary magnets), which are also microsystems in our sense.

It is simplest to use quantum physics from the beginning, so that the state of each microsystem is defined by a set of numbers, called its *quantum numbers*. We confine our attention to stationary states, by which we mean states which are solutions of the time-independent Schrödinger equation. We shall see in Chapter 7 that even "continuous" quantum states, such as those of a free particle, can be treated as discrete states whose separation goes to zero as the volume within which the particle is confined goes to infinity. Some examples of microsystems and their quantum states are given in Appendix B.

A macrosystem is a collection of many such microsystems interacting with each other. At any instant, the macrosystem is in a particular *microstate*, which is a many-particle state defined by the quantum numbers of all the microsystems of which it consists. For example, if the positions and velocities of *all* the atoms in a monatomic gas were specified, we would have a particular microstate. In general, a very large number of microstates will correspond to a particular macrostate; in the case of the monatomic gas, there is an enormous number of possible combinations of molecular positions and velocities that are consistent with a given pressure and volume. It is neither necessary, possible, nor even desirable to know the quantum numbers of the microstates; all we need is certain average quantities. For any given macrostate, the most important of these quantities are its energy and its *multiplicity* Ω, which is the number of distinct microstates corresponding to a given macrostate.[1] In general, Ω is an extremely large number, but it is finite and can, in principle, be calculated. We shall find that Ω plays a fundamental role in statistical mechanics, and that the most important concept in thermal physics after energy, entropy, is proportional to its logarithm. In the next section we calculate Ω for a simple case. In most real systems, a direct calculation of Ω is beyond our powers; nevertheless, the number exists and can usually be derived by indirect means which we will discuss in later chapters.

The multiplicity Ω is important for the following reason. The fundamental assumption of statistical mechanics is that *a macrosystem in a given macrostate is equally likely to be in any of the Ω microstates which constitute that macrostate*. When we observe a particular macrostate in an experiment, we are averaging over all the microstates that it comprises, giving equal weight (often called the *a priori* probability) to each. In quantum theory, the assumption that all states of the same energy are equally likely to be occupied is best formulated, and can be justified, in terms of density matrix theory, which is beyond the scope

---

[1] Degeneracy is sometimes used instead of multiplicity, but the two words are not exactly synonymous. Degeneracy in quantum mechanics means the number of discrete quantum states with the same energy, while macrostates may be distinguished by properties other than energy (such as, in the case discussed in this chapter, their magnetic moment). In this book, we restrict the use of the word degeneracy further, to the number of discrete quantum states *of a single microsystem* that have the same energy. If we make this restriction, degeneracy is usually a small number, while multiplicity can be unimaginably large.

of this book.[2] In classical statistical mechanics, this assumption follows from the *ergodic hypothesis*. The ergodic hypothesis has gone through a number of metamorphoses,[3] but in essence states that a macrosystem passes rapidly through all the Ω possible microstates, spending an equal amount of time in each. Thus, a time average is the same as an average over the microstates. This hypothesis used to be called the ergodic theorem, until it was shown to be not strictly true. However, in a large system, the ergodic hypothesis is an excellent approximation to the truth and we will assume it. It follows that the statistical weight of a macrostate is its multiplicity Ω; by this we mean that, *other things being equal, the probability of observing a particular macrostate is proportional to Ω for that macrostate*.

4.3 The Spin 1/2 Magnet

To clarify the meaning of these rather abstract terms, we consider a concrete and simple example. It turns out that the workhorse of classical thermodynamics and kinetic theory, the ideal gas mentioned in Chapter 2, is not the easiest system to understand by statistical mechanics, and we postpone considering it until Chapter 10. Instead, we calculate the multiplicity Ω for a collection of weakly interacting spin 1/2 magnets (called spins; see Appendix B for a more detailed description), such as the atomic nuclei in a chunk of solid phosphorus. It should be understood that the spin 1/2 system is merely an illustration, chosen because it is one of the few systems in which the calculation of Ω is elementary. We shall find that the same results apply to a variety of physically different but mathematically analogous systems. While Ω is not directly calculable in most systems, in later chapters we shall see that such a calculation is in fact unnecessary. The purpose of this chapter is to demonstrate the physical meaning of Ω and to show that it can be expressed as a function of U, the internal energy of the system.

We assume a model in which each spin is at a fixed location, as is approximately true in a solid, but is free to orient in a magnetic field. We also assume that the energy of any one spin is independent of the state of any of the others,[4] but that they can exchange energy and so come into thermal equilibrium with each other. Each spin has only two quantum states, conventionally called up and down (see Appendix B), which form the simplest possible non-trivial quantum system.

Let us first consider the states of a single spin. The phosphorus nucleus, like the electron, has intrinsic angular momentum of $\hbar/2$. In atomic physics, it is convenient to measure angular momentum in units of $\hbar$, so that the nucleus is said to have spin 1/2. Associated with the spin is a magnetic moment μ, which is much less than that of the electron, but is readily measurable. The rules of quantum theory (see Appendix B) require that the component of the angular momentum along any particular axis be quantized. For one nucleus in a magnetic field B (whose direction defines this axis), this component can have only two values, $\pm 1/2$, corresponding to magnetic moments of $+\mu$ and $-\mu$, respectively, and there are two states, with energies $E_- = -\mu B$ and $E_+ = +\mu B$, as shown in Figure B.2. These up and down states will be denoted $\uparrow$ and $\downarrow$, respectively. Note that if $B = 0$, the two states have the same energy; that is, the $\uparrow$ and $\downarrow$ states are degenerate.

A microstate of a collection of such spins is defined by its particular configuration of spin orientations; for example, first spin $\uparrow$, second spin $\downarrow$, third $\downarrow$, and so on. Macrostates can be defined by their different values of the total magnetic moment M (which from now on we simply call the "moment"), which can be measured in the laboratory. In other words, all microstates with the same value of M belong to the same macrostate.

---

[2] See, for example, J. J. Sakurai, *Modern Quantum Mechanics*, 2nd ed. (Addison-Wesley, 1994), pp. 182–187.
[3] See S. G. Brush, *The Kind of Motion We Call Heat* (North Holland, 1976), Section 10.10.
[4] This is usually a valid assumption for atomic nuclei except at extremely low (microkelvin) temperatures.

Consider two spins. There are four possible configurations or microstates, shown in Figure 4.1, with the corresponding values of the total moment M.

Note that the two microstates with zero moment ($M = 0$) have the same energy even when $B \neq 0$. If we measured M, we would not be able to distinguish between these two microstates; they belong to the same macrostate, which has multiplicity $\Omega = 2$. Thus, the macrostates can be classified by their moment M and multiplicity Ω, as shown in Figure 4.2.

For three spins there are eight microstates, shown, with the corresponding values of M, in Figure 4.3. These microstates can be grouped into four macrostates, as in Figure 4.4.

A pattern is beginning to emerge; the values of Ω are the binomial coefficients (see Appendix D). We will now prove that this is true in general for the spin 1/2 system.

Suppose that we have N spins and that we want to know the multiplicity of a macrostate with $M = n\mu$, where n is the net number of up spins. To produce this value of M there must be $N_\uparrow$ spins up and $N_\downarrow$ down, where $N_\uparrow + N_\downarrow = N$ and $N_\uparrow - N_\downarrow = n$. The number of ways that $N_\uparrow$ up spins can be obtained is the number of different ways of selecting $N_\uparrow$ objects from N, which is given by the binomial coefficient $c(N, N_\uparrow)$ defined in Appendix D:

$$\Omega(N, n) = c(N, N_\uparrow) = \frac{N!}{N_\uparrow! N_\downarrow!} = \frac{N!}{[(N + n)/2]![(N - n)/2]!}. \tag{4.1}$$

$\Omega(N, n)$ is the multiplicity of the macrostate with moment $M = n\mu$, given that we have N spins. The total number of microstates, corresponding to all possible spin configurations, is [from Equation (D.2)]:

$$\Omega_t = \sum_n \Omega(N, n) = \sum_{N_\uparrow} c(N, N_\uparrow) = 2^N. \tag{4.2}$$

Physically, Equation (4.2) follows directly from the fact that each spin can be chosen to be ↑ or ↓, independently of all the others, so that adding one spin doubles the total number of microstates.

$$
\begin{array}{ccccc}
 & \uparrow\uparrow & \uparrow\downarrow & \downarrow\uparrow & \downarrow\downarrow \\
M \quad = & 2\mu & 0 & 0 & -2\mu
\end{array}
$$

FIGURE 4.1   Possible microstates of two spins.

$$
\begin{array}{cccc}
M \quad = & 2\mu & 0 & -2\mu \\
\Omega \quad = & 1 & 2 & 1
\end{array}
$$

FIGURE 4.2   Possible macrostates of two spins.

$$
\begin{array}{ccccccccc}
 & \uparrow\uparrow\uparrow & \uparrow\uparrow\downarrow & \uparrow\downarrow\uparrow & \downarrow\uparrow\uparrow & \uparrow\downarrow\downarrow & \downarrow\uparrow\downarrow & \downarrow\downarrow\uparrow & \downarrow\downarrow\downarrow \\
M \quad = & 3\mu & \mu & \mu & \mu & -\mu & -\mu & -\mu & -3\mu
\end{array}
$$

FIGURE 4.3   Possible microstates of three spins.

$$
\begin{array}{ccccc}
M \quad = & 3\mu & \mu & -\mu & -3\mu \\
\Omega \quad = & 1 & 3 & 3 & 1
\end{array}
$$

FIGURE 4.4   Possible macrostates of three spins.

In the absence of a magnetic field, all these 2^N microstates have the same energy, but one can in principle measure M in an infinitesimally small field, and thus distinguish between the macrostates of different n by their different magnetic moments.

According to our fundamental assumption that each microstate is equally probable, the probability[5] $P(N, n)$ of observing a given macrostate of moment $n\mu$ is proportional to the multiplicity $\Omega(N, n)$. Since the system must be in *some* microstate, $\sum_n P(N, n) = 1$. It follows that the probability of observing a moment $M = n\mu$, in the absence of a magnetic field, is given by the binomial distribution:

$$P(N, n) = \frac{\Omega}{\Omega_t} = \frac{N!}{2^N[(N+n)/2]![(N-n)/2]!}. \tag{4.3}$$

Equation (4.3) is identical to Equation (D.13), and there are many other physical systems which are mathematically equivalent to the spin 1/2 system. Some examples are given later in this chapter.

It is shown in Appendix D that $P(N, n)$ has a maximum at $n = 0$ and that when N is large, $P(N, n)$ is only significantly different from zero if $n \ll N$. Since N and n are large numbers, it is convenient to treat n as a continuous variable. It is shown in Appendix D that the probability density is then given to a good approximation by the Gaussian (Equation D.18):

$$P(N, n) \approx P(N, 0) \exp\left(-\frac{n^2}{2N}\right), \tag{4.4}$$

where $P(N, 0) = (2\pi N)^{-1/2}$. The standard deviation of this distribution is $\langle \Delta n^2 \rangle^{1/2} = N^{1/2}$ (Equation D.20), so that if N is a very large number the probability of finding a value of n, and hence a magnetic moment, that is significantly different from zero (relative to N) is vanishingly small. For example, in 1 gram of a typical solid there are $\sim 10^{22}$ atoms. If we take $N = 10^{22}$, we see that the magnetic moment of 1 gram of atoms with magnetic moment μ is most unlikely to deviate from zero by more than a few times $10^{11}\mu$, or about 10^{-11} of the maximum moment that would be achieved if all the spins were aligned, which is $N\mu$. The probability of observing a deviation of 1 part in 10^{10} is $\sim 10^{-22}$. Only for very small samples do fluctuations from the most probable value of M become at all significant.

Unless we are specifically concerned with fluctuations, we can assume that in a macroscopic system, the vast majority (essentially all) of the macrostates with the same energy have the same (zero) value of n and hence, zero magnetic moment. They are thus indistinguishable by a laboratory measurement and can be combined into one macrostate, with $\Omega \approx \Omega_t = 2^N$. We shall find in the next chapter that the significant quantity is $\ln \Omega$ rather than Ω itself. Note that $\ln \Omega$ is a number of order N. Hence, if N is large, an error in Ω by a factor much smaller than 2^N produces a negligible error in $\ln \Omega$.

Example 4.1: The Drunkard's Walk

The spin 1/2 problem is mathematically identical to the random walk (or "drunkard's walk") problem in one dimension. The random walk can be visualized as follows. A drunk starts from a lamp post and takes steps of length ℓ to the left or to the right at random with equal probability, using the curb as a guide, so that all his steps fall on a line. What is the probability that after N steps, he will (a) be back at the lamp post or (b) be at a distance $n\ell$ from it? Assume that N and n are even. Give approximate values of the two probabilities for $N = 60$, $n = 20$.

---

[5] Note that this is a probability, not a probability density [see the paragraph following Equation (D.8) in Appendix D].

(a) The different microstates of the spin 1/2 problem correspond to different orderings of the steps, and $|n|$, the net number of up or down spins, corresponds to the distance from the lamp post after N steps. (Note that the direction the drunk has gone is not specified, only the distance.) To get back to the lamp post, the drunk must make an equal number of steps to the right and to the left; that is, $n = 0$. The probability that $n = 0$ if N is even is from Equation (D.17):

$$P(N, 0) = 2^{-N}\Omega(N, n) \approx \left(\frac{2}{\pi N}\right)^{1/2}.$$

(b) In order to have reached a point at a distance $n\ell$ from the lamp post, the drunk must have made $|n|$ more steps in one direction than the other, so that when N is large the probability is:

$$P(N, |n|) = 2\frac{\Omega(N, n)}{2^N} \approx 2\left(\frac{2}{\pi N}\right)^{1/2} \exp\left(-\frac{n^2}{2N}\right),$$

from Equation (D.17).

For $N = 60$, $n = 20$, the probabilities are $(1/30\pi)^{1/2} \approx 0.1$ and $2(1/30\pi)^{1/2}e^{-400/120} \approx 0.007$, respectively.

A good way to get a feel for the distribution $P(N, n)$ is to run a random walk program, which sums N values of a number which can be either $+1$ or -1 at random, for various values of N (see Appendix D). Of course, these values of N are always going to be much smaller than those that occur in reality.

The result we have obtained for our simple spin 1/2 system is typical; whatever property is measured, if the sample contains a large number of atoms, the deviation of that property from its most probable value is likely to be very small. This does not mean that fluctuations can always be neglected, since some observable properties are sensitive to what happens on a small scale. An example familiar to everyone (whether they recognize it or not) is Rayleigh scattering, which is responsible for the fact that the sky is blue and not black. Light is scattered by refractive index fluctuations in the upper atmosphere. The refractive index depends on the air density averaged over a volume of the order of a cubic wavelength. Such a small volume contains relatively few molecules and its density fluctuations are significant (see Problem 6.13).

The calculation of Ω for the simple spin system discussed above is relatively elementary. For more complex systems, even for an ideal gas, the direct calculation is much more difficult. Fortunately, as pointed out above, we shall find in later chapters that such a calculation is usually unnecessary, since Ω can be obtained by indirect means.

So far we have only considered the case of zero field, when all the macrostates of different M have the same energy. In a non-zero field, the assumption of equal *a priori* probability for all the microstates can no longer be made, since the energies of the different macrostates are no longer equal. However, we can still assume equal probability for microstates of a given magnetic moment; that is, of a given n, since they all have the same energy. The energy of a single spin in a magnetic field B is $\pm\mu B$, where the sign is positive for a down spin and negative for an up spin. Hence,

$$U = (N_\downarrow - N_\uparrow)\mu B = -n\mu B. \tag{4.5}$$

For a given n, the multiplicity is given by Equation (4.1), so that one can combine Equations (4.1) and (4.5) to eliminate n and hence obtain $\Omega(U)$; that is, the multiplicity as a function of the internal energy.

If n/N is small, so that $|U| \ll -N\mu B$, Stirling's approximation, which was used to derive Equation (4.4), is valid and Ω is given by the Gaussian (Equation 4.4). However, one can imagine a situation where all the spins are lined up with the field; that is, $N_\downarrow = 0$, $n = N$, and $U = -N\mu B$. Equation (4.4) is then no longer valid and we must use the exact formula (Equation 4.1), which gives $\Omega = 1$.

To find the actual magnetization in a non-zero field, we need to formally introduce the concept of temperature. This is the topic of the next chapter.

4.4 Multiplicity and Disorder

Another problem that is mathematically identical to the spin 1/2 case is to find the probability of obtaining $N_\uparrow = \frac{N+n}{2}$ heads in N tosses of a fair coin; the probability is calculated exactly as before, and is given by Equation (4.3). If N pennies are laid out on a table with all heads up (so that $N_\uparrow = N$), and N is large, it is highly unlikely that such an arrangement would be achieved by random tosses of the coins. For some reason, the pennies must have been deliberately laid out that way; that is, the array is *ordered*. From Equation (4.1), $\Omega = 1$ for this arrangement. If, on the other hand, the number of heads in any arbitrarily chosen region of the array is close to the number of tails, one would say that the array is *disordered*, and the arrangement could well be the result of random tosses. This arrangement, like the spin system with $n \approx 0$, has a large multiplicity: $\Omega \sim 2^N$. This example illustrates the connection between multiplicity and disorder. In general we can say that small multiplicity implies order, while large multiplicity implies disorder in the sense that an arrangement with large Ω could be achieved by a random process, while one with small Ω is very unlikely to be.

The same connection between multiplicity and disorder can also be seen in the case of the magnetic system discussed earlier. Figure 4.5a shows a two-dimensional array of spins all pointing in one direction (as will happen in a sufficiently large magnetic field). They are perfectly ordered, and $\Omega = 1$, since there is only one way this can be done. Figure 4.5b, on the other hand, shows a random array of spins with net spin close to zero.[6] The spins are completely disordered and the orientation of one spin is entirely uncorrelated with the orientation of any other. The vast majority of possible spin arrangements have the same, or nearly the same, net magnetic moment as the arrangement shown, and $\Omega \sim 2^N$.

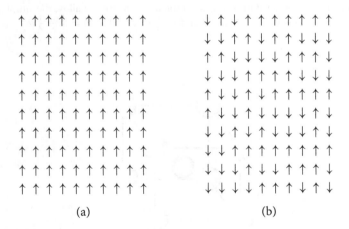

(a)   (b)

FIGURE 4.5   (a) Ordered and (b) disordered arrays of spins.

---

[6] Figure 4.5b was generated by randomly flipping (inverting) the spins in (a). The decision whether to flip or not was determined by whether a computer-generated random fraction was greater or less than 0.5.

Example 4.2: The Random Binary Alloy

To further illustrate the connection between multiplicity and disorder, we consider another system, physically more complex than the spin 1/2 magnet, but mathematically almost identical to it. This is a binary alloy—for instance, CuZn ("β-brass")—in which equal numbers of Cu (copper) and Zn (zinc) atoms are distributed over two possible types of atomic site. We call the two types of atom A and B. Their distribution may be ordered, A atoms occupying only one type of site and B atoms the other type; it may be completely disordered, A and B atoms being distributed at random; or it may be partially ordered. The crystal structure of β-brass is body-centered cubic (see Figure 4.6), each atom being at the center of a cube formed by eight others. In the disordered phase, the eight corners are occupied randomly by A or B atoms. In the ordered phase, every A atom is entirely surrounded by eight B atoms and vice versa. There can also be partial ordering, in which a fraction f of the sites is occupied by the "correct" atom. In the disordered phase, the fraction $f = 1/2$, while in the perfectly ordered phase, $f = 1$ or 0. These values of f, 1, or 0, are equivalent, since there is no physical distinction between the "a" sites and the "b" sites. Note that in this model the atomic sites are fixed in position; only the atom occupying each site varies.

We consider a sample with N sites, containing $\frac{1}{2}N$ A atoms and $\frac{1}{2}N$ B atoms. If an A atom is on an a site, or a B atom on a b site, we say that the site occupancy is correct. Let the fraction of correctly occupied sites be f. The multiplicity Ω is the product of two factors: the number of different ways that the $\frac{1}{2}fN$ sites occupied by A atoms can be chosen from the $\frac{1}{2}N$ a sites, and the number of ways that the $\frac{1}{2}fN$ sites occupied by B atoms can be chosen from the $\frac{1}{2}N$ b sites. The two factors are the same, so that the multiplicity is, from Equation (D.1):

$$\Omega(f) = \left[\frac{(N/2)!}{(fN/2)![(1-f)N/2]!} \right]^2. \tag{4.6}$$

We shall find in Chapter 15, where we discuss transitions between ordered and disordered phases in more general terms, that it is convenient to define an *order parameter* ψ such that $\psi = 0$ in the disordered phase and $\psi = \pm 1$ if ordering is complete. In the binary alloy, the simplest function of f with these properties is $\psi = 2f - 1$, so that $f = \frac{1}{2} + \frac{1}{2}\psi$, $(1-f) = \frac{1}{2} - \frac{1}{2}\psi$.

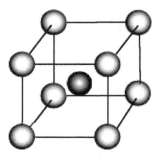

FIGURE 4.6 Perspective view of part of the crystal structure of β-brass (body-centered cubic). The structure is repeated indefinitely in all three dimensions. The two sets of atomic sites, a and b, are shaded light and dark, respectively. Each b (dark) site is at the center of a cube of a (light) sites. Similarly each a site is at the center of a cube of b sites. In the perfectly ordered phase, all the a sites are occupied by one type of atom and all the b sites by the other. In the disordered phase, each type of atom is randomly distributed over the a and b sites.

Substituting for f in Equation (4.6) and using Stirling's approximation (Equation D.4) to the factorial gives, after some algebra:

$$\Omega(\psi) = \frac{2^{N+2}}{\pi(1 - \psi^2)N}$$
$$\times \exp\left\{-\frac{N}{2}[(1 + \psi)\ln(1 + \psi) + (1 - \psi)\ln(1 - \psi)]\right\}. \tag{4.7}$$

If the alloy is only weakly ordered, so that $\psi \ll 1$, we can expand the logarithms, using Equation (A.5):

$$\ln(1 + \psi) \approx \psi(1 - \psi/2), \quad \ln(1 - \psi) \approx -\psi(1 + \psi/2),$$

so that:

$$\Omega(\psi) \approx \Omega(0)\exp\left\{-\frac{N}{2}\left[\psi(1 + \psi)\left(1 - \frac{\psi}{2}\right) - \psi(1 - \psi)\left(1 + \frac{\psi}{2}\right)\right]\right\}$$
$$= \Omega(0)\exp\left[-\frac{1}{2}N\psi^2\right], \tag{4.8}$$

which is identical to the expression for the multiplicity of the spin 1/2 magnet, with ψ substituted for n/N; compare Equations (4.8) and (4.4).

We shall need these results when we discuss phase transitions in Chapter 15.

4.5 Envoi: Multiplicity in Other Systems

We have introduced the concepts of macrostate and microstate, and of multiplicity Ω, which is the number of microstates contained in a given macrostate. We have used the spin 1/2 system as a model to illustrate just what is meant by Ω, and found that there are other systems that are mathematically identical to it. Another relatively simple system is a collection of identical harmonic oscillators, for which Ω is given by Equation (D.24). However, the majority of systems in which we are interested are much more complicated, and in general it is difficult, even impossible, to calculate Ω directly. Fortunately, as we shall see in the next two chapters, this calculation is rarely necessary, and when we need Ω we can find it to a good approximation by indirect means.

One class of systems which we discuss at length in later chapters is the perfect gas. A *perfect gas* is one in which the interactions between molecules can be neglected; to be more precise, a gas in which these interactions do not significantly affect the energy of a molecule. The ideal gas, mentioned in the last chapter, is a special case of the perfect gas. A molecule in a gas has a continuum of states, rather than the two available to a single spin. We shall see in Chapter 7 that the number of such states with a given energy increases rapidly with increasing energy, and also with increasing volume in which the molecule is free to move. The actual value of Ω in a gas is not easy to calculate, and we will return to this problem in Chapters 10 through 13.

4.6 Problems

4.1 What is meant by the terms *macroscopic* and *microscopic* in statistical physics? What are the orders of magnitude of the macroscopic and microscopic length scales (i.e., the physical sizes of the macrosystem and microsystem) in the following cases:

(a) A liter of air

(b) A living cell, made up primarily of macromolecules containing $\sim 10^4$ atoms each

(c) A galaxy, considered as a collection of stars whose internal structure is ignored?

4.2   Calculate the approximate energies, in eV and in joules, of the n'th level of:
      (a) An electron in a one-dimensional potential well of width 1 nm (10^{-9} m),
      (b) A carbon atom (mass 12 M_H) in the same well, and (c) a dust particle of mass 1 picogram
          (10^{-15} kg) in a well of width 1 μm (10^{-6} m). Assume infinitely high barriers to the well.

4.3   Consider two non-interacting hydrogen atoms in zero field as a system. What is the degeneracy
      (including electron spin) of (a) the electronic ground state and (b) the first excited state of the
      combined system? Neglect spin-orbit interaction and the relativistic separation of the 2s and 2p
      states. Note that in (b), only one of the two atoms is excited. Ignore the spin degeneracy of the
      proton.

4.4   Consider a system of three independent harmonic oscillators of equal frequency v. The macrostates
      of this system have energies (taking the energy of the ground state, $i = 0$, as zero) $E_i = ihv$, where
      i is a non-negative integer. Find the multiplicity $\Omega(3, i)$ of each of the four lowest macrostates
      ($i = 0, 1, 2, 3$) by enumerating all possible microstates with energy E_i. Compare your results with
      the general formula for the multiplicity of the i'th energy level of a system of N such oscillators,
      which is given in Equation (D.24):

$$\Omega(N, i) = \frac{(N + i - 1)!}{i!(N - 1)!}.$$

4.5   (a) In N successive tosses of a coin, what is the probability that a particular sequence of heads and
          tails, chosen in advance, will occur?
      (b) What is the probability that the number of heads and number of tails will be equal after 100
          tosses?
      (c) Suppose that you win 1 cent for a head and lose 1 cent for a tail, and that at some point in
          the game you have won 50 cents. What is the probability that after another 50 tosses your
          opponent will have got exactly his or her money back? Why is this probability different from
          the result in (b)?

4.6   Many atoms (and some molecules: e.g., O_2) have a spin of 1, so that there are three possible spin
      states, corresponding to $S_z = +1, 0, -1$, where S_z is the component of the spin along some chosen
      axis (see any textbook of quantum mechanics). Consider a collection of N such atoms, where
      $N \gg 1$, and assume that interactions between these atoms can be neglected, so that the orienta-
      tion of the spin of any one atom is independent of the orientation of all the others. Let the number
      of atoms in the states $S_z = +1, 0, -1$ be N_+, N_0, N_-, respectively, where $N = N_+ + N_0 + N_-$
      (a) Starting from the multinomial distribution (Equation D.2) and using Stirling's formula in its
          crude form (Equation D.6), show that in zero magnetic field the most probable value of each
          number, N_+, N_0, and N_-, is $N/3$
      (b) If we put $N_+ = \frac{N}{3} + x, N_- = \frac{N}{3} + y$, where $x, y \ll N$, show that $\Omega(x, y) \approx \Omega(0, 0)e^{-\phi(x,y)}$,
          where ϕ is a positive definite quadratic function of x and y, and find $\phi(x, y)$. Make a rough
          perspective sketch of $\phi(x, y)$.

4.7   (Unbranched polymer.) A polymer such as rubber can be regarded as an unbranched chain of
      atoms. The length of the chain is determined by the way the bonds joining the atoms are oriented
      relative to each other. The following one-dimensional model,[7] while very much oversimplified,
      accounts qualitatively for the essential mechanical and thermal properties of such a polymer. The
      bonds are assumed to have equal length, and the angle between them to be restricted to 0° or 180°,
      so that each bond can only be oriented to the right or to the left. The energy is assumed to be
      independent of bond angle. The polymer molecule is assumed to be embedded in a matrix so that
      it cannot rotate

---

[7] See L. R. G. Treloar, *The Physics of Rubber Elasticity*, 3rd ed. (Oxford, UK, 1975). A somewhat more realistic model is
    considered in Problem 7.15.

(a) If there are N links of length ℓ_0, and N is even, find the multiplicity of the macrostate of the polymer with length ℓ, where ℓ/ℓ_0 is an even integer. For simplicity, take ℓ as the distance between the unattached ends of the two end links

(b) Hence, show that if N is large and $\ell \ll N\ell_0$,

$$\Omega(\ell) = 2\Omega(0) \exp\left[-\frac{\ell^2}{2N\ell_0^2}\right]. \tag{4.9}$$

We will use Equation (4.9) in the next chapter.

4.8 (a) Starting from Equation (D.15), find Ω for a system of N spins in which all the spins but one are aligned, so that $n = N - 2$. Compare your result with the exact value

(b) Starting from Equation (4.7), find Ω for an almost perfectly ordered binary alloy in which just one A and one B atom are wrongly sited, and compare your result with the exact value.

4.9 I make a bet with my students that if any two of them have the same birthday then I win, but if anyone in the class has the same birthday as me, then they win: (a) Show the probabilities for these two events are different and favor the instructor and (b) determine the number of students such that the instructor has better than 50/50 chance of winning

4.10 An instructor in a 50 seat classroom tells you that there 45 students in the class. What does that tell you about how the students are arranged in the classroom? What does that tell you about the nearest neighbors of any randomly selected student in the class? How do the answers change if the room is set up at 5 rows of 10 chairs versus a large circle of 50 chairs?

4.11 A coin is tossed repeatedly with a probability of heads, p, and a probability of tails, $1 - p$: (a) What is the probability that I can flip N heads in a row? (b) What is the probability that the next flip after N tosses is tails? (c) An "undoctored" coin has an equal probability of heads or tails. If the coin is weighted such that $p = 0.55$, ten percent better than random, evaluate the difference in probabilities for the outcomes in (a) and (b).

4.12 Make a table of coin flips for $N = 2, 4, 6, 10, 20, 40, 60, 100$ coin flips. Record the measured probability of heads and compare it to the theoretical probability of 0.5. Compare your results with a classmate by keeping track of the sequence HTHTTH.. etc. Notice that if two of you find the same probability as N becomes large you are very unlikely to have recorded the identical sequence. Comment on the difference between a microstate and a macrostate.

5

Entropy, Free Energy, and the Second Law of Thermodynamics

First law: You can't win, you can only break even.
Second law: You can break even only at absolute zero.
Third law: You can't reach absolute zero.

An anonymous student

Perhaps the greatest triumph of probability theory within the frame work of nineteenth century physics was Boltzmann's (statistical) interpretation of the irreversibility of thermal processes.

Ernst Nagel

5.1 Gedanken

I recently was teaching my young son that every square is a special sort of rectangle, but not every rectangle is a square. We need to step back carefully from our macroscopic understanding of entropy that erroneously leads us to think of it synonymously as heat. Heat flow carries with it entropy, but not every entropy change involves heat. Recall our Carnot engine as a set of processes that allow us to extract work from (at least) two heat baths of different temperatures. We reverse the Carnot cycle and can instead just as well think of doing work to return ourselves to the same initial state of the system described by the state variables $U, P, V, S, \& T$. Work must be done to change the entropy of the cycle back to its initial value. I walk into a room of temperature T with a box of argon atoms, also at temperature T. The box of argon atoms (gas) is already in thermal equilibrium with the room. I open the box and allow the argon atoms to mix with the air in the room. I close the box—the system (of the room plus the box) is *not* in its initial state. To restore the system to its initial state in phase space, I would have to walk through the room with molecular tweezers and capture every argon atom and air molecule and re-order the system with all the argon in the box and all the air in the room. Approximating both gases as ideal gases, the energy of the system never changed because the whole process occurred at temperature T. So the work I did of capturing and sorting the atoms didn't go into changing the energy of the system, but to restoring the entropy of the system to its initial state value. There was never a flow of heat in the system, but there was a change in entropy.

5.2 The Condition for Thermal Equilibrium

A macrosystem, or system for our present purposes, is a collection of microsystems which can exchange energy between each other on a time scale that is short compared with the time taken to make a measurement. An example is a gas, in which molecules (the microsystems) are continually colliding and exchanging energy. Suppose that we have two such macrosystems, Σ_1 and Σ_2, initially isolated from each other, each in a well-defined macrostate. The macrostates of the two systems have energy U_1 and U_2 and multiplicity Ω_1 and Ω_2, respectively. If the energy of a system is changed, so is its multiplicity, and there exist functional relationships $\Omega_1 = \Omega_1(U_1)$, $\Omega_2 = \Omega_2(U_2)$. For the moment, we need not calculate these functions, but merely recognize that they exist, and we assume that they are analytic. We assume that the combined system remains isolated from all outside influences, and that no work is done on or by either system (this means that the volumes remain fixed), so that the total energy $U = U_1 + U_2$ is constant.

What happens when we bring the two systems into thermal contact with each other, as shown schematically in Figure 5.1? In this figure, the thick line represents an insulating wall isolating the two systems from their surroundings. The thin line represents a diathermal (i.e., heat conducting) wall that permits them to exchange heat, which is the name that we give energy that is transferred in this manner (see Chapter 2). We assume that the thermal contact has no effect on the microstates of either system; this assumption is reasonable since we know that energy can be exchanged through a diathermal wall over macroscopic distances, while the interactions that determine microstates operate on the atomic scale of distance. Then, for each microstate of Σ_1, there are Ω_2 microstates of Σ_2, so that the multiplicity Ω of the combined system is the product of the separate multiplicities:

$$\Omega = \Omega_1 \Omega_2. \tag{5.1}$$

The fundamental hypothesis of statistical mechanics, stated in the previous chapter, is that the probability of observing a given macrostate, relative to other macrostates of the same energy, is proportional to its multiplicity Ω. It follows that *if one starts with the system in an improbable macrostate, it will evolve to the most probable one, while the reverse process cannot occur, so that in equilibrium the multiplicity is maximized.* This is, in essence, the second law of thermodynamics. The final (equilibrium) macrostate of the combined system is the state that maximizes the product in Equation (5.1), subject to the condition that the total energy U is constant.

For Ω to be a maximum, $d\Omega = 0$. By the product rule,

$$d\Omega = \Omega_1 \, d\Omega_2 + \Omega_2 \, d\Omega_1.$$

Dividing by $\Omega = \Omega_2\Omega_1$ and using $d\ln(x) = \frac{dx}{x}$, we see that equilibrium is reached when:

$$d\ln\Omega = d\ln\Omega_2 + d\ln\Omega_1 = 0. \tag{5.2}$$

Equilibrium is established by transfer of energy (in the form of heat) from one system to the other, and is reached when the rate of energy transfer from Σ_1 to Σ_2 is equal, on average, to the rate of transfer from Σ_2 to Σ_1, so that the net transfer is zero.

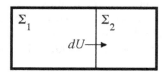

FIGURE 5.1 Two macrosystems in thermal contact with each other but isolated from their surroundings. An amount of energy dU is transferred from one to the other. The thick line represents an insulating wall and the thin line a diathermal wall.

Because, by our fundamental hypothesis, Ω is a maximum in equilibrium, if we make a small change—for example by transferring an infinitesimal amount of energy dU from System 1 to System 2 (see Figure 5.1)—Ω does not change to first order in dU. Ω_1 and Ω_2 will (in general) change, but the changes are related by Equation (5.2). Energy is conserved, so that the increase in the energy of Σ_2, due this transfer, must be equal to the decrease in the energy of Σ_1; that is $dU_2 = -dU_1 = dU$. These changes in U_1 and U_2 produce corresponding changes in $\ln \Omega_1$ and $\ln \Omega_2$, so that Equation (5.2) can be written:

$$d \ln \Omega = \frac{\partial \ln \Omega_2}{\partial U_2} dU_2 + \frac{\partial \ln \Omega_1}{\partial U_1} dU_1 = \left[\frac{\partial \ln \Omega_2}{\partial U_2} - \frac{\partial \ln \Omega_1}{\partial U_1} \right] dU = 0, \quad (5.3a)$$

where the partial derivatives are taken at constant volume. Hence the condition for thermal equilibrium is:

$$\left(\frac{\partial \ln \Omega_2}{\partial U_2} \right)_V = \left(\frac{\partial \ln \Omega_1}{\partial U_1} \right)_V. \quad (5.3b)$$

We will examine the significance of this result in Section 5.4, where we shall see that it enables us to formally define an absolute scale of temperature.

5.3 Entropy and the Second Law

We see from Equations (5.2) and (5.3) that the quantity of interest is $\ln \Omega$ rather than Ω itself, so we need a name for it. Further reasons for using $\ln \Omega$ rather than Ω will become clear later. Because $\ln \Omega$ is a very large number ($\ln \Omega \simeq N$, the number of particles in the system, as we saw in Chapter 4), it is convenient to multiply it by a small constant called *Boltzmann's constant* k_B, whose numerical value we will determine later. Following Boltzmann,[1] we define the *entropy* of a system in a macrostate of multiplicity Ω by:

$$\boxed{S \equiv k_B \ln \Omega.} \quad (5.4)$$

Besides convenience, there is a more fundamental reason for including the constant k_B in Equation (5.4), rather than simply writing $S = \ln \Omega$. We shall see later that the *thermodynamic limit* of statistical mechanics, in which all fluctuations vanish, corresponds to simultaneously letting $N \to \infty$ and $k_B \to 0$, Nk_B remaining finite. Thus, we can distinguish an expression for a statistical quantity such as the average fluctuation, which vanishes in the thermodynamic limit, from one for a thermodynamic quantity such as U, which remains finite, by the presence of k_B in the expression without a compensating factor N. For example, we saw in Chapter 4 that $\ln \Omega \sim N$, so that $S \sim Nk_B$ and is a thermodynamic quantity.

We saw in Section 5.2 that for systems that do not affect each other's microstates, Ω is multiplicative, so that S is additive. It follows that entropy S is proportional to the size of the system; that is, it is an extensive property like energy or volume. This is another reason for using $\ln \Omega$ rather than Ω itself.

Clausius'[2] aphoristic formulation of the second law of thermodynamics, "die Entropie striebt einem maximum zu;" that is, "The entropy (of an isolated system) tends to a maximum," is a concise expression of the fundamental assumption that equilibrium is achieved when the most probable macrostate of an isolated system is reached. We shall see in later chapters that all the other forms of the second law, which are expressed in terms of the impossibility of certain processes, follow from this formulation.

---

[1] Ludwig Boltzmann (Austrian physicist, 1844–1906) proposed this statistical definition of entropy in 1872. He rightly considered this the most important of his many achievements and asked that the formula "$S = k \log W$" (W was his symbol for multiplicity) be engraved on his tombstone.

[2] Rudolf Clausius (German physicist, 1822–1888) is generally regarded as the founder of thermodynamics as a rigorous discipline [See W. H. Cropper, "Rudolf Clausius and the road to entropy," *Am. J. Phys.* **54**, 1069–1074 (1986)].

Example 5.1: A Small, Countable Spin System

Consider two magnetic systems of four spins each, in a magnetic field. The spins in System 1 are completely aligned; in the notation of Chapter 3, $N_\uparrow - N_\downarrow = 4$. The spins in System 2 are completely random, so that $N_\uparrow = N_\downarrow$. The two systems are brought together so that energy is exchanged between them. Energy conservation requires that the magnetic moment, and therefore the net spin, is unchanged, so that in equilibrium we have a system of eight spins with $N_\uparrow - N_\downarrow = 4$. What is the entropy of the combined system before and after they are brought into contact? Is this process permitted by the second law of thermodynamics?

The numbers are too small to use the Gaussian approximation discussed in Chapter 4, and instead we use the exact expression Equation (4.1) for the multiplicity. For System 1, $\Omega_1 = \frac{4!}{4!} = 1$, while for System 2, $\Omega_2 = \frac{4!}{2!2!} = 6$. Hence, from Equation (5.1), the multiplicity of the combined system before contact is $\Omega_{\text{init}} = \Omega_1\Omega_2 = 6$ and the initial entropy is $S_{\text{init}} = k_B \ln 6$.

After equilibrium is established, we have a single system with $N = 8$ and $n = 4$, so that $\Omega_{\text{final}} = \frac{8!}{6!2!} = 28$.

Hence, the entropy $S_{\text{final}} = k_B \ln 28$. Since $S_{\text{final}} > S_{\text{init}}$, this process is permitted by the second law.

5.4 Absolute Temperature

We will now show that the zero'th law of thermodynamics, which states that if two bodies are separately in thermal equilibrium with a third they are in equilibrium with each other, follows from the condition for thermal equilibrium (Equation 5.3b).

With S defined by Equation (5.4), Equation (5.3a) becomes:

$$dS = \left[\left(\frac{\partial S_2}{\partial U_2}\right)_V - \left(\frac{\partial S_1}{\partial U_1}\right)_V\right] dU, \tag{5.5}$$

and Equation (5.3b) becomes:

$$\left(\frac{\partial S_2}{\partial U_2}\right)_V = \left(\frac{\partial S_1}{\partial U_1}\right)_V. \tag{5.6}$$

Furthermore, if we were to bring each system separately into contact with a third, System 3, and reach equilibrium in each case, we would have corresponding relationships:

$$\left(\frac{\partial S_1}{\partial U_1}\right)_V = \left(\frac{\partial S_2}{\partial U_2}\right)_V = \left(\frac{\partial S_3}{\partial U_3}\right)_V. \tag{5.7}$$

Equation (5.7) shows that equality of $\left(\frac{\partial S}{\partial U}\right)_V$ is a transitive relation, so that this equation is equivalent to the zero'th law of thermodynamics (see Section 2.1). It follows that $\left(\frac{\partial S}{\partial U}\right)_V$, or any monotonic single-valued function of it, defines a scale of temperature, which is absolute in the sense that the definition is independent of the properties of any particular substance.

What function of $\left(\frac{\partial S}{\partial U}\right)_V$ should we choose for our "absolute temperature" scale? We note that for dU positive in Equation (5.5) (i.e., when energy in the form of heat is transferred from System 1 to System 2), S increases if and only if $\frac{\partial S_1}{\partial U_1}$ is *less* than $\frac{\partial S_2}{\partial U_2}$. Since, by our fundamental hypothesis, change can only occur in the direction that increases S, we see that heat is transferred from the system with the lower $\frac{\partial S}{\partial U}$ to the higher one. Since we want our temperature scale to be such that higher temperature means hotter, in the sense that heat moves from the hotter to the colder system, we define our scale of temperature T (the Kelvin[3] scale) by:

$$T \equiv \left(\frac{\partial S}{\partial U}\right)_V^{-1} = \left(\frac{\partial U}{\partial S}\right)_V.$$ (5.8)

Besides the volume, the number of particles in the system is kept constant in Equation (5.8), as is any parameter (such as the magnetic field) whose change can cause work to be done on or by the system. T is an intensive quantity, like pressure; it does not depend on the size of the system. We denote the absolute scale of temperature by T to distinguish it from the empirical scales denoted by θ in Chapter 2, whose definition depends on the properties of the particular substance chosen.

Let us pause to consider the question of dimensions. If k_B is regarded as merely a dimensionless conversion factor (like the factor $\pi/180$ that converts degrees to radians), S is also dimensionless and T has the dimensions of energy. However, it is often convenient to think of temperature as a dimension itself, in which case k_B and S have the dimensions of energy divided by temperature. For an illuminating discussion of when and why temperature can be treated as a dimension of its own, see the discussion of the "Rayleigh-Riabouchinsky Paradox," by G. B. West in Cooper and West, page 2. West, following Rayleigh, shows that temperature can be usefully treated as a separate dimension if we confine our attention to macroscopic phenomena and ignore the statistical nature of thermodynamic quantities; that is, if the thermodynamic limit $k_B \to 0$ is valid [see the discussion of Equation (5.4) above]. Thus, it is essential to write S in the form of Equation (5.4), rather than simply as $\ln \Omega$ (which corresponds to taking $k_B = 1$), if we are to retain the ability to use dimensional analysis to the fullest.

Example 5.2: Temperature and Entropy of a Paramagnet

The choice of Equation (5.8) to define temperature makes sense in terms of our experience of the effect of temperature on the occupancy of energy levels (these ideas will be made quantitative in Chapter 6). At low temperature, we expect only a few low lying energy levels to be occupied, while at high temperature, the available energy should be spread more or less equally over many levels. To see that this is the case for our definition of T, consider the collection of N elementary magnets (spins) discussed in the last chapter. Suppose that we apply a magnetic field B, so that a macrostate of moment $M = n\mu$ has energy $U = -n\mu B$, where we use the notation of Chapter 4: $n \equiv N_\uparrow - N_\downarrow$. First, consider the case where all energy levels are equally occupied, so that a spin is equally likely to be up or down and $n = 0$. $\Omega(n)$, and hence $S(U)$, is at its maximum when $n = 0$, so that $\left(\frac{\partial S}{\partial U}\right)_B \to 0$ and $T = \left(\frac{\partial U}{\partial S}\right)_B \to \infty$ as $n \to 0$.

Now consider the case where only the lowest energy levels (illustrated in Figure 5.2) are significantly occupied, so that practically all the spins are up; that is, $N_\uparrow \to N, N_\downarrow \to 0, n \to N$. Hence, the macrostates have large n, and we confine our attention to the lowest two: the macrostate with $n = N$ (all spins aligned) and the next, with $n = N - 2$ (one spin flipped). When $n = N$,

n	U
$N - 4$	$(4 - N)\mu B$
$N - 2$	$(2 - N)\mu B$
N	$-N\mu B$

FIGURE 5.2 Lowest states of an N-spin system in a magnetic field B.

---

[3] Lord Kelvin (Scottish physicist, 1824–1907; before his ennoblement, William Thomson) formulated the laws of thermodynamics independently of Clausius, but from an engineering perspective (see Chapter 3).

$U = -N\mu B$, and $\Omega = c(N, N) = 1$, so that $S = 0$ in the lowest macrostate. When $n = N - 2$, $U = (2 - N)\mu B$, $\Omega = c(N, N - 1) = N$, $S = k_B \ln N$. The change in U on going from the lowest macrostate to the next lowest is $\Delta U = 2\mu B$, while the change in S is $\Delta S = k_B \ln N$. Both of these changes are small, so that $\left(\frac{\partial U}{\partial S}\right)_B \approx \frac{\Delta U}{\Delta S} = \frac{2\mu B}{k_B \ln N}$. Putting in numbers for nuclear spins ($\mu \sim 5 \times 10^{-27}$ J/T, $B \sim 10$ T, $k_B = 1.38 \times 10^{-23}$ J/K, $N \sim 10^{22}$), we find $T \sim 10^{-5}$ K, a very low temperature, as expected. In a one-dimensional array of spins, it should be noted, that up and down are established by the direction of the spin relative to the magnetic field and there is, by this, a natural symmetry in the system that we can take advantage of in the next section.

5.5 Negative Temperature

Suppose that we can somehow make $n < 0$ in our spin system. Such a situation cannot occur spontaneously, since it requires higher energy levels to be more occupied than lower ones, but it can be created artificially in the laboratory. The simplest method conceptually is as follows. We put the sample (i.e., the material containing the spins) into a large magnetic field at a very low temperature, so that the spins line up parallel to the field. We then remove it from the field, turn it over, and put it back so that the spins are now aligned anti-parallel to the field; that is, n is negative. The operation must be carried out in a time short compared with the spin relaxation time (the time that the spins take to exchange energy with their surroundings) so that the spins do not have time to re-orient. For nuclear spins at low temperature, this relaxation time can be very long (seconds to hours), while transfer of magnetic energy from one spin to its neighboring spins is relatively fast, so that equilibrium is quickly established amongst the spins and their entropy and temperature can be defined. It is easy to see from the form of $\Omega(n)$ that $\left(\frac{\partial S}{\partial U}\right)_B < 0$ if $n < 0$, so that T is now negative. There is nothing unphysical about this, although it cannot happen in a system (such as a gas) with no upper limit to its internal energy, since infinite energy would be required. Substituting $\frac{\partial S_1}{\partial U_1} > 0$, $\frac{\partial S_2}{\partial U_2} < 0$ in Equation (5.5), we see that the entropy would decrease if energy were transferred from System 1 to System 2, so that the second law requires that heat will always flow from an object with $T < 0$ to one with $T > 0$. Hence, negative temperatures are *higher* than positive ones. Since, as we shall see, it is impossible to actually reach absolute zero (the third law of thermodynamics), one can only go from a negative temperature to a positive one by cooling through $T = \infty$. In a sense, T^{-1} is a more fundamental quantity than T, since it goes smoothly through zero and cannot be infinite, $T = 0$ being inaccessible.

5.6 Entropy and Heat: Isentropic and Isothermal Processes

The macrostate of any system (here abbreviated to "state") is defined by a certain small number of variables. For instance, the state of a non-magnetic fluid of fixed mass and composition can be defined by two variables p and V. S and U can be written as functions of p and V:

$$U = U(p, V) \tag{5.9}$$

$$S = S(p, V). \tag{5.10}$$

Let us solve Equation (5.10) for p (i.e., we express p as a function of S and V), and substitute for p in Equation (5.9). We obtain U as a function of S and V, $U(S, V)$, whose differential can be written:

$$dU = \left(\frac{\partial U}{\partial S}\right)_V dS + \left(\frac{\partial U}{\partial V}\right)_S dV. \tag{5.11}$$

By the second law of thermodynamics, the entropy of an isolated system can never decrease. It follows that the change of entropy of such a system in a reversible process must be zero; if it were positive when the process goes in one direction, it would have to be negative in the other. Hence, a reversible process in an isolated system must be at constant entropy (i.e., $dS = 0$) and is called an *isentropic* process.[4]

We can now identify the two partial derivatives in Equation (5.11). By definition, $\left(\frac{\partial U}{\partial S}\right)_V = T$ (Equation 5.8). We found in Section 2.3 that for a reversible process in an isolated system, $dU = -p\,dV$. Since such a process is isentropic, this means that:

$$\left(\frac{\partial U}{\partial V}\right)_S = -p. \tag{5.12}$$

Hence, for a general process in which work is done and heat is transferred, Equation (5.11) can be written:

$$\boxed{dU = T\,dS - p\,dV.} \tag{5.13}$$

In a magnetic system, we add a term $-\mu_0 M\,dH$ to the right-hand side of Equation (5.13); see Equation (2.16).

Equation (5.13) is a statement of the first law of thermodynamics entirely in terms of state functions and is perhaps the most important equation in thermodynamics. It is worthwhile to step back and consider what it means. We first consider a reversible process. Think of the differentials as very small increments; when V changes by so small an amount that p does not change appreciably, the work done on the gas in a reversible process is $đW = -p\,dV$. Since the first law requires that the change in internal energy is equal to the work done on the gas plus the heat put in; that is, $dU = đW + đQ$, it follows that $T\,dS = đQ$, or:

$$\boxed{dS = \frac{đQ}{T},} \tag{5.14}$$

in a reversible process. This is the thermodynamic definition of entropy, which Clausius[5] introduced in 1854, long before Boltzmann gave his statistical definition $S \equiv k_B \ln \Omega$.

Now consider an irreversible process. We derived Equation (5.13) by considering only reversible processes, but because the equation contains only state functions, it must hold for any process whatever. However, as pointed out in Chapter 2, only in a reversible process can $-p\,dV$ be identified with work $đW$, so that *only in a reversible process can $T\,dS$ be identified as the heat input $đQ$*, and only then does Equation (5.14) hold. The advantage of our definition of entropy over the thermodynamic one (Equation 5.14) is that the physical meaning of entropy is now clear; it is defined by Equation (5.4) and is proportional to the logarithm of the number of microstates in the given macrostate. In thermodynamics, entropy appears as a mathematical necessity, needed if we are to write the first law in terms of state functions, but whose physical meaning is by no means clear.[6]

To illustrate the distinction between the state function S and the heat transferred Q, consider a reversible process in which a cylinder with perfectly diathermal walls, and containing an ideal gas, is immersed in a heat bath at temperature T. (A *heat bath* is a system so large that its temperature is unaffected by transfer of energy to and from the object in question; see Chapter 6 for a more precise definition.) The gas expands isothermally and reversibly from volume V_1 to V_2, doing work on a piston as it does so. The internal energy U of an ideal gas is a function only of temperature, so that $dU = 0$ in an isothermal change, and Equation (2.6) gives:

$$đQ = -đW = p\,dV.$$

---

[4] For reasons that will become clear in the next section, a reversible process in an isolated system can also be (and usually is) called an *adiabatic* process.

[5] See the reference in Footnote 2 in this chapter.

[6] See Appendix E, in particular the reference in its Footnote 4.

Hence, $dS = \frac{p\,dV}{T}$ in, and the entropy change of the gas in this reversible process is:

$$\Delta S = \int_{V_1}^{V_2} \frac{p(V, T)}{T}\, dV.$$

The relation $p = p(V, T)$ is called the *equation of state* of the gas. It will be shown in Chapter 10 that for an ideal gas, the equation of state is $p = nk_B T = \frac{Nk_B T}{V}$, where N is the number of gas molecules in volume V, and n is the particle density N/V. Hence,

$$\Delta S = Nk_B \int_{V_1}^{V_2} \frac{dV}{V} = Nk_B \ln\left(\frac{V_2}{V_1}\right). \tag{5.15}$$

Note that ΔS depends, as it must since it is a state function, *only* on the initial and final states. During the expansion, the heat bath has transferred an amount of heat $Q = T\,\Delta S$ to the gas in order to keep the temperature constant, so that the bath's entropy has decreased by ΔS. Thus, entropy ΔS has been supplied to the gas by the heat bath, and the total entropy of the gas plus heat bath is unchanged. This is a necessary condition for reversibility; if the piston is pushed back to its original position, entropy ΔS is returned to the bath. In reversible processes, we can think of entropy as a conserved quantity.

Now consider the irreversible process illustrated in Figure 2.5, with the modification that the cylinder has a diathermal rather than an insulating wall. The gas is initially confined to V_1 and is allowed to expand to V_2 by the breaking of a diaphragm. Since the temperature is constant, the initial and final states of the gas are identical to the previous case, so that the entropy change of the gas, ΔS, is still given by Equation (5.15). However, since there is no change in U and no work has been done, no heat has been transferred, so that it is no longer true that the entropy of the heat bath has decreased. The total entropy has increased; that is, *entropy is created in an irreversible process*. Furthermore, we can no longer write $dS = \frac{dQ}{T}$; Equation (5.14) holds *only* for a reversible process. However, because ΔS depends only on the initial and final states, we can calculate ΔS for a reversible process connecting these states, and then use this calculated value to find the change of entropy in an irreversible process with the same initial and final states.

The thermodynamic definition of entropy, $dS = \frac{dQ}{T}$, is very useful, since we often know dQ and can hence obtain dS without having to go through the difficult task of calculating Ω. For example, if we heat an object whose heat capacity is $C(T)$, $dQ = C(T)\,dT$, so that S can be found as a function of T. While the relation $dS = \frac{dQ}{T}$ is valid only in a reversible process, it can nevertheless be used when irreversible processes occur, so long as we are careful to make sure that the part of the process that we are using to find dS can be modelled by a reversible process. For example, consider two objects, respectively, at temperatures T_1 and T_2 ($T_1 > T_2$). We put them in thermal contact, but let us suppose that the thermal conductivity of the material separating them is finite. The process of coming into equilibrium is, of course, irreversible, but nevertheless we can calculate the change of entropy of each object when a small amount of heat dQ is transferred from object 1 to object 2 as follows. The removal of energy dQ from object 1 can be modelled as a reversible process, in which the energy is transferred to another object at the *same* temperature, and this is also true of the supply of energy dQ to object 2. Hence, the total change of entropy when dQ is transferred is:

$$dS = \frac{dQ}{T_2} - \frac{dQ}{T_1},$$

which is greater than zero, as it must be because the conduction process is irreversible. Note that the second law, in the form "entropy cannot decrease," forbids flow of heat from a cooler to a hotter body if they are isolated from their surroundings, since dS would be negative if T_2 were greater than T_1.

5.7 Entropy of Mixing

The fact that $dS = \frac{dQ}{T}$ in a reversible transfer of heat might suggest that entropy change is exclusively associated with heat transfer, but the irreversible processes illustrated in Figure 2.2 and discussed in the previous section show that this is not the case. In general, there are contributions to the entropy that have nothing directly to do with heat. For example, suppose we have two pure crystals[7] of different elements A and B at a very low temperature, where thermal vibrations can be neglected. Every atom is in its place, so that there is only one microstate and the entropy is zero. Now suppose that we melt these two crystals (containing N_a and N_b atoms, respectively) together and allow the resulting alloy to recrystallize, forming a crystal of $N = N_a + N_b$ atoms. We then return the crystal to its original temperature. In many cases, it is found that the different atoms are distributed at random over the possible sites in the new crystal, so that the probability of finding an A atom in any arbitrarily chosen site is $x = \frac{N_a}{N}$, and the probability of finding a B atom is $1 - x$. If we neglect thermal vibrations, the multiplicity of this macrostate is, from Equation (D.1):

$$\Omega = \frac{N!}{(xN)![(1 - x)N]!}.$$

If $N \gg 1$ the entropy is, from Stirling's formula in the form given in Equation (D.6):

$$S = -Nk_B[x \ln x + (1 - x) \ln(1 - x)], \tag{5.16}$$

(note that $0 < x < 1$, so that both logarithms are negative and the entropy is positive).

If $N_a = N_b$, so that $x = \frac{1}{2}$, S has its maximum value:

$$S = Nk_B \ln 2. \tag{5.17}$$

Thus, the entropy is greater than that of the separate pure crystals by an amount of order Nk_B. This entropy increase is called the *entropy of mixing*, and an increase of this order occurs whenever two distinct substances mix irreversibly; see Problem 5.5c for the case of two ideal gases mixing.

The entropy of mixing is not small; for $x \sim 0.5$, it is of the same order as the increase in the entropy of an ideal gas with N molecules when its absolute temperature is doubled (see Chapter 10).

5.8 Free Energy

In an infinitesimal reversible compression of a gas, the work done on the gas is $dW = -p\,dV$, from Equation (2.4). If the gas is isolated from its surroundings, so that no heat is transferred in or out, this energy must go entirely into internal energy. Since the compression is reversible, and $dQ = 0$, it is isentropic. Then, from Equation (5.13) $dW = dU$, as follows directly from the conservation of energy.

What is the corresponding expression in an infinitesimal *isothermal* compression; that is, one in which heat is exchanged with the surroundings in order to keep the temperature constant? The first law of thermodynamics in the form Equation (2.3) tells us that the work done on the gas is:

$$dW = dU - dQ.$$

If the process is reversible, this can be written:

$$dW = dU - T\,dS.$$

---

[7] In the language of solid state physics, a crystal is an ordered three-dimensional array of atoms. Here we ignore the possible presence of defects in this crystal.

We define a quantity F, called the *free energy*,[8] by

$$\boxed{F \equiv U - TS.}$$ (5.18)

Since U, T, and S are state functions, so is F, and its differential is:

$$dF = dU - T\, dS - S\, dT.$$ (5.19)

Substituting dU from Equation (5.13) gives:

$$\boxed{dF = -S\, dT - p\, dV.}$$ (5.20)

In an isothermal process, $dT = 0$, so that the work done on the gas in a reversible isothermal process is:

$$đW = -p\, dV = dF.$$ (5.21)

Hence, in a reversible isothermal expansion, the work obtained $(-đW)$ is equal to the decrease in F; in an irreversible expansion, such as that discussed in Section 2.4, the work obtained will be less than this because $|đW| < p\, dV$.

Equation (5.21) shows that the maximum work we can get from an isothermal process in which the system goes from state 1 to state 2 is $F_1 - F_2$, just as $U_1 - U_2$ is the maximum work we can get from an isentropic process. This is why F is called the "free" energy. However, we shall see that F has a much wider application, playing the same role in isothermal processes that U plays in isentropic ones.

For example, from Equation (5.12), $p = -\left(\frac{\partial U}{\partial V}\right)_S$, while from Equation (5.20):

$$p = -\left(\frac{\partial F}{\partial V}\right)_T.$$ (5.22)

Thus, if we know the free energy of a system, we can use Equation (5.22) to find the equation of state.

To illustrate the significance of the free energy, think of two gases, Gas 1 and Gas 2, contained in a closed cylinder with insulating walls, with an insulating piston dividing them (see Figure 5.3). Suppose that initially the gases are at different pressures. Since there is a net force on the piston, it will move until they are at the same pressure. For a small reversible change in the position of the piston about its equilibrium position, the change in the total internal energy is (remembering that no heat is transferred):

$$dU = dU_1 + dU_2 = -p_1\, dV_1 - p_2\, dV_2 = 0,$$ (5.23)

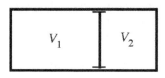

FIGURE 5.3   Two gases separated by an insulating piston and isolated from their surroundings.

---

[8] Usually called the Helmholtz free energy (although Helmholtz was not the first to introduce the concept; see Bailyn, p. 246) to distinguish it from the Gibbs free energy. It is more logical to call the latter the *free enthalpy*, as is done in this book (see Section 5.10), so that no confusion can arise. In the chemical literature, the free energy is sometimes called the *work function*. This term has a different meaning in physics (see Glossary).

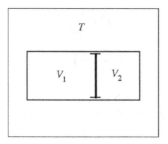

FIGURE 5.4 Two gases separated by a piston and in thermal contact with a heat bath at temperature T.

since the total volume is fixed and the pressures are equal. As is obvious from what we know about stable mechanical equilibrium, if no heat can be transferred into or out of the system, the equilibrium position of the piston corresponds to minimum U.

Now suppose that the cylinder, while still closed, has diathermal walls and is immersed in a heat bath that maintains the system at a constant temperature, so that all processes are isothermal (Figure 5.4). Now, from Equation (5.23), $dF_1 = -p_1\, dV_1$ and $dF_2 = -p_2\, dV_2$, so that in equilibrium:

$$dF = dF_1 + dF_2 = -p(dV_1 + dV_2) = 0.$$

Hence, when the temperature is fixed, stable equilibrium corresponds not to minimum U but to minimum F. Problem 5.4 illustrates the same principle for the case of a simple magnetic system.

In the next chapter, it will be shown by a statistical argument that the free energy F is minimized in equilibrium for *any* process at constant volume and temperature (this can also be shown by a purely thermodynamic argument).[9] The reason why U is not minimized in isothermal equilibrium is that the total entropy of the system *plus* the heat bath has to be maximized, and F takes this into account. The fact that entropy must be maximized while U is minimized shows that F is a minimum, not a maximum, in stable equilibrium. Note, however, that at absolute zero ($T = 0$), the entropy term $-TS$ in Equation (5.18) vanishes, as does the distinction between isothermal and isentropic processes. In general, the entropy term in F becomes larger relative to U as the temperature rises, a fact of great importance in understanding phase transitions, as we shall see in Chapter 15.

5.9 The Gibbs-Helmholtz Equation

Another important result provides a differential equation for the free energy. From Equation (5.20), we see that:

$$S = -\left(\frac{\partial F}{\partial T}\right)_V, \tag{5.24}$$

where not only N and V, but all variables (such as magnetic field) whose change can cause work to be done, are kept constant.

Substituting Equation (5.24) in Equation (5.18) gives us a differential equation for $F(V, T)$ which can be written in three equivalent ways:

$$\boxed{U = F - T\left(\frac{\partial F}{\partial T}\right)_V;} \tag{5.25a}$$

[9] Goodstein, p. 23.

$$U = -T^2 \left(\frac{\partial (F/T)}{\partial T} \right)_V ; \qquad (5.25b)$$

$$= \left(\frac{\partial (\beta F)}{\partial \beta} \right)_V , \qquad (5.25c)$$

where we define $\beta \equiv \frac{1}{k_B T}$ (this definition will be found convenient in subsequent chapters). Equation (5.25) in any of its three forms is called the Gibbs-Helmholtz equation. Given $F(T)$, one can use this equation to find $U(T)$. Alternatively, given $U(T)$, one can obtain $F(T)$, to within an undetermined function of V, by integrating with respect to T.

5.10 Free Enthalpy (Gibbs Free Energy)

As pointed out in Section 2.6, constant pressure (isobaric) processes are far more common in practice than constant volume processes. The quantity that is minimized in an isobaric isothermal process is the *free enthalpy*[10] G, defined by:

$$G \equiv H - TS, \qquad (5.26)$$

where H is the enthalpy ($H \equiv U + pV$; see Equation 2.16).

Differentiating Equation (5.26) gives:

$$dG = dH - T\,dS - S\,dT$$
$$= dU + p\,dV + V\,dp - T\,dS - S\,dT.$$

Substituting for dU from Equation (5.13), we find:

$$\boxed{dG = V\,dp - S\,dT.} \qquad (5.27)$$

As an illustration of the minimization of the free enthalpy G, consider the condensation of a gas. The enthalpy H is lower in the condensed phase (solid or liquid) than it is for the same material in gaseous form, because of the interatomic attraction and the smaller volume. On the other hand, the gas has higher entropy, because there are more states available for the particles. At low temperature, the enthalpy difference ΔH dominates and the condensed phase has a lower G than the gas. At high temperature, on the other hand, the entropy term $-T\,\Delta S$ dominates and the gas has the lower G, so that it is the stable phase. If the pressure is kept constant and the temperature is reduced, the transition occurs when the free enthalpies of the two phases are equal. Phase transitions are discussed in more detail in Chapter 15.

If a quantity X has an extremum, we can express this mathematically by writing $dX = 0$. For example, if T and V are kept constant, so that $dT = 0$, and $dV = 0$, then Equation (5.20) shows that $dF = 0$ when the system is in equilibrium; that is, the free energy $F \equiv U - TS$ is a minimum. As pointed out above, F is a minimum, not a maximum, since we are trying to simultaneously *maximize S* and *minimize U*. Similarly for an isothermal isobaric process (one in which T and p are constant), Equation (5.27) gives $dG = 0$ in equilibrium, so that the free enthalpy G is a minimum. For an isentropic process, on the other hand, $dS = 0$. If the volume is kept constant, Equation (5.13) gives $dU = 0$, so that the internal energy U is minimized for an isentropic constant volume process, while in an isentropic isobaric process, $dH = 0$, so that the enthalpy H is minimized. The conditions under which each of the four quantities U, H, F, and G are minimized in equilibrium are summarized in Table 5.1.

---

[10] Often called the Gibbs free energy (or Gibbs function) after J. Willard Gibbs (1833–1903), the American chemical physicist who put chemical thermodynamics and classical statistical mechanics on a sound mathematical basis (see the quotation from Bohr at the beginning of Chapter 8).

TABLE 5.1 Quantities Minimized in Equilibrium When Different Parameter Pairs are Held Constant. Example: If T and V Are Constant for a Fluid, or T and B for a Magnetic System, F Is Minimized

Parameters kept constant	V, B	p, M
S	U	H
T	F	G

In a magnetic system, the same reasoning that led to Equation (5.22) and Equation (5.23) gives for the magnetic moment [see Equation (5.41) in Problem 5.8]:

$$M = -\mu_0^{-1} \left(\frac{\partial U}{\partial H} \right)_S \tag{5.28}$$

$$= -\mu_0^{-1} \left(\frac{\partial F}{\partial H} \right)_T . \tag{5.29}$$

What we usually measure is the magnetic susceptibility χ. Susceptibility can be measured (in principle) either isentropically or isothermally. The isentropic and isothermal susceptibilities are given, respectively, by:

$$\chi^S = \left(\frac{\partial M}{\partial H} \right)_S = -\mu_0^{-1} \left(\frac{\partial^2 U}{\partial H^2} \right)_S , \tag{5.30}$$

$$\chi^T = \left(\frac{\partial M}{\partial H} \right)_T = -\mu_0^{-1} \left(\frac{\partial^2 F}{\partial H^2} \right)_T . \tag{5.31}$$

These two susceptibilities may differ substantially. In practice, it is usually easier to make measurements at constant temperature than at constant entropy, so that F and G are more useful quantities than U and H.

Example 5.3: The Elastic Polymer

In Problem 4.7, we described a simple one-dimensional model of a polymer, consisting of N links of length ℓ_0, each of which can be oriented to the left or to the right. We now calculate the response of this model polymer to an applied force. We assume that the length ℓ is much less than $N\ell_0$, the fully stretched length of the polymer, so that the Gaussian approximation (Equation 4.9) can be used for the multiplicity. Suppose that the polymer is stretched to a length ℓ by a force f. We can use Equation (2.7) to obtain differential expressions for the energy and free energy analogous to Equations (5.13) and (5.20):

$$dU = T\,dS + f\,d\ell, \tag{5.32}$$

$$dF = -S\,dT + f\,d\ell. \tag{5.33}$$

so that:

$$f = \left(\frac{\partial F}{\partial \ell} \right)_T .$$

The model assumes that U is independent of the relative orientations of the links, so that $\left(\frac{\partial U}{\partial \ell}\right)_T = 0$. Using $F = U - TS$, we find that the force is given by:

$$f = -T\left(\frac{\partial S}{\partial \ell}\right)_T. \tag{5.34}$$

There are two main contributions to the entropy. One comes from the vibrations of the atoms comprising the polymer; that is, from the internal energy U. Since we have assumed that U is independent of ℓ, this contribution to S is also independent of ℓ. The other contribution comes from the different ways that the links can be oriented; we call this the *conformational* entropy, and its dependence on ℓ can be obtained from Equation (4.9). Substituting Ω from Equation (4.9) in $S = k_B \ln \Omega$, we find:

$$S(\ell) = S(0) - \frac{k_B}{2N\ell_0^2}\ell^2, \tag{5.35}$$

where $S(0)$ includes the vibrational contribution that is independent of ℓ.

From Equations (5.34) and (5.35), we have

$$f = -T\left(\frac{\partial S}{\partial \ell}\right)_T = \frac{k_B T}{N\ell_0^2}\ell. \tag{5.36}$$

Thus $f \propto \ell$; that is, Hooke's law is obeyed, with a force constant f/ℓ proportional to temperature. For a given force, the length *decreases* with temperature; that is, the expansion coefficient is negative. As T increases the entropy term in the free energy becomes more important; the polymer "wants" to twist up and increase its entropy.

If the stretching is done isentropically, rather than isothermally, the decrease in conformational entropy must be balanced by an increase in vibrational entropy, so that the polymer warms up. This is, of course, required by the conservation of energy; the work done on the polymer by the stretching force must increase the internal energy, as pointed out in Example 2.3.

This model is too crude to account quantitatively for the mechanical properties of polymers, but it gives a good qualitative account of the behavior of rubber at normal temperatures.[11]

5.11 Adiabatic Processes

The above reasoning—in particular, the derivation of Equation (5.13) and the consequent identification of $T\,dS$ with dQ in a reversible process—is somewhat abstract, and we will now consider how it applies to the system of spin 1/2 nuclear magnets discussed in Chapter 4. In particular, we show for this model that a process that is *adiabatic* in the quantum mechanical sense (which will be explained) is necessarily isentropic.

Consider first a single isolated spin in a magnetic field B, in a particular state (say, spin up), with energy $-\mu B$. Now we *slowly* change B. In order to produce a transition (a *spin flip*) to the spin down state, whose energy is $+\mu B$, quantum mechanics tells us that we need a component of the field oscillating at the frequency corresponding to the energy separation of the states; that is, $\frac{2\mu B}{h}$. If B were to change rapidly, it could have a Fourier component at this frequency, but we will assume that B changes sufficiently slowly that no such high frequency field is present. Then the spin must stay in whatever state it began in, even though the energy of this state is changing. The energy needed is supplied (or taken up) by the electromagnetic field. Such a slow change, in which the occupation of quantum states does not change, is called an *adiabatic* change.

---

[11] See L. R. G. Treloar, *Physics of Rubber Elasticity* (Oxford University Press, 1958) for a fuller treatment.

$$\uparrow\downarrow \Rightarrow \downarrow\uparrow$$

FIGURE 5.5 Mutual spin flip, conserving total spin.

Now suppose that we have a collection of spins isolated from its surroundings. Although in our model we neglected the interaction between them, the spins must interact weakly if they are to come into equilibrium with each other. The interaction occurs because the motion of one spin in the field generates a high frequency magnetic field of just the right frequency to cause a neighboring spin to flip; this spin-spin interaction ensures that the whole assemblage of spins reaches its equilibrium condition of maximum Ω. However, the quantum mechanics of this interaction tells us that if one spin flips $\uparrow$ to $\downarrow$, the other has to flip from $\downarrow$ to $\uparrow$ (see Figure 5.5), so that the total spin is conserved. Although there is a change of microstate, the macrostate does not change, so that Ω (and hence S) does not change. While this is a very specific model, it illustrates the general result, which we justify in Chapter 6, that an adiabatic change is isentropic.

In general, a process is adiabatic in the quantum mechanical sense if it does not change the average occupancy of the quantum states. Ω is unchanged, so that such a process does not change the entropy and is called an isentropic process. An adiabatic process in this sense is always possible in principle. Here we have shown this for a spin system, which has discrete energy levels; the case of continuous levels, as in a gas, will be dealt with later (see Example 10.2).

5.12 Irreversible Adiathermal Processes: Joule-Thomson Expansion and Enthalpy Conservation

In Chapter 2, it was pointed out that the rapid expansion of a gas into a vacuum conserves internal energy, but being irreversible, it necessarily increases entropy. In practice, a more important type of irreversible process is the continuous flow process, of which Joule-Thomson expansion is an example. This process is widely used in refrigerators (see Section 3.7) and is illustrated schematically in Figure 5.6.

A gas flows at a steady rate through a throttle valve (or more usually, a porous plug) which reduces its pressure from p_1 to p_2. The walls are insulating, so that no heat is transferred in or out. If the gas is not ideal, the temperature will, in general, change. We now show that it is the enthalpy H rather than the internal energy U that is conserved in this process. This is easiest to see if we imagine the gas being pushed through the valve by a piston, and pushing against another piston after throttling. These imaginary pistons are shown by the thin lines. Suppose that a certain amount of gas, of initial volume V_1, is passed through the valve, its volume changing to V_2. The net work done on the gas by the two pistons is $p_1 V_1 - p_2 V_2$. Since no heat is transferred, energy conservation requires that the internal energy be changed by this amount; that is:

$$U_2 - U_1 = p_1 V_1 - p_2 V_2.$$

Since $H \equiv U + pV$, this equation can be written $H_1 = H_2$; enthalpy is conserved in this irreversible process.

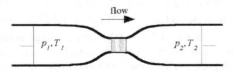

FIGURE 5.6 Joule-Thomson expansion through a porous plug. The "pistons" indicated at the input and output are purely imaginary.

Example 5.4: A Gas Cycle Refrigerator

In a refrigerator, we are interested in the temperature change that is produced by Joule-Thomson expansion, and we define the *Joule-Thomson coefficient* as the rate of change of temperature with pressure at constant enthalpy, $\left(\frac{\partial T}{\partial p}\right)_H$. Derive an expression for the Joule-Thomson coefficient in terms of the equation of state and the heat capacity. Find the condition for the Joule-Thomson coefficient to be positive, so that a drop in pressure produces a drop in temperature. Joule-Thomson expansion can then be used in refrigeration.

Substituting $T\,dS$ for $\dbar Q$ in Equation (2.17) we find:

$$dH = T\,dS + V\,dp. \tag{5.37}$$

The derivative of H with respect to p at constant T is thus:

$$\left(\frac{\partial H}{\partial p}\right)_T = T\left(\frac{\partial S}{\partial p}\right)_T + V.$$

From the triple product rule (Equation E.2),

$$\left(\frac{\partial H}{\partial T}\right)_p \left(\frac{\partial T}{\partial p}\right)_H = -\left(\frac{\partial H}{\partial p}\right)_T.$$

From Example 2.6(a), $\left(\frac{\partial H}{\partial T}\right)_p = C_p$, the heat capacity at constant pressure (we have substituted T for θ).

Hence,

$$\left(\frac{\partial T}{\partial p}\right)_H = -C_p^{-1}\left(\frac{\partial H}{\partial p}\right)_T = -C_p^{-1}\left[T\left(\frac{\partial S}{\partial p}\right)_T + V\right].$$

We can derive the following Maxwell relation from Equation (5.27) (see Appendix E):

$$\left(\frac{\partial S}{\partial p}\right)_T = -\left(\frac{\partial V}{\partial T}\right)_p.$$

Since C_p is defined as the *kmolar* heat capacity at constant pressure, we must put the volume equal to the kmolar volume V_m, and:

$$\left(\frac{\partial T}{\partial p}\right)_H = C_p^{-1}\left[T\left(\frac{\partial V_m}{\partial T}\right)_p - V_m\right]$$

$$= \frac{V_m}{C_p}\left[\alpha_b T - 1\right]. \tag{5.38}$$

where $\alpha_b \equiv \frac{1}{V}\left(\frac{\partial V}{\partial T}\right)_p$ is the bulk expansion coefficient (which is always positive for a gas).

In an ideal gas, $V_m = \frac{R_0 T}{p}$, so that $\alpha_b = T^{-1}$ and the right-hand side of Equation (5.38) vanishes. However, in an imperfect gas it does not, and the sign of the Joule-Thomson coefficient depends on whether $\alpha_b T$ is greater or less than 1. If $\alpha_b T > 1$, $\left(\frac{\partial T}{\partial p}\right)_H > 0$, so that Joule-Thomson expansion cools the gas and can be used for refrigeration. Real gases have Joule-Thomson coefficients which are positive at sufficiently low temperature, but at a certain inversion temperature T_{inv}, which decreases with pressure, the Joule-Thomson coefficient changes sign. If T_{inv} is below room temperature (as it is for hydrogen and helium but not for nitrogen and oxygen), the gas must be cooled to T_{inv} by some other means, such as isentropic expansion, before Joule-Thomson expansion can be used to cool it further.

The reason for the existence of an inversion temperature can be qualitatively understood as follows. The equation of state of an imperfect gas deviates from the ideal gas law because of interaction between molecules. We shall see in Chapter 15 that this interaction is weakly attractive at large distances and strongly repulsive at short, as shown by the full line in Figure 15.9. The effect of the interaction can be expressed formally by expanding the equation of state as a "virial series" in powers of the particle density n:

$$p = k_B T \left[n + B_2(T)n^2 + B_3(T)n^3 + \dots \right]. \tag{5.39}$$

The B_i are functions of temperature only and are called the *virial coefficients*.

It can be shown (see Problem 5.9) that at low pressure, where $B_2 n$ is small and all higher terms in Equation (5.39) are negligible, the Joule-Thompson coefficient is approximately:

$$\left(\frac{\partial T}{\partial p} \right)_H = \frac{N_A}{C_p} \left[T \frac{dB_2}{dT} - B_2 \right] \tag{5.40}$$

$$= \frac{N_A}{C_p} T^2 \frac{d(B_2/T)}{dT}.$$

One can make a qualitative guess at the temperature dependence of B_2 by considering the magnitude of the intermolecular interaction relative to the kinetic energy of the molecules. At low temperatures, the kinetic energy is small and the weak attractive interaction reduces the pressure below its ideal gas value; hence $B_2 < 0$. Joule-Thomson expansion cools the gas because work must be done against this attraction. The effect of the attractive interaction decreases as the kinetic energy increases, so that $\frac{d|B_2|}{dT} < 0$ and (since $B_2 < 0$) $\frac{dB_2}{dT}$ is positive. Hence, $\left(\frac{\partial T}{\partial p} \right)_H$ is positive at low temperature. At high temperatures, on the other hand, the weak attraction is negligible relative to the kinetic energy, and the strong "hard sphere" repulsion between molecules at short distances dominates; this increases the pressure above its ideal gas value, so that B_2 becomes positive and almost independent of temperature. Hence, $\left(\frac{\partial T}{\partial p} \right)_H$ is negative. This argument can be made more quantitative by considering a particular equation of state, such as van der Waals' equation (derived in Chapter 15); see Problem 5.10.

5.13 The Third Law of Thermodynamics

The alert reader will have noticed that while the statistical definition of entropy, $S = k_B \ln \Omega$, is consistent with the thermodynamic one, $dS = \frac{dQ}{T}$, the two definitions are not equivalent. The thermodynamic definition only specifies differences of entropy, while the statistical definition specifies its absolute value, at least in principle. The two definitions can be made to coincide by hypothesizing that for any system in equilibrium, $S \to 0$ and $\Omega \to 1$ as $T \to 0$. This hypothesis is called the third law of thermodynamics, a somewhat grandiose title that implies greater generality than is really the case. A large body of data[12] on the heat capacities of molecular systems at low temperature support the hypothesis, which in quantum terms simply follows from the fact that any system in true internal equilibrium at $T = 0$ is in its ground state, which is usually non-degenerate. However, there are many systems, such as glasses, that retain significant disorder and, therefore, non-zero entropy, as $T \to 0$. Thus, the third law in this form does not have the universal validity of the other three laws of thermodynamics. An alternative version of the third

---

[12] Principally due to W. F. Giauque (1895–1982) and his collaborators. These data are reviewed in R. H. Fowler and E. A. Guggenheim, *Statistical Thermodynamics* (Cambridge, UK, 1939), Sections 532 and 533, where possible ambiguities in the definition of Ω (due, e.g., to the nuclear spin degeneracy) are also discussed.

law is an impossibility statement[13]: "It is impossible by any means whatsoever to reach the absolute zero of temperature," and no exceptions to this have been found. This impossibility statement implies that any two systems *which can be interconverted by a reversible process* have the same entropy at 0 K, since if they did not, it would be possible to start with the lower entropy system at non-zero temperature and cool to 0 K by isentropic conversion to the higher entropy system. Systems such as glasses, which have non-zero entropy at 0 K, cannot be obtained from a lower entropy system (such as the crystalline form of the material) by a reversible process.[14]

5.14 Extensive versus Intensive Variables

The first law is an application of conservation of energy where dU equals a sum of terms, each of which are a pairing of the extensive variable and the intensive variable that correspond to how the energy of the system changes with respect to the extensive variable. Hence, $dU = TdS - pdV$ where $T = dU/dS$ (at constant volume) and $-p = dU/dV$ (at constant entropy). Not only does this allow us to then add any contribution to the energy of a system as a pairing of the extensive and corresponding intensive variable defined by the gradient of energy with respect to the extensive variable, but it also bridges the application of the first law to include both the macroscopic picture (extensive variable) with the microscopic picture (intensive variable). By this, very powerful modifications can be made to the first law with very little intimate knowledge of the system. For instance, at the surface of a fluid, one can assert an energy scale that is dependent upon the surface area of the fluid. So $dU = \sigma dA$ where A is the area of the fluid surface and $\sigma = dU/dA$ (with all other extensive variables held constant). As a consequence, thermodynamics holds court at both microscopic and macroscopic scales and must always reconcile to both.

5.15 Envoi

After energy, entropy is the most important concept in thermal physics. The physical meaning of entropy, which has confused generations of thermodynamics students (and sometimes, I fear, their teachers[15]), is straightforward when defined statistically in terms of the multiplicity; that is, the number of microstates (quantum states) in a macrostate. When the second law of thermodynamics is expressed in terms of entropy defined in this way, it becomes intuitively almost obvious, while the third law is a natural consequence of what we know about quantum systems. Having a satisfactory definition of entropy also enables us to define absolute temperature without the appearance of mathematical legerdemain from which the conventional thermodynamic definition as an integrating factor and the traditional statistical definition as a Lagrangian multiplier (see Problem 6.10) both suffer.

Once the concept of entropy has been grasped, we can proceed to develop all of thermodynamics in terms of state functions; that is, quantities whose values depend only on the current state of the system in question, and not on how that state was reached. Besides entropy, the most important of these state functions are the free energy F and free enthalpy G, which play the same role in isothermal processes that the internal energy U and enthalpy H play in isentropic ones. Furthermore, we shall see in the next chapter that F has a useful statistical interpretation.

---

[13] See *No Way: the Nature of the Impossible*, eds. Philip J. Davis and David Park (Freeman, 1987) for an extended discussion of the importance of impossibility statements in physics and other disciplines. The first epigraph to this chapter makes the same point more snappily.
[14] See J. S. Dugdale, *Entropy and its Physical Meaning* (Taylor & Francis, 1996), Chapter 16.
[15] When I was an engineering student, we were taught that entropy means "buzzing aboutness."

5.16 Problems

5.1 For Equation (5.5) to correspond to maximum rather than a minimum entropy, and hence to thermal equilibrium, $\left(\frac{\partial^2 S}{\partial U^2}\right)_V$ must be negative. Show that this implies that the heat capacity at constant volume of a system in thermal equilibrium, $C_V \equiv \left(\frac{\partial U}{\partial T}\right)_V$, must be positive[16]

5.2 (a) A body of mass M with heat capacity (per unit mass) C, initially at temperature $T_0 + \Delta T$, is brought into thermal contact with a heat bath at temperature T_0. Show that if $\Delta T \ll T_0$, the increase ΔS in the entropy of the entire system (body plus heat bath) when equilibrium is reached is proportional to ΔT^2, and find the constant of proportionality

   (b) Find ΔS, given that the body in (a) is a bacterial cell of mass 1 pg (10^{-15} kg) consisting mostly of water, its initial temperature is 300.03 K, and the heat bath temperature is 300.00 K. Take the heat capacity of water as 4 kJ kg^{-1} K^{-1} and ignore the contribution of the other components of the cell

   (c) Make an order of magnitude estimate of the probability of subsequently finding the system in (b) back in its original condition (i.e., with the cell at 300.03 K) for 1 psec (10^{-12} sec) during the lifetime of the universe ($\sim 10^{18}$ sec). What does your result tell you about the probability of detectable fluctuations on a macroscopic scale?

   (d) Repeat (b) for a virus of mass 10^{-20} kg (for simplicity, assume that the heat capacity per kg is the same as for water), with initial temperature 300.3 K. Estimate the probability of subsequently finding the virus at 300.3 K in one observation

5.3 Consider the system of N spin 1/2 magnets of moment μ discussed in Chapter 4, in a magnetic field B. The energy of a macrostate of moment $M = n\mu$ in a field B is $U = -n\mu B$. Assume that N is large number

   (a) Using the Gaussian approximation (Equation 4.9), find S as a function of U

   (b) Hence, find $U(T, B)$

   (c) The isothermal susceptibility χ^T is defined by Equation (5.31). Show that in the Gaussian approximation $\chi^T \propto T^{-1}$. This is known as Curie's law and is found to hold for most non-metallic paramagnets at moderate temperatures

   (d) For a given temperature, roughly for what range of B is the Gaussian approximation valid? If the elementary magnets are electrons with $\mu = 9 \times 10^{-24}$ J/T (here "T" stands for "Tesla"), and $B = 10$ T, in what temperature range do you expect the approximation to fail?

5.4 This is an another derivation of Curie's law, designed to illustrate the point that in a constant temperature environment equilibrium corresponds to minimum F. For the spin system of N spins discussed in Chapter 4, use the Gaussian approximation (Equation 4.4) to obtain the entropy S. Hence find the free energy F at a given T as a function of B and M, where $M = n\mu$, n being the net number of up spins, as before. Minimize F with respect to M at constant T and B to find M, and hence find χ^T

---

[16] Note the essential qualifier, "in thermal equilibrium." The heat capacity of a self-gravitating ideal gas is negative (the temperature *rises* as the gas loses energy by radiation), but the gas is not in equilibrium. A main sequence star, which approximates to such a gas, is only saved from collapse by the continuous generation of energy by thermonuclear fusion, and collapses to a white dwarf when the star has no hydrogen left to fuel this reaction (see Example 13.3). For a thorough discussion of this fascinating and astrophysically important system, and analogous ones which occur at phase transitions, see D. Lynden-Bell, "Negative specific heat in astronomy, physics and chemistry," *Physica* **A263**, 293–304 (1999). For a formal treatment of some simple model systems exhibiting a negative heat capacity when in a steady non-equilibrium state, see R. K. P. Zia, E. L. Preastgaard and O. G. Mouritsen, "Getting more from pushing less: negative specific heat and conductivity in non-equilibrium steady states," *Am. J. Phys.* **70**, 384–392 (2002).

5.5   The equation of state of ν kmoles of an ideal gas is $pV = Nk_B T$, where $N = \nu N_A$
      (a) Starting from Equation (E.6), show that for an ideal gas:

$$\left(\frac{\partial S}{\partial V} \right)_T = \frac{Nk_B}{V}$$

      (b) Hence, derive Equation (5.15)
      (c) 0.5 kmole of one ideal gas is allowed to mix with 0.5 kmole of another (chemically different)
           ideal gas. The initial temperature and pressure of each gas is the same, and the temperature and
           total pressure remain constant throughout the mixing; that is, the mixing is both isothermal
           and isobaric. Use the result of (b) to find the increase in entropy, assuming that the gases do
           not react or interfere with each other in any way, so that their entropies are additive. Compare
           your result with Equation (5.17). Explain why these results agree

5.6   In the model of rubber given in Example 5.3, what is the difference between the kmolar heat capac-
      ity of rubber at zero length and at length ℓ (a) at constant ℓ and (b) at constant force f? Assume
      that $\ell \ll N\ell_0$

5.7   Starting from the fact that all the variables in Equation (5.13) are state functions, prove the Maxwell
      relation:

$$\left(\frac{\partial S}{\partial p} \right)_V = - \left(\frac{\partial V}{\partial T} \right)_S .$$

Hence, show that:

$$\left(\frac{\partial S}{\partial V} \right)_p = \left(\frac{\partial p}{\partial V} \right)_S \left(\frac{\partial V}{\partial T} \right)_S = \left(\frac{\partial p}{\partial T} \right)_S ,$$

so that by measuring the rate of change of T with p as a material is compressed isentropically, one
can deduce the entropy as a function of volume

5.8   Starting from Equation (2.15), show that in an incompressible magnetic system, Equation (5.20)
      should be replaced by:

$$dF = -S\, dT - \mu_0 M\, dH. \tag{5.41}$$

Hence, derive Equation (5.29)

5.9   Derive Equation (5.40) from Equations (5.38) to (5.39), assuming that $B_2 n$ is small and that all
      higher terms in Equation (5.39) are negligible

5.10 The van der Waals equation of state of an imperfect gas is (see Chapter 15):

$$p = \frac{nk_B T}{1 - bn} - an^2, \tag{5.42}$$

where a is a constant representing the effect of long-range attraction between molecules, which
reduces the actual pressure p below its kinetic value, and b is a constant representing the effect of
hard-sphere repulsion, which reduces the volume available and hence raises the effective particle
density above its actual value of n

(a) Assuming that $an^2 \ll p$ and $bn \ll 1$, put Equation (5.42) into the virial form (Equation 5.39) and find $B_2(T)$ to first order in a and b

(b) Show that with these approximations, the Joule-Thomson inversion temperature is[17]:

$$T_{\text{inv}} = \frac{2a}{k_B b}.$$

5.11 (Difference between the heat capacities of a gas.)
In Example 5 of Chapter 2, the heat capacities at constant pressure and constant volume, C_p and C_V, were shown to be $\left(\frac{\partial H}{\partial T}\right)_p$ and $\left(\frac{\partial U}{\partial T}\right)_V$, respectively

(a) Show that these can be written $C_p = T\left(\frac{\partial S}{\partial T}\right)_p$, $C_V = T\left(\frac{\partial S}{\partial T}\right)_V$

(b) Hence, show that for a system whose entropy is a function of T and V alone,

$$C_p - C_V = T\left(\frac{\partial V}{\partial T}\right)_p \left(\frac{\partial S}{\partial V}\right)_T,$$

(c) Use a Maxwell relation and the triple product rule to show that this can be written as

$$C_p - C_V = V_m T \alpha_b^2 B_T, \tag{5.43}$$

where $\alpha_b \equiv \frac{1}{V}\left(\frac{\partial V}{\partial T}\right)_p$ is the bulk expansion coefficient and $B_T \equiv -V\left(\frac{\partial p}{\partial V}\right)_T$ is the isothermal bulk modulus

(d) Show that for a gas obeying the equation of state given by Equation (5.39), if terms quadratic in B_2 and all higher virial coefficients are negligible, Equation (5.43) becomes:

$$C_p - C_V \approx R_0 \left(1 + nT\frac{dB_2}{dT}\right) \tag{5.44}$$

(e) Show that Equation (5.43) applies to the elastic polymer (see Example 5.3), if we interpret C_p as the heat capacity at constant force, C_V as the heat capacity at constant length, α_b as the linear expansion coefficient at constant force $\frac{1}{\ell}\left(\frac{\partial \ell}{\partial T}\right)_f$, and B_T as the force constant (force per unit extension). Explain physically why $C_p - C_V$ is still positive even though the polymer shrinks on heating

5.12 Show that:

$$H = G - T\left(\frac{\partial G}{\partial T}\right)_P, \tag{5.45}$$

analogous to the Gibbs-Helmholtz Equation (5.25)

5.13 (a) Derive Equation (6.16)

(b) For the case $x = \frac{1}{2}$, estimate how small N must be for the neglected terms in the "crude" Stirling approximation (Equation D.6) to make a 10% contribution to the entropy of mixing

5.14 Suppose that you put a slightly soluble solid into a beaker of water. It requires a certain energy ϵ to remove an atom from the solid and put it into solution ($N_A\epsilon$ is called the heat of solution). Use Equation (5.16) to find the free energy (relative to the solid) of the solution, to the lowest order in the concentration x, when $x \ll 1$. If the limit of solubility at temperature T is x_{max} ($x_{\text{max}} \ll 1$), find ϵ

---

[17] In Chapter 15, we shall find that the critical temperature T_c, above which the gas-liquid distinction vanishes, is $T_c = \frac{8a}{27bk_B}$ for a van der Waals gas. Thus $\frac{T_{\text{inv}}}{T_c} = \frac{27}{4} \approx 7$. In real gases of cryogenic interest, $\frac{T_{\text{inv}}}{T_c}$ is found to lie between 5 and 10 at low pressure. However, experimentally T_{inv} depends on pressure, and higher virial coefficients must be included in the calculation if one is to obtain the pressure dependence.

5.15 Two identical blocks of metal at 300K and 310K are placed in contact. What is the entropy increase of the universe? What changes (and what does not) if the blocks are 300K and 320K?

5.16 Integrate Equation (5.13)—a state function, to produce a function of U. Differentiate this state function to show that a necessary condition of Equation (5.13) is that $SdT - VdP = 0$

5.17 Determine all of the Maxwell relationships for the Gibbs Free Energy as given in Equation (5.26). See Problem 5.7

5.18 The energy of a bubble can be written as $dU = TdS - pdV + \sigma dA$. Here, σ is a surface tension cost for the bubble. Calculate the free energy F necessary to construct such a spherical bubble. Assuming a spherical bubble, the free energy can be reduced to a function of F(T,r), where r is the radius of the bubble. Show that the free energy predicts an unstable bubble for a radius below a critical radius of $2\gamma/p$. What happens to any bubble one attempts to make of this size?

5.19 In a closed system, show that entropy maximization is equivalent to minimizing the Helmholtz free energy

5.20 What is the behavior of the Helmholtz free energy as T approaches absolute zero?

6

The Canonical Distribution: The Boltzmann Factor and the Partition Function

We avoid the gravest difficulties when, giving up the attempt to frame hypotheses concerning the constitution of matter, we pursue statistical inquiries as a branch of rational mechanics.

J. Willard Gibbs

6.1 Gedanken

You learn at a very young age that when something is dropped and is broken, it is impossible to put it back together again *exactly* the way it was. There is something deeply profound about this knowledge. So profound, that it becomes a natural part of how you see the universe around you, and if you were to see it violated you'd know something was wrong with the universe. Imagine someone showing you a short movie. The movie starts with a view of the cracked and broken pieces of a dish on the floor. The pieces come together to form a complete porcelain dish that leaps up from the floor and lands gently on a table. You know on an innate level what you have just seen: the person showing you the movie is running it in reverse. Through the concept of entropy, the physics of thermodynamics and statistical mechanics has something fundamental to say about any physical process that evolves in time. Whether it is charging a capacitor or putting a mass on a spring or any other physical process that takes a system from one state of equilibrium to another, the concept of entropy holds court. We may think (incorrectly) of time as a scalar quantity. There is a direction to time, determined by entropy in the second and third laws, so once again, statistical mechanics fits just as nicely into thinking of spacetime as a four dimensional vector space.

6.2 A System in Contact with a Heat Bath

We are now in a position to quantify the relationship between temperature and the probability of occupation of energy levels. We consider a microsystem in equilibrium with a heat bath at temperature T. Our purpose is to find a mathematical expression for the probability that a state of the microsystem with energy E is occupied, as a function of E and T.

First, we must make the concept of "heat bath" precise. Suppose that we have two systems, Σ_0 and Σ, where Σ_0 is extremely large while Σ is much smaller (Figure 6.1a). We assume that both have fixed

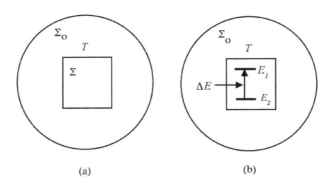

FIGURE 6.1   (a) A system Σ in thermal contact with a heat bath Σ_0 at temperature T. (b) Energy ΔE is transferred to Σ from Σ_0, raising Σ from state 2 to state 1.

volume and are isolated from the surroundings. Initially, Σ_0 is at temperature T, while Σ is at a different temperature. The two systems are then brought into contact. They come into equilibrium by transferring a certain amount of energy, in the form of heat, in one direction or the other. However, because Σ_0 is much larger than Σ, the effect of this energy transfer on the properties of the large system Σ_0 is very small compared with its effect on the small system Σ. It follows that in equilibrium, both systems are at temperature T. A large system like Σ_0, whose temperature is essentially unaffected by heat transfer, is called a heat bath.

6.3 The Boltzmann Factor and the Canonical Distribution

Now let us suppose that Σ is a microsystem (in the sense of Chapter 4) and has only two states, labeled 1 and 2 (Figure 6.1b), which are nondegenerate. These differ in energy by $\Delta E = E_1 - E_2$. An example of such a system is a single spin in a magnetic field (see Figure B.2). It is important to remember that even when two systems such as Σ_0 and Σ are in thermal equilibrium with each other, energy is still continually being transferred back and forth between them, although the *average* flow of energy is zero. We ask what happens to the entropy of the combined system when Σ changes from state 2 to state 1. When this happens, conservation of energy requires that an amount of energy equal to ΔE be transferred from the heat bath Σ_0 to Σ. If the energy of Σ_0 is U_0 when Σ is in state 2, it is $U = U_0 - \Delta E$ when Σ is in state 1.

The entropy S of the heat bath is a function of its internal energy U and its volume V, and in this case V is kept constant. Since the heat bath is very large, ΔE is a very small amount of energy relative to U, so that we can expand $S(U)$ as a Taylor series in $U - U_0$. We take the first (linear) term to find the change ΔS in the entropy of the heat bath when it goes from a macrostate of energy U_0 to one of energy U:

$$\Delta S = \left(\frac{\partial S}{\partial U} \right)_V (U - U_0),$$

$$= -\frac{\Delta E}{T}, \tag{6.1}$$

from the definition of T (Equation 5.8). It can be shown (see Problem 6.2) that the next and all higher terms in the Taylor expansion of ΔS vanish in the limit of an infinitely large heat bath.

Whichever state of Σ is occupied, $\Omega = 1$ for this microsystem, so that its entropy is always zero. Hence, the change in the entropy of the combined system is ΔS. The fundamental hypothesis of statistical mechanics (see Section 4.2) is that the probability of observing a particular macrostate is proportional to its multiplicity Ω. From Equation (5.4), $\Omega = e^{S/k_B}$, so that the ratio of the probability that the small system Σ is in state 1 to the probability that it is in state 2 is:

$$\frac{P_1}{P_2} = \frac{\Omega_1}{\Omega_2} = e^{\Delta S/k_B}. \tag{6.2}$$

Substituting for ΔS, we find:

$$\frac{P_1}{P_2} = e^{-\Delta E/k_B T} = e^{(E_2 - E_1)/k_B T}$$

$$= \frac{e^{-E_1/k_B T}}{e^{-E_2/k_B T}}. \tag{6.3}$$

The analysis can be extended to the case where the system Σ has any number of states as follows. Since Σ is in equilibrium internally, we can separate out any pair of states and regard this pair as a two-state system, including all the other states in the heat bath Σ_0. If we do this successively for every pair of states, we find a relation like Equation (6.3) for each pair. This can only be the case if, for a single state i of energy E_i, the probability of occupation is given by:

$$\boxed{P_i = Z^{-1} e^{-E_i/k_B T},} \tag{6.4}$$

where Z^{-1} is a normalizing factor to be determined. Equation (6.4) is called the *canonical distribution*. It is probably the most important formula in statistical physics. The factor $e^{-E_i/k_B T}$ in Equation (6.4) is known as the *Boltzmann factor*.

The canonical distribution (Equation 6.4) was first established as a general result by Gibbs,[1] and we follow his nomenclature. It is often called the Boltzmann distribution,[2] but this terminology is not only historically inaccurate but is confusing, since the same name is often given to the "Maxwell-Boltzmann distribution" which is derived in Chapter 9. The Maxwell-Boltzmann distribution has the same mathematical form as the canonical distribution but applies *only* to an ideal gas, while the canonical distribution is much more general. The Boltzmann factor and the canonical distribution apply to *any* system with a fixed number of particles which is in thermal equilibrium.

The canonical distribution gives quantitative expression to the intuitive idea, illustrated in Example 5.2, that at low temperature, only low energy states are populated, while at sufficiently high temperature, all states are populated almost equally. Note that so long as the temperature is finite and positive, the probability of occupation of any one non-degenerate state always decreases with the energy of the state; constant probability corresponds to infinite temperature. If states of higher energy are more populated than those of lower, T must be negative (see Section 5.5).

---

[1] J. W. Gibbs, *The Elementary Principles of Statistical Mechanics* (1902); reprinted in 1981 by Oxbow Press. Since the canonical distribution is so important, it is worth your while to look at other derivations. Algorithmic derivations that do not invoke the concept of entropy are given in Chapter 7 of C. A. Whitney, *Random Processes in Physical Systems* (Wiley, 1990), and in *Am. J. Phys.* **63**, 876–878 (1995); ibid. **64**, 13–14 (1996). Most of these derivations are restricted to a particular system (such as a collection of harmonic oscillators, or the ideal gas). The derivation in most textbooks involves the maximization of the entropy of a collection of identical microsystems by the method of Lagrangian multipliers (see Problem 6.10). The clearest *experimental* demonstration of the canonical distribution in a model gas, using readily available equipment, is described by Jeffrey J. Prentis in Section IV of "Experiments in statistical mechanics," *Am. J. Phys.* **68**, 1073–1083 (2000). For a simple experiment that demonstrates the canonical distribution of electrons in a semiconductor, see Footnote 12 later in this chapter.

[2] Or sometimes Boltzmann's law (e.g., by Feynman in his lectures).

6.4 The Partition Function

The normalizing factor Z can easily be found. Since the microsystem must be in *some* state, the sum of the probabilities given by Equation (6.4) is 1; that is:

$$\sum_{i=1}^{n} P_i = Z^{-1} \sum_{i=1}^{n} e^{-\frac{E_i}{k_B T}} = 1, \tag{6.5}$$

where n is the total number of states and $i = 1$ labels the lowest energy state (often called the "ground" state). For simplicity, we normally write $\sum_i$ for the summing operation $\sum_{i=1}^{n}$ unless this causes ambiguity.

Hence,

$$\boxed{Z = \sum_i e^{-\frac{E_i}{k_B T}}.} \tag{6.6}$$

Z is usually called the *partition function* in English (a term due to Gibbs), although the German "Zustand-summe" (state sum) is more descriptive. It is often convenient to write β for $\frac{1}{k_B T}$, so that Equation (6.6) becomes:

$$Z = \sum_i e^{-\beta E_i}. \tag{6.7}$$

Note that every state must be included separately in this sum. A degenerate energy level contributes g_j identical terms $e^{-\beta E_j}$ to the sum, where E_j is the energy of the level and g_j is its degeneracy. Hence, we can lump all states with the same energy together and write Z as a sum over energy levels rather than over states:

$$\boxed{Z = \sum_j g_j e^{-\beta E_j},} \tag{6.8}$$

since the sum over g_j states with energy E_j is simply $g_j e^{-\beta E_j}$.

The mean (average) energy $\langle E \rangle$ of the microsystem can be found directly from Z:

$$\boxed{\langle E \rangle = \sum_i P_i E_i = Z^{-1} \sum_i E_i e^{-\beta E_i} = -Z^{-1} \frac{\partial Z}{\partial \beta} = -\frac{\partial \ln Z}{\partial \beta},} \tag{6.9}$$

where in the third step we differentiate Z term by term, keeping all the energy levels fixed, and the fourth step follows by the chain rule from $\frac{d \ln Z}{dZ} = Z^{-1}$. Since the energy levels, which depend in general on volume (and possibly on external influences such as magnetic field), are fixed during the differentiation, the partial derivative in Equation (6.9) is taken with V, H, and any other external influence kept constant.

Consider a macrosystem consisting of N identical and weakly interacting microsystems. By "weakly interacting" we mean that the energy levels of one microsystem are assumed to be unaffected by the state of any other, so that the energy U of the macrosystem is the sum of the energies of the component microsystems. Each microsystem has a mean energy $\langle E \rangle$, so that the internal energy is given by:

$$U = N \langle E \rangle = -N \frac{\partial \ln Z}{\partial \beta} = N k_B T^2 \frac{\partial \ln Z}{\partial T}, \tag{6.10}$$

where the energies E_i are kept constant in the differentiation.

We shall find that all the important thermodynamic quantities can be derived from Z. When one applies statistical mechanics to a particular system, the first and most important task is usually to determine Z.

6.5 The Many-Particle Partition Function

There is nothing in the derivation of the partition function or the Boltzmann factor that restricts it to a microsystem. The only condition is that it be much smaller than the heat bath Σ_0, which we can make as large as we please. We shall often find it convenient to use the partition function for an entire macrosystem, which we call Z_N to distinguish it from the partition function of the constituent microsystems. We call Z_N the *many-particle partition function* to distinguish it from the *single-particle partition function Z* (or Z_1 when we need to emphasize the distinction). The mean energy of a macrosystem is simply U, so that Equation (6.9) becomes:

$$U = -\frac{\partial \ln Z_N}{\partial \beta}. \tag{6.11}$$

To find Z_N in terms of Z_1, we limit ourselves here to the case where the microsystems are weakly interacting, in the sense that, while they can exchange energy, the energy levels of any one microsystem are unaffected by the state of any other. We furthermore assume for now that the microsystems are *distinguishable*. What this means is that it is possible in principle to determine which microsystem is in which state. For example, the spins in Chapter 4 are distinguishable, since each spin is a certain position in the solid, and one can imagine doing an experiment in which one determines whether a particular spin is up or down.[3] Identical particles in motion, on the other hand, are indistinguishable; their combined wave function is either symmetric or antisymmetric in exchange of particles, and it is impossible even in principle to determine, and meaningless to ask, which particle is in which state (see Footnote 4 of Chapter 10).

We begin by considering two identical, but distinguishable, spins, labelled A and B. Each has two states, ↑ and ↓, with energies $E_\uparrow$ and $E_\downarrow$. The partition function for an individual spin is:

$$Z_1 = e^{-\beta E_\uparrow} + e^{-\beta E_\downarrow}.$$

The combined system of two spins has the four states illustrated in Figure 4.1, and its partition function is:

$$\begin{aligned} Z_2 &= e^{-2\beta E_\uparrow} + 2e^{-\beta(E_\uparrow + E_\downarrow)} + e^{-2\beta E_\downarrow} \\ &= \left(e^{-\beta E_\uparrow} + e^{-\beta E_\downarrow}\right)^2 \\ &= Z_1^2. \end{aligned}$$

Similarly, a system of three spins has the eight states illustrated in Figure 3.2, and its partition function is:

$$\begin{aligned} Z_3 &= e^{-3\beta E_\uparrow} + 3e^{-\beta(2E_\uparrow + E_\downarrow)} + 3e^{-\beta(E_\uparrow + 2E_\downarrow)} + e^{-3\beta E_\downarrow} \\ &= \left(e^{-\beta E_\uparrow} + e^{-\beta E_\downarrow}\right)^3 \\ &= Z_1^3. \end{aligned}$$

---

[3] Although such an experiment has not yet been done, it is not beyond the bounds of possibility. In fact, an experiment has already been proposed to detect the change of state of a single spin at a well-defined point in space, using modern techniques of scanning microscopy.

The pattern is obvious, in fact,

$$Z_N = Z_1^N. \tag{6.12}$$

Equation (6.12) holds for any number of weakly interacting identical but *distinguishable* microsystems, which may have any number (even infinite) of states, but its formal proof requires a tricky combinatorial argument and is rather difficult to make rigorous.

We only sketch the proof of Equation (6.12) here. Each microsystem has a set of states, labelled by i, with energies E_i, and its partition function Z_1 is given by Equation (6.7). Z_N is the sum of the Boltzmann factors of all possible combinations of these states. It can be written $Z_N = \sum_I e^{-\beta E_I}$, where I labels a many-particle state. Since the microsystems are weakly interacting, the energy E_I of this state is the sum of the energies of the N individual microsystems, which we label by k ($k = 1$ to N); that is, $E_I = \sum_{k=1}^{N} E_k$, where E_k is the energy of the k'th microsystem in some particular state. Hence the Boltzmann factor for the many-particle state is $\prod_{k=1}^{N} e^{-\beta E_k}$, and $Z_N = \sum \prod_{k=1}^{N} e^{-\beta E_k}$, where the sum is over all possible products.

Now consider the product Z_1^N. This is a product of sums, and each term in the product has the form $\prod_{k=1}^{N} e^{-\beta E_k}$, just as each term in Z_N has. Each possible combination of the individual states gives a separate term in the sum, so that Equation (6.12) follows. Combining Equations (6.11) and (6.12) gives $U = -\frac{\partial \ln Z_N}{\partial \beta} = -N \frac{\partial \ln Z_1}{\partial \beta} = N\langle E \rangle$, in agreement with Equation (6.10).

Note that Equation (6.12) only applies to *distinguishable* microsystems. In Chapter 9, we use the more powerful method of the grand partition function to show that for indistinguishable microsystems, such as the particles in a gas, the right-hand side of Equation (6.12) must be divided by a factor $N!$. It is possible, and customary, to derive this factor by eliminating the terms in the sum which are impermissible for indistinguishable particles. However, the argument is not trivial,[4] and we do not use it. The derivation in Chapter 9 does not use Equation (6.12).

6.6 Examples of the Use of the Canonical Distribution

In this section, we discuss some examples of the calculation and use of the single-particle partition function. We write Z for Z_1.

Example 6.1: Excited States of An Atom

The ground level of the neutral lithium atom is doubly degenerate (i.e., $g = 2$). The first excited level is 6-fold degenerate and is at an energy 1.2 eV above the ground level. In the outer atmosphere of the sun, which is at a temperature of about 6000 K, what fraction of the neutral lithium is in the excited level? Since all the other levels of Li are at a much higher energy, it is safe to assume that they are not significantly occupied.

If the ground level energy is defined as zero and E is the energy of excited level,

$$Z = 2 + 6e^{-\beta E}.$$

The probability that the atom is in its excited level is, from Equation (5.8):

$$P(E) = 6Z^{-1}e^{-\beta E} = \frac{3}{e^{\beta E} + 3}.$$

---

[4] See, for example, Reif, Section 7.3. The derivation has been discussed critically by R. Baierlein in "The fraction of 'all different' combinations: Justifying the semi-classical partition function," *Am. J. Phys.* **65**, 314 (1997).

$E = 1.2$ eV and $T = 6000$ K, whence $\beta E = 2.32$, $e^{\beta E} \approx 10$.

$$P(E) = \frac{3}{10 + 3} = 0.23.$$

Example 6.2: DNA Unwinding: The Zipper Model

A simple model of the separation of two-stranded DNA into single strands has been proposed by Kittel.[5] The double strand is modeled as a zipper of N links. Opening one link requires an energy E_0 and increases the degeneracy by a factor g_0, since opening a link gives the subunits of the strands more ways to orient themselves than when it is closed. A link can open only if all the links between it and one end are already open. Let the number of open links in a particular double strand be r. We want to find the average number of links open at temperature T, for the cases $g_0 = 1$ and $g_0 > 1$, and thus find the temperature at which complete unwinding (called *denaturation*) occurs. To simplify the calculation, we assume that the unwinding can only occur from one end, as in a real zipper, that g_0 does not depend on the position of the link in the zipper, and that the completely linked state is non-degenerate.

(a) Find the partition function.

The energy of the state with r links open is $E(r) = rE_0$, and its degeneracy $g(r) = g_0^r$, so the partition function is:

$$Z = \sum_{r=0}^{N} g(r)e^{-\beta E(r)} = \sum_{r=0}^{N} g_0^r e^{-r\beta E_0} = \sum_{r=0}^{N} \left(g_0 e^{-\beta E_0} \right)^r.$$

The sum from $r = 0$ to N of the geometric series a^r is $\sum_{r=0}^{N} a^r = \frac{1-a^{N+1}}{1-a}$. Hence,

$$Z = \frac{1 - g_0^{N+1} e^{-(N+1)\beta E_0}}{1 - g_0 e^{-\beta E_0}}.$$

(b) Taking $g_0 = 1$, find the average number $\langle r \rangle$ of open links under the following conditions:
(i) $k_B T \ll E_0$; (ii) $E_0 \ll k_B T \ll NE_0$.

$E(r) = rE_0$, so that:

$$\langle r \rangle = \frac{\langle E(r) \rangle}{E_0} = -E_0^{-1} \frac{\partial \ln Z}{\partial \beta}.$$

States with $\beta E(r) \gg 1$ are not significantly occupied, so in both cases we can put $N = \infty$ in the equation for Z without error.

Hence, $Z = \frac{1}{1-e^{-\beta E_0}}$, and $\ln Z = -\ln \left(1 - e^{-\beta E_0} \right)$.

(i) $k_B T \ll E_0$; that is, $\beta E_0 \gg 1$. Then $e^{-\beta E_0} \ll 1$ and $\ln Z \approx e^{-\beta E_0}$, since $\ln(1 + x) \approx x$ for small x.

$$\langle r \rangle = -E_0^{-1} \frac{\partial \left(e^{-\beta E_0} \right)}{\partial \beta} = e^{-\beta E_0}.$$

The probability of one link being open is $e^{-\beta E_0}$, which is small, and the probability of two or more being open is negligible.

---

[5] C. Kittel, "Phase transition of a molecular zipper," *Am. J. Phys.* **37**, 917 (1969).

(ii) When $E_0 \ll k_B T \ll N E_0$, the term $e^{-(N+1)\beta E_0}$ can still be neglected, but $\beta E_0 \ll 1$ so that $1 - e^{-\beta E_0} \approx \beta E_0$. Now we have:

$$Z \approx \frac{1}{\beta E_0} \quad \text{and} \quad \ln Z \approx -\ln \beta - \ln E_0.$$

$$\langle r \rangle = -E_0^{-1} \frac{\partial \ln Z}{\partial \beta} = E_0^{-1} \frac{\partial (\ln \beta)}{\partial \beta} = \frac{k_B T}{E_0}.$$

This has a simple interpretation: the thermal energy $k_B T$ is shared over $\langle r \rangle$ links.

(c) Taking $g_0 = 1$, find $\langle r \rangle$ when $k_B T \gg N E_0$.

Since $k_B T \gg N E_0$ implies $N \beta E_0 \ll 1$, one might be tempted to put $e^{-(N+1)\beta E_0} \approx 1 - (N+1) \beta E_0$, giving $Z \approx \frac{(N+1)\beta E_0}{\beta E_0} = N + 1$. However, since this is a constant, taking its derivative gives no information, and we must include the next term in the Taylor expansion:

$$e^{-(N+1)\beta E_0} \approx 1 - (N+1)\beta E_0 + \frac{1}{2} [(N+1)\beta E_0]^2 .$$

$$Z \approx (N+1) + \frac{1}{2}(N+1)^2 \beta E_0 = (N+1) \left[1 + \frac{1}{2}(N+1)\beta E_0 \right].$$

$$\ln Z \approx \ln(N+1) + \ln \left[1 + \frac{1}{2}(N+1)\beta E_0 \right]$$

$$\approx \ln(N+1) + \frac{1}{2}(N+1)\beta E_0,$$

since $N \beta E_0 \ll 1$.

Hence,

$$\langle r \rangle = -E_0^{-1} \frac{\partial (\ln Z)}{\partial \beta} = \frac{1}{2}(N+1).$$

In this (unrealistic) version of the model, the strands never completely unwind.

(d) Taking $g_0 > 1$, find the temperature at which complete unwinding (denaturation) occurs.

So long as $g_0 e^{-\beta E_0} < 1$, the sum to infinity for Z converges. If $g_0 e^{-\beta E_0} \geq 1$, this is not the case. The probability of r links being open then increases as r increases, so that the strands separate completely. At low temperature $e^{-\beta E_0}$ is small so that $g_0 e^{-\beta E_0} < 1$, but as the temperature is raised we reach a point where $g_0 e^{-\beta E_0} = 1$ and the strands unwind. The temperature where this happens is $T_t = \frac{E_0}{k_B \ln g_0}$. Note that so long as g_0 is a constant (i.e., it is the same whichever link is broken), this temperature is independent of the length of the strand, an important fact biologically since DNA comes in many different lengths. According to Kittel,[5] $g_0 \sim 10^4$ for DNA, and $E_0 \approx 0.3$ eV, so that $T_t \approx 350$ K. If N is large, the transition is quite abrupt: at the temperature of the human body (310K), $\langle r \rangle \approx 0.15$, so that only about 15% of the DNA molecules have even one link open. DNA and RNA will be discussed further in Chapter 18.

Example 6.3: The Harmonic Oscillator: Einstein's Model of a Solid

In 1907 Einstein, in the first application of quantum theory to a problem other than radiation, modelled a solid body containing N atoms as a collection of $3N$ harmonic oscillators (since each atom can vibrate in three directions). For simplicity, he assumed that all these oscillators have the same frequency and are only weakly coupled to each other. Find the partition function for a single oscillator, and the internal energy and heat capacity of this model solid.

We know from quantum mechanics (see Appendix B) that the states of a harmonic oscillator of frequency ν have energies given by:

$$E_n = \left(n + \frac{1}{2}\right) h\nu, \tag{6.13}$$

where h is Planck's constant and n is any non-negative integer. Substituting Equation (6.13) into Equation (6.6), we obtain for the partition function of a single oscillator:

$$Z = \sum_n e^{-\beta E_n} = e^{-\beta h\nu/2} \sum_n e^{-n\beta h\nu} = e^{-\beta h\nu/2} \sum_n \left(e^{-\beta h\nu}\right)^n.$$

The sum is a geometric series of the form $\sum_{n=0}^{\infty} a^n = (1-a)^{-1}$ $(a < 1)$, so that:

$$Z = \frac{e^{-\beta h\nu/2}}{1 - e^{-\beta h\nu}} = \frac{1}{2}\text{cosech}\left(\frac{\beta h\nu}{2}\right). \tag{6.14}$$

Either expression for Z may be used, as convenient. The hyperbolic form is particularly useful when taking derivatives of Z, since the algebra is simpler.

The mean energy of a single oscillator is obtained by substituting Equation (6.14) into Equation (6.9):

$$\langle E \rangle = -\frac{\partial \ln Z}{\partial \beta} = \frac{1}{2}h\nu \coth\left(\frac{\beta h\nu}{2}\right) = h\nu\left(\frac{1}{2} + \frac{1}{e^{\beta h\nu} - 1}\right), \tag{6.15a}$$

while for Einstein's model of $3N$ weakly interacting identical oscillators:

$$U(T) = 3N\langle E \rangle = \frac{3}{2}Nh\nu \coth\left(\frac{\beta h\nu}{2}\right) = 3Nh\nu\left(\frac{1}{2} + \frac{1}{e^{\beta h\nu} - 1}\right). \tag{6.15b}$$

The internal energy is not a directly measurable quantity, and instead we measure the heat capacity, which is the temperature derivative of U: $C_V = \frac{dU}{dT}$ (we need not distinguish C_p from C_V in this case, since the expansion of the solid is negligible). This is usually referred to one kmole of material, for which $N = N_A$. Differentiating Equation (6.15b) with respect to T and remembering that $\beta \equiv \frac{1}{k_B T}$, we find for the kmolar heat capacity (the heat capacity per kmole):

$$C_V(T) = \frac{dU}{dT} = -k_B \beta^2 \frac{dU}{d\beta}$$

$$= 3R_0 \left(\frac{\beta h\nu}{2}\right)^2 \text{cosech}^2\left(\frac{\beta h\nu}{2}\right) = 3R_0 \left(\beta h\nu\right)^2 \frac{e^{\beta h\nu}}{\left(e^{\beta h\nu} - 1\right)^2} \tag{6.16}$$

where $R_0 = N_A k_B$.

It is convenient to define the "Einstein temperature" by $\theta_e \equiv \frac{h\nu}{k_B}$, and to look at the limiting behavior of Z and $C(T)$ at temperatures that are either high or low relative to θ_e. At high T $(T \gg \theta_e)$, $\beta h\nu$ is small. For small x, $\text{cosech}(x) \approx 1/x$, and substituting this in Equation (6.16) gives the high T limit:

$$C_V \rightarrow 3R_0 \tag{6.17}$$

The lowest order fractional correction to C_V is shown in Appendix A to be $-\frac{1}{12}\left(\frac{\theta_e}{T}\right)^2$, so that C_V deviates significantly from $3R_0$ only when T is close to or below θ_e.

At low T, $(T \ll \theta_e)$ $\beta h\nu$ is large, and we find:

$$C_V(T) \approx 3R_0 \left(\frac{\theta_e}{T}\right)^2 e^{-\theta_e/T}. \tag{6.18}$$

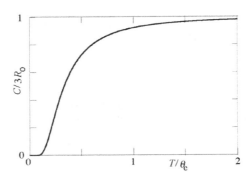

FIGURE 6.2   Heat capacity of an Einstein solid as a function of temperature, from Equation (6.16).

Because of the exponential, C_V at low T is much less than at high T. The full dependence of C_V on temperature in Equation (6.16) is shown in Figure 6.2, in which $\frac{C_V(T)}{3R_0}$ is plotted against $\frac{T}{\theta_e} = \frac{1}{\beta h v}$.

A collection of classical oscillators, whose energy levels are continuous, corresponds to $h v \to 0$ in Equation (6.16). The high T limit (Equation 6.17) then applies at all temperatures, and C_V should be independent of T. It was established by experiment early in the nineteenth century that Equation (6.17) applies to most solid elements at ordinary temperatures,[6] and is called the Dulong and Petit law. This fact was subsequently explained as being a direct consequence of the classical equipartition theorem (see Section 7.8 and Appendix G). What the classical theory could not explain was that the law fails in certain solids (e.g., diamond) at room temperature, and fails in all solids at sufficiently low temperature. Einstein showed that this apparent anomaly could be qualitatively understood in terms of the then-new quantum theory, and that the quantum statistical argument given above leads to the curve shown in Figure 6.2. This is called the Einstein heat capacity curve, and a model solid with one discrete vibrational frequency is known as an Einstein solid. The reason why the deviation from the Dulong and Petit law is particularly noticeable in diamond is that this is a hard material (indicating strong interatomic forces) made up of light atoms. As a result, the average frequency of vibrations is high compared with that of softer, heavier elements, just as the frequency of the oscillations of a light mass on a stiff spring is higher than that of a heavy mass on a soft spring.

Einstein's assumption of a single frequency is clearly incorrect for most solids, since it does not take into account the interactions between the atoms, which produce a continuum of frequencies. A more accurate theory was put forward by Debye in 1912; it bears a close relation to the problem of thermal radiation and is discussed in Section 7.8. Einstein's model, which can easily be extended to take account of more than one discrete frequency, applies to the contribution of internal molecular vibrations to the heat capacity of a simple molecular solid or gas, since their frequencies are few and discrete (see Section 10.6).

6.7 Fluctuations

A system in thermal contact with a heat bath is continually exchanging energy with it. Hence, unlike the isolated system considered in Chapter 4, the energy of such a system is not exactly constant: its instantaneous value fluctuates about the mean energy $\langle E \rangle$. A measure of this fluctuation is the mean square deviation from the average (the variance): $\langle \Delta E^2 \rangle \equiv \langle (E - \langle E \rangle)^2 \rangle$ (see Appendix D). The variance can be calculated from the partition function as follows.

---

[6] See Bailyn, pp. 32–33.

From Equation (D.10), we have $\langle \Delta E^2 \rangle = \langle E^2 \rangle - \langle E \rangle^2$, where $\langle E^2 \rangle \equiv \sum_j P(E_j) E_j^2$. Here $P(E_j)$ is the probability of the system having an energy E_j. From Equation (6.8), $P(E_j) = Z^{-1} g_j \exp(-\beta E_j)$ so that:

$$\langle E^2 \rangle = Z^{-1} \sum_j g_j E_j^2 e^{-\beta E_j}$$

$$= Z^{-1} \frac{\partial^2 Z}{\partial \beta^2}, \tag{6.19}$$

where, as before, we differentiate Z term by term, keeping the E_j constant.

From Equation (6.9):

$$\langle E \rangle^2 = Z^{-2} \left(\frac{\partial Z}{\partial \beta} \right)^2,$$

so that:

$$\langle \Delta E^2 \rangle = Z^{-1} \frac{\partial^2 Z}{\partial \beta^2} - Z^{-2} \left(\frac{\partial Z}{\partial \beta} \right)^2$$

$$= \frac{\partial^2 (\ln Z)}{\partial \beta^2}. \tag{6.20}$$

From Equation (6.9), $\langle E \rangle = -\frac{\partial (\ln Z)}{\partial \beta}$, so we can write Equation (6.20) in the more transparent form:

$$\langle \Delta E^2 \rangle = -\frac{\partial \langle E \rangle}{\partial \beta}$$

$$= k_B T^2 \frac{\partial \langle E \rangle}{\partial T}, \tag{6.21a}$$

where everything that can change the energy levels (e.g., volume and magnetic field) is kept constant when taking the derivative.

For a macrosystem consisting of weakly interacting microsystems, for which Z is replaced by Z_N, $\langle E \rangle$ in Equation (6.21a) is replaced by U as in Equation (6.11), and $\frac{\partial U}{\partial T} = C_V$, so that Equation (6.21a) becomes

$$\langle \Delta U^2 \rangle = k_B T^2 C_V. \tag{6.21b}$$

If the heat capacity diverges, as it often does at a second order phase transition, Equation (6.21b) shows that the fluctuations become very large. Phase transitions are discussed in Chapter 15.

The fractional root mean square fluctuation in the internal energy of a macrosystem consisting of N identical weakly interacting microsystems is, from Equations (6.9) and (6.20):

$$\left\langle \left(\frac{\Delta U}{U} \right)^2 \right\rangle^{1/2} = \frac{\left(\frac{\partial^2 \ln Z_N}{\partial \beta^2} \right)^{1/2}}{\frac{\partial \ln Z_N}{\partial \beta}}$$

$$= N^{-1/2} \left\langle \left(\frac{\Delta E}{E} \right)^2 \right\rangle^{1/2}, \tag{6.22}$$

where E is the energy of the microsystem. The fractional fluctuation in U drops off as $N^{-1/2}$ as N increases. Equation (6.22) is the generalization of the result obtained for a particular model in Chapter 4. This inverse square root dependence on the size of the sample is a general statistical result for the fractional root mean square fluctuation of the sum of independent random events.[7] Note that it only applies to weakly interacting microsystems. However, Equations (6.21a) and (6.21b) apply to any system.

Example 6.4: Energy Fluctuations of a Harmonic Oscillator

What is the mean square fluctuation of the energy of a harmonic oscillator of frequency v in contact with a heat bath at temperature T?

The energy fluctuations of the harmonic oscillator are of considerable historical interest, since they were instrumental in leading Planck to his radiation law and Einstein to the photon hypothesis (see the reference in Footnote 18 of Chapter 11). Applying Equation (6.20) to the partition function (Equation 6.14), we find:

$$\langle \Delta E^2 \rangle = \left(\frac{hv}{2} \right)^2 \operatorname{cosech}^2 \left(\frac{\beta hv}{2} \right) = \left(\frac{hv}{2} \right)^2 \left[\coth^2 \left(\frac{\beta hv}{2} \right) - 1 \right].$$

From Equation (6.15a):

$$\langle E \rangle = \left(\frac{hv}{2} \right) \coth \left(\frac{\beta hv}{2} \right)$$

so that:

$$\langle \Delta E^2 \rangle = \langle E \rangle^2 - \left(\frac{hv}{2} \right)^2.$$

Substituting $E = hv \left(n + \frac{1}{2} \right)$ from Equation (6.13), noting that $\langle \Delta E^2 \rangle = (hv)^2 \langle \Delta n^2 \rangle$, that $\langle \Delta n \rangle = 0$ from the definition of a mean, and canceling the factor $(hv)^2$, we obtain:

$$\langle \Delta n^2 \rangle = \langle n \rangle^2 + \langle n \rangle. \tag{6.23}$$

Einstein obtained a result equivalent to Equation (6.23) from Planck's radiation law (see Chapter 10) and recognized that the $\langle n \rangle^2$ term in Equation (5.23) is what is expected for the fluctuations of classical waves. The term linear in $\langle n \rangle$ has the form of the density fluctuations of a gas, and implies that an oscillator in its n'th energy level can be regarded as a collection of n particles of energy hv; that is, photons. We shall return to this in Chapter 11, where we discuss the photon gas.

6.8 Obtaining the Entropy and Free Energy from the Partition Function

If we know the partition function Z, the free energy F and entropy S can be obtained with the help of the Gibbs-Helmholtz Equation (5.25). Eliminating U from Equations (6.11) and (5.25c) gives:

$$\left(\frac{\partial (\beta F)}{\partial \beta} \right)_V = - \left(\frac{\partial \ln Z_N}{\partial \beta} \right)_V. \tag{6.24}$$

---

[7] See, for example, C. M. Grinstead and J. L. Snell, *Introduction to Probability* (American Mathematical Society, 1997), Theorem 6.9.

This can be integrated at constant V to give:

$$\beta F = - [\ln Z_N + \phi(V)],$$

or

$$F = -k_B T [\ln Z_N + \phi(V)], \qquad (6.25)$$

where $\phi(V)$ is an as yet undetermined function that does not depend on T.

Substituting Equation (6.25) in Equation (5.24), we obtain:

$$S = - \left(\frac{\partial F}{\partial T} \right)_V = k_B \left[T \frac{\partial \ln Z_N}{\partial T} + \ln Z_N + \phi(V) \right]. \qquad (6.26)$$

To find $\phi(V)$, we note that when $T \to 0$, only the lowest energy level of the macrosystem is occupied. If we define the energy of this level as $U = 0$, and it has multiplicity Ω_0, $Z_N \to \Omega_0$ as $T \to 0$. From the original definition of entropy, Equation (5.3):

$$S(T \to 0) = k_B \ln \Omega_0. \qquad (6.27)$$

Substituting $T = 0$, $Z_N = \Omega_0$, and $S = k_B \ln \Omega_0$ in Equation (6.26) shows that $\phi(V) = 0$, so that:

$$F = -k_B T \ln Z_N, \qquad (6.28a)$$
$$= -N k_B T \ln Z_1 \qquad (6.28b)$$

for distinguishable microsystems for which Equation (6.12) holds.

Equation (6.28a) can be written:

$$\boxed{Z_N = e^{-\beta F}.} \qquad (6.29)$$

Since $\phi(V) = 0$, and $\beta \propto T^{-1}$, the entropy is:

$$S = k_B \left(\ln Z_N - \beta \frac{\partial \ln Z_N}{\partial \beta} \right). \qquad (6.30)$$

Algebraic manipulation (see Problem 6.1a) of Equation (6.30) gives

$$\boxed{S = -k_B \sum_i P_i \ln P_i,} \qquad (6.31)$$

where $P_i = Z_N^{-1} e^{-\beta E_i}$ is the probability that the macrosystem is in the i'th microstate. The expression for the entropy of mixing (Equation 5.16) can be regarded as a special case of Equation (6.31), the sum being taken over the N atomic sites in the crystal.

Entropy is often defined by Equation (6.31). This is more general than $S = k_B \ln \Omega$, our original definition of entropy (Equation 5.4), since $S = k_B \ln \Omega$ follows from Equation (6.31) if we take all the P_i's equal to $1/\Omega$ (see Problem 6.1b). Equation (6.31) refers to a system in contact with a heat bath, while Equation (5.4) is for an isolated system of fixed energy. Although the physical significance of Equation (6.31) is less clear than Equation (5.4), it has the advantage as a definition of S that it makes an immediate connection with information theory,[8] in which the missing information of a garbled message is $\mathcal{M} \equiv - \sum_i P_i \ln P_i$,

---

[8] See, for example, R. Baierlein, *Atoms and Information Theory, an Introduction to Statistical Mechanics* (W. H. Freeman, 1971), Chapter 3, and H. B. Callen, *Thermodynamics and an Introduction to Thermostatistics*, 2nd ed. (Wiley, 1985), Section 17.1.

where P_i is the probability that the message has the i'th possible meaning. We do not in general know these probabilities, but their most likely values are those that maximize $\mathcal{M}$. $\mathcal{M}$ is analogous (in fact, mathematically identical) to entropy and is often called *informational entropy*. It can be shown (see Problem 6.10) that S (or $\mathcal{M}$) is maximized when the P_i's are given by the canonical distribution Equation (6.4). This interpretation is implicit in the commonly quoted loose definition of entropy as the "degree of disorder" of a system. However, we shall not pursue this connection here.

There is another important result implicit in Equation (6.31). A system undergoes an adiabatic process if the occupancies of its quantum states do not change. The occupancies are simply NP_i, and N is fixed, so that the P_i do not change; then, from Equation (6.31), the entropy does not change. Thus, an adiabatic process is necessarily isentropic. Simple cases illustrating this fact are described in Chapter 5 (Section 5.11) and Example 10.2.

6.9 Scaling of Energy and Temperature

Many simple microsystems can be described by a single characteristic energy ϵ (e.g., $h\nu$ for the harmonic oscillator or μB for the spin in a magnetic field), so that Z is a dimensionless function of $\beta\epsilon$ only. It follows from Equations (6.12) and (6.30) that if the interactions are weak, the dimensionless quantity S/Nk_B, and hence S, is also a function of $\beta\epsilon$ only. Similarly, it follows from Equation (6.28) that F, and hence all quantities that have the dimensions of energy, such as U, G, and H, must have the form $\epsilon f(\beta\epsilon)$ where $f(x)$ is a dimensionless function.

The temperature dependence of any thermodynamic property of such a system can be reduced to a single universal curve, such as Figure 6.2. Another useful consequence of the fact that S is a function of $\beta\epsilon$ alone is that in an isentropic process, $\beta\epsilon$ cannot change. If the energy levels are changed (e.g., by varying a magnetic field), T must change in proportion. An example of this scaling is given in Example 6.5.

6.10 Free Energy and Equilibrium

A closed system of fixed energy comes into equilibrium when its entropy is a maximum. However, this cannot be the case for a system in thermal equilibrium with a heat bath, since (at positive temperature) the entropy of the system can always be increased by increasing its internal energy. What matters is the entropy of the system *plus the heat bath*, and at the beginning of this chapter we found that the entropy of the heat bath decreases as the energy of the system increases. We saw in Chapter 5, by considering some particular cases, that equilibrium is achieved when the free energy is minimized. We now show that this is a general result for any isothermal process at constant volume; minimization of F determines the optimum compromise between maximizing entropy and minimizing energy.

Suppose that the system Σ is a macrosystem that can be in a variety of macrostates of different energy U and multiplicity Ω. One of these is the equilibrium state, which by our fundamental hypothesis is the most probable state.[9] The probability of Σ being in any particular macrostate is proportional to the Boltzmann factor for that state; that is, $\Omega e^{-\beta U}$. Since $\Omega = e^{S/k_B}$,

$$\Omega e^{-\beta U} = e^{(S/k_B - \beta U)} = e^{\beta(TS-U)} = e^{-\beta F}, \tag{6.32}$$

---

[9] The other states may be metastable; that is, while their free energy is higher than the equilibrium state, some constraint prevents them from reaching true equilibrium (see Section 6.11).

where $F = U - TS$ is the free energy. Hence, the most probable state is the one with minimum F; if at a fixed temperature and volume a system is initially in some other state, it will change (if this is permitted by the constraints) until F is minimized.

We can apply these ideas to find an expression for the fluctuation in any quantity that appears in the expression for the free energy F. Let the instantaneous value of this quantity be x, where F is a function of x, as well as of other variables, and its mean value be $\langle x \rangle$. The probability $P(x)\, dx$ that the quantity has a value between x and $x + dx$ is, from Equation (6.32):

$$P(x)\, dx = \frac{e^{-\beta F(x)}\, dx}{\int e^{-\beta F(x)}\, dx}.$$

In a macrosystem, $P(x)$ is sharply peaked about the mean value $\langle x \rangle$, so we can expand the free energy about $\langle x \rangle$, putting $\xi \equiv x - \langle x \rangle$ and using the fact that $F(x)$ is a minimum for $x = \langle x \rangle$, so that the linear term in the Taylor expansion in ξ vanishes:

$$F(x) = F(\langle x \rangle) + \frac{1}{2}F''\xi^2 \ldots \tag{6.33}$$

We have written F'' for $\frac{\partial^2 F}{\partial x^2}$, the second derivative being taken at $x = \langle x \rangle$, with all variables except x kept constant.

Hence, $P(\xi)$ has a Gaussian distribution, given by:

$$P(\xi) \approx \frac{\exp\left(-\frac{1}{2}\beta F''\xi^2\right)}{\int \exp\left(-\frac{1}{2}\beta F''\xi^2\right)\, d\xi},$$

where the constant factor $e^{-\beta F(\langle x \rangle)}$ has been cancelled, and we have used the fact that $d\xi = dx$.

The mean square fluctuation in x is, since $\xi \equiv x - \langle x \rangle$,

$$\langle \Delta x^2 \rangle = \langle \xi^2 \rangle = \frac{\int_{-\infty}^{\infty} \xi^2 \exp\left(-\frac{1}{2}\beta F''\xi^2\right)\, d\xi}{\int_{-\infty}^{\infty} \exp\left(-\frac{1}{2}\beta F''\xi^2\right)\, d\xi}$$

$$= \frac{k_B T}{F''}. \tag{6.34}$$

The equation for the energy fluctuation (Equation 6.21) is a special case of this general relation (see Problem 6.13).

The physical significance of Equation (6.34) is that if F'' (i.e., $\frac{\partial^2 F}{\partial x^2}$ at $x = \langle x \rangle$, with every parameter but x kept constant) is small, so that the curve $F(x)$ is very flat near its minimum, large fluctuations are to be expected. An extreme case of this occurs at a second order phase transition (see Chapter 15).

Example 6.5: Susceptibility and Magnetic Heat Capacity of a Spin 1/2 System; The Schottky Bump

The simplest possible magnetic system is the spin 1/2 system discussed in Chapter 4. The energy levels are $\pm \mu B$, where B is the magnetic field, and are non-degenerate, so that the partition function for a single spin is:

$$Z = e^{\beta \mu B} + e^{-\beta \mu B}. \tag{6.35}$$

At high temperature, where $\beta \mu B \ll 1$, Z can be expanded to give:

$$Z \approx 2 + (\beta \mu B)^2 = 2\left[1 + \frac{1}{2}(\beta \mu B)^2\right]. \tag{6.36}$$

From Equation (6.12), $Z_N = Z^N$, since the spins are distinguishable by their location, so that from Equation (6.28):

$$F = -Nk_BT \ln Z. \tag{6.37}$$

Expanding $\ln Z$ as a series in $\beta \mu B$, using Equation (A.5), we have:

$$\ln Z \approx \ln \left[2 + (\beta \mu B)^2 \right] = \ln 2 + \ln \left[1 + \frac{1}{2} (\beta \mu B)^2 \right]$$
$$\approx \ln 2 + \frac{1}{2} (\beta \mu B)^2 .$$

so that:

$$F \approx -Nk_BT \left[\ln 2 + \frac{1}{2} (\beta \mu B)^2 \right]. \tag{6.38}$$

From Equation (5.31) the isothermal susceptibility at high temperature ($k_B T \gg \mu B$) is:

$$\chi^T = -\left(\frac{\partial^2 F}{\partial B^2} \right)_T = N\beta \mu^2 = N \frac{\mu^2}{k_B T}, \tag{6.39}$$

which is Curie's law (see Problem 5.3c). Another way to derive χ is to calculate the average magnetization per spin from Z. (see Problem 6.3).

We now calculate the magnetic contribution $C_m(T)$ to the heat capacity, at both high and low temperature. We first find U from Z, using Equation (6.10):

$$U = -N \frac{\partial \ln Z}{\partial \beta} = -N Z^{-1} \frac{\partial Z}{\partial \beta}$$
$$= -N\mu B \frac{e^{\beta \mu B} - e^{-\beta \mu B}}{e^{\beta \mu B} + e^{-\beta \mu B}}$$
$$= -N\mu B \tanh(\beta \mu B). \tag{6.40}$$

Since μ is at most of the order of a few Bohr magnetons, $\frac{\mu B}{k_B}$ is only a few degrees Kelvin at ordinary laboratory fields of a few Tesla, so that $\beta \mu B \ll 1$ except at very low temperatures. If this the case,

$$U = -N\beta (\mu B)^2 = -N \frac{(\mu B)^2}{k_B T},$$

so that:

$$C_m(T \to \infty) = N \frac{(\mu B)^2}{k_B T^2}. \tag{6.41}$$

Thus, C_m varies as T^{-2} at high T.

As $T \to 0$, $\beta \mu B \to \infty$ and $U \to -N\mu B$, so that at $C_m(T \to 0) = 0$ low T.

The full expression for C_m can be obtained from Equation: (6.40), using

$$C_m(T) = \left(\frac{\partial U}{\partial T} \right)_B = -k_B \beta^2 \left(\frac{\partial U}{\partial \beta} \right)_B$$
$$= -R_0 (\beta \mu B)^2 \text{sech}^2(\beta \mu B), \tag{6.42}$$

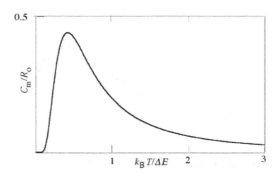

FIGURE 6.3 Heat capacity C_m of a two-level system (Schottky bump).

where we have put $N = N_A$, $R_0 = N_A k_B$, so that C_m in Equation (6.42) is the magnetic contribution to the kmolar heat capacity. $C_m(T)$ is plotted (in units of R_0) in Figure 6.3, where we have put $\Delta E = 2\mu B$. This curve is characteristic of any system with two energy levels separated by ΔE, and is called the Schottky bump. It has a maximum at $T = 0.42 \, \Delta E/k_B$, and the maximum value of C_m is $0.44R_0$, independent of ΔE. If ΔE is very small (e.g., the case of a nuclear spin in a magnetic field), the peak occurs at a very low temperature, but it is just as high as when ΔE is large. A similar curve is obtained for any system with a small number of approximately equally spaced levels.

The physical reason for the general shape of the curve is simple. At low temperature, most of the population is in the ground state. The population of the excited state, and therefore the internal energy U, increases exponentially with temperature, so that the heat capacity, being the derivative of U, also increases exponentially from zero. At high temperature, on the other hand, the populations are almost equal and little change can take place, so that the heat capacity tends to zero as $T \to 0$. Since the slope of $C_m(T)$ is positive at low T and negative at high T, there must be a maximum in between.[10]

Example 6.6: Magnetic Anisotropy

Crystalline solids are in general anisotropic; that is, their properties depend on direction. An elementary magnet in such a crystal has an energy that depends on its orientation relative to the crystal axes. This causes *magnetic anisotropy*, which is a very important property of magnetic materials, particularly those used in permanent magnets. In this example we find the free energy and the torque on such a crystal exerted by a magnetic field in an arbitrary direction, using a simple model and assuming that the temperature is not too low.

The simplest model of an anisotropic magnet assumes that it contains N independent spin 1/2 magnets whose moment μ depends on orientation. If μ is to have the same sign at all orientations, the simplest possible dependence is $\mu = \mu_m \cos^2 \theta$, where θ is the angle that the magnetic field B makes with the z-axis of the crystal and μ_m is a constant (assumed positive in this example). Do not confuse the angle θ with empirical temperature. We first find the free energy F as a function of θ, B, and T, assuming that $\mu_m B \ll k_B T$. By a process analogous to the derivation of pressure (Equation 5.22), we then find the magnetic torque on the crystal from F.

---

[10] The fact that the maximum value of C_m is independent of ΔE can also be understood qualitatively. $U = -\frac{1}{2}N_A \, \Delta E$ at $T = 0$ and $U = 0$ at $T = \infty$. The integral under the curve is just the change in U, so that $\int_0^\infty C_m \, dT = \frac{1}{2}N_A \, \Delta E$. The scale of temperature, and hence the width of the curve, is also proportional to ΔE (see Section 6.9). Since the integral under a curve of a given shape is proportional to the product of its height and width, it follows that the height must be independent of ΔE.

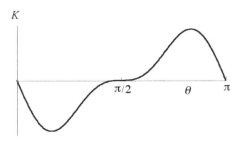

FIGURE 6.4   Torque on an anisotropic magnetic crystal (Equation 6.43).

The form of the partition function and free energy is the same as in Example 6.5, with $\mu_m \cos^2 \theta$ substituted for μ. Since $\beta \mu_m B \ll 1$, we can use Equation (6.38), giving:

$$F = -Nk_B T \left[\ln 2 + \frac{1}{2} \left(\beta \mu_m B \cos^2 \theta \right)^2 \right].$$

Note that if F has minima when $\theta = 0$ or π, and maxima when $\theta = \pm \pi/2$.

Now suppose that the crystal is suspended in a large magnetic field. The spins tend to line up with the field, but if $\theta \neq 0$ the crystal exerts a torque K on its suspension, trying to pull the $\theta = 0$ axis parallel to the field, since this maximizes μ and hence minimizes F. If the crystal turns through an angle $d\theta$, the work done is $K\, d\theta$. It is important to pay careful attention to the sign of K. If $K\, d\theta > 0$, work is done *by* the crystal *on* the suspension, so that the contribution to the internal energy is $-K\, d\theta$. Hence, the first law of thermodynamics for an anisotropic magnet can be written:

$$dU = T\, dS - K\, d\theta,$$

and the free energy is:

$$dF = -S\, dT - K\, d\theta.$$

Hence,

$$K = -\left(\frac{\partial F}{\partial \theta} \right)_B = -2N\beta \mu_m B^2 \cos^3 \theta \sin \theta. \tag{6.43}$$

K is plotted as a function of θ in Figure 6.4. The sign is such that if $K > 0$, θ tends to increase. Equilibrium corresponds to $K = 0$. Note how the equilibrium is stable at $\theta = 0$ and π, small deviations being subject to a restoring torque proportional to $-\theta$ near $\theta = 0$ and to $-(\pi - \theta)$ near $\theta = \pi$, but is weakly unstable at $\theta = \pi/2$.

In a real anisotropic ferromagnet, such as iron containing cobalt, B can be thought of as the Weiss molecular field (see Chapter 15) which is very large, so that the assumption $\mu B \ll k_B T$ is not usually justified. However, the dependence of torque on angle is qualitatively similar to that shown in Figure 6.4.

6.11 Thermally Activated Processes: Arrhenius' Law

Suppose that a system is in a metastable state; that is, it is not in its true equilibrium state, but can only reach equilibrium by passing over an energy barrier. A simple analogy is a car on a roller coaster track, illustrated schematically in Figure 6.5a. If the car is stationary at the point M, it cannot reach the true equilibrium point O unless it is somehow provided with enough energy to overcome the barrier B.

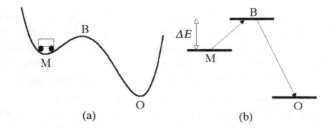

FIGURE 6.5 Model of a metastable system.

A schematic energy level diagram of an analogous chemical system is shown in Figure 6.5b. If it is initially in the metastable state M, it must acquire enough thermal energy to reach the state B if it is to reach the equilibrium state O. An example that is all around us is paper, which is predominantly made of carbon and hydrogen. In the presence of oxygen, the material is not in its lowest free energy state, which it would reach if it could combine with the oxygen (i.e., burn) to form carbon dioxide and water. However, there is an energy barrier to this chemical reaction, which can only be overcome by raising the temperature, (e.g., with a lighted match). Once this temperature is reached, in the absence of a heat sink such as water, the reaction is self-sustaining, since it releases sufficient energy to maintain the needed high temperature.

Like most real chemical reactions, the burning reaction is rather complicated. We simplify it by considering one molecule, which can either be in a metastable state M or in an equilibrium state 0, with an intermediate state B (known in chemistry as the *transition state*) at an energy ΔE above M. ΔE is usually called the *activation energy*. For simplicity, we assume all the states to be non-degenerate and ignore the fact that transition states above B also contribute. We also assume that the molecule remains in the state M long enough to come into "local" thermal equilibrium, so that the canonical distribution applies. The rate R_c at which molecules cross the barrier is proportional to the probability that a molecule will find itself in state B, having started from M. At temperature T, this probability is proportional to the Boltzmann factor $e^{-\beta \Delta E}$. Hence,

$$R_c \propto e^{-\beta \Delta E}. \tag{6.44}$$

This is called Arrhenius' law.[11] The "constant" of proportionality (the *prefactor*) in Equation (6.44) is difficult to calculate, and in general, varies as some power of temperature. In most chemical reactions $\Delta E \gg k_B T$, so that the exponential dominates the temperature dependence and the effect of variation of the prefactor with temperature can often be neglected. The contribution of states above B to the transition rate only affects the prefactor, as will be shown in the next chapter. Note that if $\Delta E \gg k_B T$, a small relative change in ΔE can change R_c by orders of magnitude. This is the function of a catalyst, whose effect is to reduce the activation energy while leaving the initial and final states unaffected.

Arrhenius' law (Equation 6.44) applies to a wide range of processes, from cooking spaghetti to thermonuclear fusion, and can be used to determine the activation energy ΔE from the temperature dependence of the rate R_c. This is usually done by plotting $\log_{10} R_c$ against T^{-1}. Such a graph is called an *Arrhenius plot* and gives a straight line if Equation (6.44) holds. The slope of the line is $-\Delta E/k_B \ln 10$, where $\ln 10 = 2.3$. However, if ΔE depends on temperature, as is usually the case, this procedure may still give a straight line, but with an erroneous value of ΔE (see Problem 6.8). If the prefactor depends on temperature, the data will not fall on a straight line, but they can be made so if one knows the temperature dependence of the prefactor. For example, if, as is often the case, the prefactor is known to vary as T^a,

---

[11] Svante August Arrhenius, Swedish physical chemist (1859–1927). Winner of the Nobel Prize for Chemistry in 1903.

where a is a small number, a plot of $\log_{10}\left(R_c T^{-a}\right)$ against T^{-1} should yield a straight line. If the temperature dependence of the prefactor is ignored, a plot of $\log_{10} R_c$ against T^{-1} over a limited temperature range may still appear to be linear within the accuracy of the data. However, if the best straight line is drawn through the data points (so that one "forces" the data to fit a straight line), the slope of this line gives an apparent value of ΔE which differs from the true value by approximately $\frac{ak_B}{\langle T^{-1}\rangle}$, where $\langle T^{-1}\rangle$ is the average value of T^{-1} over the range of measurement (see Problem 6.8c).

One application of Equation (6.44) is to the rate at which electrons cross a potential barrier (i.e., the current) in a transistor. In this case, $\Delta E = \phi_b - qV$ where q is the charge on the electron, V is the applied base-emitter voltage, and ϕ_b is a constant called the barrier height ($\phi_b/q \approx 1$ eV in silicon). In Chapter 18, we will examine an analogous situation for enzymes that catalyze biochemical reactions.

Figure 6.6 shows log-linear plots of current I against V at three different temperatures, obtained in such an experiment.[12] Applying Equation (6.44) to this situation, and including the $T^{3/2}$ dependence of the prefactor (derived in Example 7.1), gives:

$$I = AT^{3/2}e^{(qV-\phi_b)/k_B T}, \tag{6.45}$$

where A is a constant that depends on the parameters of the transistor. The full lines in Figure 6.6 are Equation (6.45), with the constant of proportionality A chosen to give the best fit to the data at 294 K. The observed dependence of I on ΔE and T follows Equation (6.45) over nearly six orders of magnitude in current, deviating at the limits for two reasons. At the highest temperature and lowest current, one cannot neglect the small generation-recombination current (see Footnote 10 of Chapter 14), which is in the opposite direction to the current given by Equation (6.45), while at the highest current, the internal resistance of the transistor reduces the voltage across the junction below that actually measured.

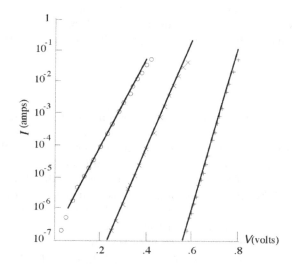

FIGURE 6.6 Log-linear plots of the current I in a silicon transistor against applied base-emitter voltage V, at three different temperatures. The measured current I is proportional to the probability that an electron has an energy exceeding ΔE, where $\Delta E = \phi_b - qV$, where ϕ_b is the barrier height. Data points: ∘, $T = 364$ K; ×, $T = 294$ K; +, $T = 200$ K. The lines are Equation (6.45). The reasons for the deviation from exponential behavior in the limits of large I or small V are given in the text. (From Sturge, M.D. and Toh, S.B., *Am. J. Phys.*, 67, 1129, 1999.)

---

[12] M. D. Sturge and S. B. Toh, "An experiment to demonstrate the canonical distribution," *Am. J. Phys.* **67**, 1129 (1999).

6.12 Envoi

The importance and universality of the canonical distribution and the partition function cannot be over-stressed. They apply to *any* system with a fixed composition which is held at a fixed temperature by contact with a heat bath, and if the many-particle partition function is known, all thermal properties of the system can be calculated. However, the actual calculation of the partition function can be a formidable task, and in this book we only attempt it in some simple systems. Even in such a simple system as an ideal gas, the calculation involves unexpected difficulties that arise from the indistinguishability of the particles. In the following chapters, we will develop methods to bypass these difficulties.

6.13 Problems

6.1 (a) Prove Equation (6.31) by substituting for P_i on the right-hand side, and using Equation (6.30)
  (b) Prove that Equation (5.4) follows from Equation (6.31) if we take all the P_i's to be equal, as is the case (according to our fundamental hypothesis; see Chapter 5) in an isolated system of fixed energy

6.2 Show that $\left(\frac{\partial^2 S}{\partial U^2}\right)_V = -\frac{1}{T^2 C_V}$, where $C_V = \left(\frac{\partial U}{\partial T}\right)_V$, and hence show that in the Taylor expansion (Equation 6.1) of the heat bath entropy, the second order term vanishes when the size of the heat bath becomes infinite

6.3 (a) The *magnetic moment per spin* μ_z is the projection of an individual magnetic moment on the z-axis, which is taken to be the direction of the field **B**. For a spin $1/2$ system, the only possible values of μ_z are $+\mu$ and $-\mu$, where μ is the elementary magnetic moment defined in Appendix B. Find the probabilities at temperature T that $\mu_z = +\mu$ and $-\mu$, and hence show that:

$$\langle \mu_z \rangle = \mu \tanh(\beta \mu B). \tag{6.46}$$

  (b) Use the result of (a) to show that the isothermal susceptibility of N spins is $\chi^T = \frac{N\mu^2}{k_B T}$ when $\mu B \ll k_B T$. Note that by using the Boltzmann factor, we have derived this result without having to calculate Ω, and in Equation (6.46), have obtained a result which is valid at all temperatures

  (c) Show that the mean square fluctuation of the contribution to the magnetic moment of an individual spin, $\langle \Delta \mu_z^2 \rangle$, is given by:

$$\langle \Delta \mu_z^2 \rangle = \mu^2 \text{sech}^2(\beta \mu B).$$

  Find its limiting behavior as $T \to 0$ and $T \to \infty$, and explain this behavior physically.

6.4 (Magnetic cooling.)
  (a) In a system consisting of N spin $1/2$ nuclei which occupy fixed sites in a solid, use the partition function to find the entropy as a function of B, (i) when $\mu B \gg k_B T$ and (ii) when $\mu B \ll k_B T$. The nuclei are distinguishable by their positions, so you may assume that Equation (5.12) holds. Compare the result in case (ii) with the multiplicity in zero field obtained in Chapter 4

  (b) Now suppose that we start with the material at a certain temperature T_0, in a field B_0 such that $\mu B_0 \gg k_B T$. Assume that the non-magnetic contributions to the heat capacity (due, for instance, to atomic vibrations) are negligible; we shall see in Chapter 7 that this is indeed the case if T_0 is low enough. What happens to the temperature if we isolate the material and reduce the field isentropically to zero? This process is called *adiabatic demagnetization*, and is generally used to obtain temperatures below a few millikelvin. It has recently been shown to be a practical method of refrigeration even at room temperature.[13]

---

[13] P. Weiss, "Magnetic refrigerator gets down and homey," *Science News* **161**, 4 (January 5, 2002); O. Tegus, "Transition-metal-based magnetic refrigerants for room-temperature applications," *Nature* **415**, 150–152 (2002).

(c) In practice, it is impossible to reduce the field seen by a spin exactly to zero, because each spin is always subject to the magnetic field of all the other spins in the material. If this residual field is $B_1 \ll B_0$, what is the final temperature? (In fact B_1 depends on $\langle \mu_z \rangle$, but you may ignore this complication.)

6.5. (A quantum mechanical rotator.) A diatomic molecule has energy levels given by $E_j = E_0 j(j + 1)$ where j is any non-negative integer and E_0 is a constant given in Appendix B. The degeneracy of the j'th level is $2j + 1$. For reasons to do with the symmetry of the nuclear wave functions,[14] this simple calculation is only valid for a heteronuclear molecule; that is, one in which the two nuclei are different

(a) Write down the partition function Z as a sum
(b) Find the separation between successive levels as a function of j, and hence find the average number of states $g(j)$ per unit energy interval at high j (this average is called the density of states, which will be defined more generally in the next chapter)
(c) Use $g(j)$ to convert Z from a sum to an integral valid at high j, and show that $Z \to \frac{k_B T}{E_0}$ as $T \to \infty$. Hence, obtain the mean energy $\langle E \rangle$ at high temperature ($k_B T \gg E_0$)
(d) Show that the rotational contribution C_{rot} to the kmolar heat capacity (i.e., the heat capacity of N_A molecules) is R_0 at high temperature
(e) Now suppose that T is so low that only the lowest two rotational levels are significantly occupied. Find $C_{rot}(T)$ in this temperature range
(f) Write a program to numerically calculate $Z(T)$ and $C_{rot}(T)$ for a heteronuclear diatomic molecule. The result of such a calculation is shown in Figure 6.7. Note how a simple interpolation, smoothly joining up the results of (d) and (e), would not have given the hump at intermediate temperature

6.6 The neutral carbon atom has a 9-fold degenerate ground level and a 5-fold degenerate excited level at an energy 0.82 eV above the ground level. Spectroscopic measurements of a certain star show that 10% of the neutral carbon atoms are in the excited level, and that the population of higher levels is negligible. Assuming thermal equilibrium, find the temperature

6.7 When an atom is raised to an excited level (e.g., in a discharge tube), it emits light as it returns to the ground state. The probability that the level is occupied at a time t after excitation is proportional to e^{-wt}, where w is the decay rate, which depends on the initial and final levels[15]. If the atom is an impurity in a solid, and the decay is slow on an atomic scale, nearby excited levels can come into thermal equilibrium with each other. The decay rate is then the average of the decay rates of the levels, weighted by their respective populations

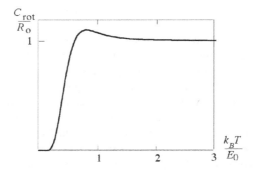

FIGURE 6.7 Molar heat capacity due to free rotation of a heteronuclear diatomic molecule.

[14] See, for example, M. Born, *Atomic Physics*, 8th ed. (Dover, 1989), p. 263.
[15] See, for example, A. P. French and E. F. Taylor, *Introduction to Quantum Physics* (Norton, 1978), Section 14.5.

(a) If there are two levels, with energies E_1 and E_2, degeneracies g_1 and g_2, and decay rates w_1 and w_2, find an expression for the average decay rate as a function of temperature, assuming thermal equilibrium between the two levels

(b) The chromium ion which is responsible for the red color of ruby has two excited levels, both doubly degenerate, 3.6×10^{-3} eV apart. The decay rate at $T = 20$ K is observed to be 262 sec^{-1} while at $T = 100$ K it is 235 sec^{-1}. Assuming thermal equilibrium, find the separate decay rates of the upper and lower levels $q/k_B = 11600$ K/eV

6.8 The activation energy ΔE in Arrhenius' law (Equation 6.44) often varies linearly with T over the range of measurement (in practice, it usually decreases with T). Suppose that we have some precise data on the reaction rate $R_c(T)$ over the range $T_1 \leq T \leq T_2$. Furthermore, suppose also that the true activation energy is given by $\Delta E = A - BT$ over this range of temperature, where $B > 0$, and that the prefactor in Equation (6.44) is independent of T

(a) Show that Equation (6.44), with a constant apparent activation energy, still fits the data on $R_c(T)$, so that if one plots $\log_{10} R_c$ against T^{-1}, one will obtain a straight line. Find the slope of this line and thus obtain a "fitted value" of the activation energy ΔE_{fit}; that is, the value of ΔE that one would obtain from the slope, if one were ignorant of the fact that ΔE depends on T. Compare ΔE_{fit} with the average value $\langle \Delta E \rangle$ of the true activation energy over the range of measurement

(b) A typical dependence of ΔE on T is shown in Figure 6.8. Suppose that measurements of R_c are made over the temperature range T_1 to T_2 and are fitted to Equation (6.44) as in (a). Make a rough copy of the figure and indicate on it the value of ΔE_{fit} that would be obtained

(c) Prove the statement in the text, that if the prefactor varies as T^a, and the plot of $\log_{10} R_c$ against T^{-1} is forced to fit a straight line, the slope of this line, which is the average slope of the true curve, gives an apparent value of ΔE which is in error by approximately $\frac{ak_B}{T^{-1}}$, where $\langle T^{-1} \rangle$ is the average value of T^{-1} over the measured range

6.9 Most of the physical constants encountered in statistical physics fall into two classes: macroscopic and microscopic quantities. Typical macroscopic quantities are the gas constant R_0, the kmolar Faraday F_q (the charge needed to electrolyze one kmole of a univalent substance), and atomic "weight" M_w. Typical microscopic quantities are Boltzmann's constant k_B, the charge on the electron q, and the actual mass of an atom. The two classes are related by Avogadro's kmolar number $N_A = 1/M_H$, where M_H is the mass of the hydrogen atom; for example, $F_q = N_A q$ and $R_0 \equiv N_A k_B$. In the thermodynamic (or continuum) limit, macroscopic quantities remain finite while microscopic quantities vanish. While the former group had been accurately determined by the middle of the nineteenth century, even at the end of that century the quantities in the latter group were only known in order of magnitude

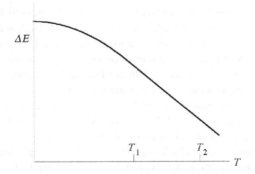

FIGURE 6.8 Typical dependence of activation energy ΔE on temperature T. Note that this is a plot of ΔE against T, not an Arrhenius plot of $\log_{10} R_c$ against T^{-1}.

In a manufacturer's catalog of laboratory equipment, an experiment in which the current I in a diode is measured as a function of applied voltage V, is described as a "measurement of the charge on the electron." Is this claim justified? Explain your answer

6.10   This problem requires knowledge of the use of undetermined (Lagrangian) multipliers in the maximization of a function of many variables, subject to constraints.[16] Prove that if one maximizes the entropy defined by Equation (6.31) with respect to the probabilities P_i, subject to the constraints $\sum P_i = 1$ and $\sum P_i E_i = \langle E \rangle$, which is fixed, one obtains the canonical distribution Equation (6.7)

6.11   (a) Show that the entropy of N independent oscillators whose energy levels are given by Equation (6.13) is

$$S = Nk_B \left[\langle n+1 \rangle \ln \langle n+1 \rangle - \langle n \rangle \ln \langle n \rangle \right] \tag{6.47}$$

(b) Use Stirling's approximation in the form of Equation (D.6) to show that the expression for the multiplicity given in Equation (D.24) is equivalent to Equation (6.47), since the total excitation $i = N\langle n \rangle$.

6.12   Show that the kmolar entropy of an Einstein solid (i.e., one in which there are $3N_A$ oscillators with a single vibrational frequency v) at a temperature $T \gg \theta_e$, where $k_B \theta_e = hv$, is:

$$S = 3R_0 \left[\ln \left(\frac{T}{\theta_e} \right) + 1 \right] \tag{6.48}$$

6.13   (a) Show from Equation (6.34) that the mean square fluctuation in the volume of a fluid containing N particles, N being fixed, is:

$$\langle \Delta V^2 \rangle = -k_B T \left(\frac{\partial V}{\partial p} \right)_T \tag{6.49}$$

(b) Hence, show that the mean square fractional fluctuation in the particle density n is:

$$\frac{\langle \Delta n^2 \rangle}{n^2} = \frac{k_B T}{N} \left(\frac{\partial n}{\partial p} \right)_T \tag{6.50}$$

(c) The equation of state of an ideal gas is $p = nk_B T$. Show that the mean square fractional fluctuation in the particle density of a fixed volume of an ideal gas containing on average N particles is

$$\frac{\langle \Delta n^2 \rangle}{n^2} = \frac{1}{N} \tag{6.51}$$

(d) When light passes through a transparent medium, it is scattered if the refractive index fluctuates on the scale of the wavelength λ of the light (Rayleigh scattering). For visible light, $\lambda \approx 5 \times 10^{-7}$ m. Fluctuations in the density of the atmosphere produce corresponding fluctuations in refractive index. Make a numerical estimate of the mean square fractional density fluctuation in a volume λ^3 of air at a height of 8 km, where the pressure is about 0.3 bar. Assume $T \approx 273$ K, where the kmolar volume at 1 bar is 22.7 m^3

This scattering is responsible for the fact that the daytime sky is not black. Why is the sky blue, rather than white like sunlight? Why is it red in the evening?

6.14   Show that Equation (6.21b) is a special case of Equation (6.34)

6.15   It is known from quantum mechanics that in a field B, an atom with spin S and magnetic moment μ has $2S + 1$ nondegenerate energy levels given by $E(m) = -m\mu B/S$, where $2m$ is an integer with

---

[16] See, for example, M. L. Boas, *Mathematical Methods in the Physical Sciences*, 2nd ed. (Wiley, 1983), pp. 174–181.

the same parity[17] as $2S$, and $-S \le m \le S$ (do not confuse S, which is spin, with the entropy S). S can be any positive integer or half-integer. Show that if $\mu B \ll k_B T$ the isothermal magnetic susceptibility of N noninteracting atoms is:

$$\chi^T = \frac{S+1}{3S} \frac{N\mu^2}{k_B T}.$$ (6.52)

Hint: It is easiest to use the high temperature expansion of the partition function as was done in Example 6.5. You will need the sum $\sum_{-S}^{S} m^2 = \frac{1}{3} S(S+1)(2S+1)$ [this formula can easily be proved by induction].

Note that as $S \to \infty$, $\chi^T \to \frac{1}{3} \frac{N\mu^2}{k_B T}$, which is the classical result for continuous energy levels (see Example 7.3)

6.16 Show that if $T \gg \theta_e$ the many-particle partition function for an Einstein solid is:

$$Z_N = \left(\frac{T}{\theta_e}\right)^{3N}$$ (6.53)

6.17 The spin 1/2 system may be defined with two different energies for the spin in an external field. For an external field in the positive z direction, the energies of the up and down spin states can be defined as $\pm\epsilon$ and or they can be defined as zero and 2ϵ, respectively. Show that these definitions result in a shift of the Helmholtz free energy that does not contribute thermodynamically (i.e., the free energies are equivalent up to a constant)

6.18 Three states in an atom are significantly separated from the others and have energies E_0, $2E_0$, and $4E_0$. (a) determine the relative occupation numbers of the three states if $E_0 = k_B T$ for N = 10 particles, 100 particles, 1000 particles. (b) Comment on the relationship between the probability and the occupation numbers for each state when N is a relatively small number

6.19 One thousand particles are allowed to occupy one of three thermodynamic states with energies $1/2 k_B$, k_B, and $3/2 k_B$. If the occupation numbers of the three states are N = 700, 300, and 100, respectively, what is the temperature of the system?

6.20 If a system of spin domains are separated by domain walls that cost an energy E_0 to create when any two adjacent spins are anti-parallel, determine the number of domains as a function of temperature in the spin system. For inspiration, see Figure D.1 and Section D.7 in Appendix D

---

[17] "Parity" means evenness or oddness.

Continuous Energy Levels, the Density of States, and Equipartition

Nothing, save the waves and I.

Lord Byron

Not everything that can be counted counts. Not everything that counts can be counted.

William Bruce Cameron

7.1 Gedanken

We imagine a large closed volume such as an empty room, with doors and windows shut and the air conditioner off. The room is quiet, the air is still. Why is the air still? On the quantum mechanical level, there are Avagadro's number of particles in each mole of the air, let's imagine, the nitrogen. Even if we model this as an ideal (non-interacting) gas, it is easy to argue that the particles themselves are not all in the same quantum state. If all of the particles were in the same quantum state, it would only be through the oddest symmetry that there was no net flow of air in a closed room, the left-going and right-going, front-going and back-going, up-moving and down-moving particles in the room exactly canceling any net momentum in the gas at temperature, T. Sillier still would be the expectation that all particles would move from one quantum state to another as the temperature was raised or lowered. No, it must be that the particles occupy a distribution of quantum states. We'd like to model (count?) which ones. At $T = 0$ we expect all of the particles to be in the ground state, but beyond that single situation, how do we determine exactly what states are occupied? Quantum mechanics would say that we cannot know the exact distribution of quantum states for all of the particles. So, we do the next best thing and want to formulate a descriptive set of quantum states that "on average" will be a good representation of what states are occupied (and by that, unoccupied) as a function of temperature.

7.2 Density of States and the Partition Function for a Continuum

So far, we have only considered systems whose energy levels are discrete, such as the harmonic oscillator or the spin 1/2 magnet. The energy levels of a particle in free space, on the other hand, are continuous, as are those of many other systems, and we now extend our treatment to cover the continuum case. We shall find that we have to replace the level degeneracy (i.e., the number of quantum states in any one energy level) with a new variable, the *density of states*, which is defined as the number of states per unit volume per unit energy interval, and which is in general a function of the kinetic energy of the particle.

We use a trick that is common in quantum mechanics, to pretend that the particle is in a large but finite box, in which the potential is zero so that the energy E is simply the kinetic energy. The energy levels in this box are discrete, although close together. At the end of the calculation, we allow the size of the box to become infinite, so that the separation of the levels tends to zero. For a given energy E, we shall find that the number of states whose energy is less than E is proportional to volume in a three-dimensional system, so that we can define $G(E)$ (not to be confused with the free enthalpy G) as the number of states per unit volume with energy less than E. In a one- or two-dimensional system, $G(E)$ is defined as the number of states with energy less than E per unit length or area respectively. The density of states is then the derivative $g(E) \equiv \frac{dG}{dE}$, the number of states per unit volume in the energy range E to $E + dE$ being $g(E)\, dE$. Note that while the degeneracy g of a discrete state is a dimensionless number, the dimensions of $g(E)$ in d dimensions are $(\text{energy})^{-1}(\text{length})^{-d}$.

The number of states between E and $E + dE$ in volume V is $V g(E)\, dE$, and from the canonical distribution Equation (6.4) the probability that one of these states is occupied is $Z^{-1}e^{-\beta E}$. Hence, the probability that the particle has an energy between E and $E + dE$ is:

$$P(E)\, dE = Z^{-1} V\, g(E)e^{-\beta E}\, dE. \tag{7.1}$$

As before, Z is the partition function, and $P(E)$ is a probability density, with dimension $(\text{energy})^{-1}$.

Since $\int_0^\infty P(E)\, dE = 1$, the partition function is:

$$Z = V \int_0^\infty g(E)e^{-\beta E}\, dE. \tag{7.2}$$

Comparing Equations (7.2) and (6.8) we see that when we go from a sum over a discrete set of energy levels to an integral over a continuum, we must replace the level degeneracy g with the number of states in the interval dE; that is, $V g(E)\, dE$.

As a simple example of this limiting process, consider the partition function at high temperature of a harmonic oscillator of frequency ν enclosed in a box of unit volume. The quantized energy levels are equally spaced and non-degenerate, being given by $E_j = (j + \frac{1}{2})h\nu$, $j = 0$ to ∞. If $k_B T \gg h\nu$, the Boltzmann factor $e^{-\beta E_j}$ changes by very little when j increases by 1, so that the energy E can be regarded as a continuous variable. The sum $\sum_j g_j e^{-\beta E_j}$ in Equation (6.8) for Z can then be approximated by an integral, with g_j replaced by $g(E)\, dE$ as in Equation (7.2). Since the separation of the energy levels is $h\nu$, there are $(h\nu)^{-1}$ states in unit energy interval, so that $g(E) = (h\nu)^{-1}$, and:

$$Z \approx (h\nu)^{-1} \int_0^\infty e^{-\beta E}\, dE = (\beta E \nu)^{-1} = \frac{k_B T}{h\nu}. \tag{7.3}$$

Hence, from Equation (6.9):

$$\langle E \rangle = -\frac{\partial \ln Z}{\partial \beta} = \beta^{-1} = k_B T. \tag{7.4}$$

For $3N_A$, oscillators $U = 3N_A \langle E \rangle = 3R_0 T$, in agreement with the high temperature limit for the Einstein solid discussed in Example 6.3.

7.3 Density of States for a Single Free Particle: One Dimension

We now calculate the density of states for a non-relativistic free particle of mass m, for which the dispersion relation is [see Equation (B.1)]:

$$E = \frac{h^2}{2m\lambda^2} = \frac{\hbar^2}{2m}k^2,$$

where λ is the de Broglie wavelength and $k = 2\pi/\lambda$. The derivation begins with a calculation of the normal modes of waves in a box with perfectly reflecting walls. This part of the calculation is valid for any dispersion relation.

We start with a one-dimensional box of length L, and calculate the density of states per unit length as $L \to \infty$. We put the origin at one end of the box and let the amplitude of the wave at a distance x from the origin at time t be $\psi(x,t)$. Since the walls are assumed to be impenetrable, the boundary conditions are the same as for a string fixed at both ends: $\psi(0,t) = \psi(L,t) = 0$. We look for the normal modes, which are simply the standing wave solutions of the wave equation with the given boundary conditions.

This problem is elementary, and we know that the solutions have the form (see Appendix B):

$$\psi(x,t) = A \sin \kappa x \, e^{i\omega t}, \tag{7.5}$$

where A is a normalizing constant, $\omega = 2\pi\nu$ (where ν is the frequency of the mode), and the boundary condition $\psi(L,t) = 0$ requires that $\kappa = n\pi/L$, where n is any positive non-zero integer. Some low κ normal modes (i.e., the solutions for $|\psi(x,t)|$ consistent with the boundary conditions) are illustrated in Figure 7.1.

We now ask how many modes there are with κ less than a certain value, say k. We call the number of these modes $N(k)$. This is the number of integers n less than $\frac{kL}{\pi}$. Since L is very large, $\frac{kL}{\pi}$ is a large number for all but the very lowest k, so that its fractional part is negligible and we can write $N(k) \approx \frac{kL}{\pi}$. Hence, the number of modes with $\kappa < k$, per unit length, is:

$$G(k) = \frac{N(k)}{L} = \frac{k}{\pi}. \tag{7.6}$$

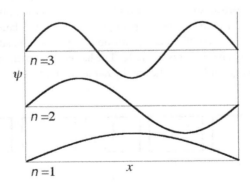

FIGURE 7.1 Normal modes in one dimension: ψ is plotted vertically (with the zero of each mode displaced for clarity) and x horizontally.

From the dispersion relation Equation (B.1) given above, $k = \frac{(2mE)^{1/2}}{\hbar}$. Substituting for k in Equation (7.6) we find for a spinless particle (which has only one quantum state for each normal mode):

$$G(E) = \frac{(2mE)^{1/2}}{\pi\hbar}. \tag{7.7}$$

If the particle has spin S, Equation (7.7) must be multiplied by the number of possible orientations of the spin, giving:

$$G(E) = (2S+1)\frac{(2mE)^{1/2}}{\pi\hbar}. \tag{7.8}$$

Differentiating Equation (7.8), we find for the density of states in one dimension:

$$g(E) = \frac{dG}{dE} = (2S+1)\frac{1}{\pi\hbar}\left(\frac{m}{2E}\right)^{1/2}. \tag{7.9}$$

In one dimension, $g(E)$ diverges as $E \to 0$, but the integrated number of states $G(E)$ is finite. Note that while the number of modes $G(k)$ does not depend on the dispersion relation, the density of states $g(E)$ does.

7.4 Density of States for a Single Particle: Two and Three Dimensions

Now let us consider a rectangular membrane (two-dimensional box) of sides L_x, L_y, with the boundary condition that ψ vanishes at the edges. The normal modes now have the form (see Appendix F):

$$\psi(x,y,t) = A \sin\kappa_x x \, \sin\kappa_y y \, e^{i\omega t}. \tag{7.10}$$

The boundary conditions are $y = 0$ for $x = 0, L$ (all y), and for $y = 0, L$ (all x). These are satisfied if and only if $\kappa_x = n_x\pi/L_x$ and $\kappa_y = n_y\pi/L_y$, where n_x and n_y are positive non-zero integers. The nodal lines (the lines on which $\psi = 0$) of some low κ modes are illustrated in Figure 7.2.

In an isotropic medium (i.e., one whose properties do not depend on direction) such as free space, the energy cannot depend on the direction of the vector κ, but only on its magnitude $\kappa = (\kappa_x^2 + \kappa_y^2)^{1/2}$. The dispersion relation must, therefore, have the form $E = E(\kappa)$. We first calculate $N(k)$, the number of states with $\kappa < k$. Such states lie within a quarter circle, of radius k, in the (κ_x, κ_y) plane, as shown in Figure 7.3. It is important to distinguish between real space, with axes (x, y), which is shown in Figure 7.2, and k-space (often called reciprocal space), with axes (κ_x, κ_y), shown in Figure 7.3.

The allowed values of κ_x and κ_y form a rectangular grid in the κ_x, κ_y plane, the points being separated by π/L_x horizontally and π/L_y vertically. Each point on the grid is associated with a rectangle of sides π/L_x and π/L_y, so that the area per point is π^2/L_xL_y. The larger the dimensions in real space, the *smaller* the corresponding grid spacings in k-space (hence the name "reciprocal space"). For large L, the number

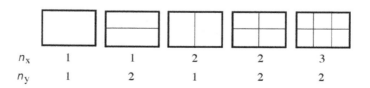

| n_x | 1 | 1 | 2 | 2 | 3 |
| n_y | 1 | 2 | 1 | 2 | 2 |

FIGURE 7.2 Nodal lines of some modes of a rectangular membrane.

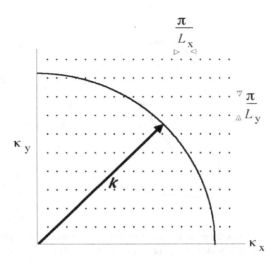

FIGURE 7.3 Reciprocal space (k-space) for a two-dimensional system. Each point represents a possible stationary state (normal mode), specified by its values of κ_x and κ_y. Note that L_x has been taken to be greater than L_y, as in Figure 7.2, so that the spacing of the points in the κ_x direction is *less* than in the κ_y direction. The curve is a quarter circle of radius k, and the number of points within the quadrant is $N(k)$; see Equation (7.11).

of points within the quarter circle is the area in the (κ_x, κ_y) plane of the quadrant, $\pi k^2/4$, divided by the area per point, so that:

$$N(k) = \frac{\pi}{4} k^2 \frac{L_x L_y}{\pi^2}. \tag{7.11}$$

Since the area in real space is $L_x L_y$, the number of states per unit area of real space is:

$$G(k) = \frac{N(k)}{L_x L_y} = \frac{k^2}{4\pi}. \tag{7.12}$$

For a particle with the dispersion relation given by Equation (B.1), and spin S, we apply the same line of reasoning to Equation (7.12) that led to Equations (7.8) and (7.9) and obtain:

$$G(E) = (2S + 1)\frac{mE}{2\pi\hbar^2}, \tag{7.13}$$

$$g(E) = \frac{dG}{dE} = (2S + 1)\frac{m}{2\pi\hbar^2}. \tag{7.14}$$

Thus, the density of states for a two-dimensional gas of non-relativistic particles is independent of energy. While this may appear to be an academic exercise, in fact the two-dimensional electron gas is of great importance in modern semiconductor physics, and we shall see in later chapters that the constancy of $g(E)$ greatly simplifies many calculations.

Finally, we come to the three-dimensional case. The particle is confined to a rectangular box of sides L_x, L_y, L_z, with the boundary condition that ψ vanish on all six faces of the box. The normal modes now have the form (see Appendix F):

$$\psi(x, y, z, t) = A \sin \kappa_x x \, \sin \kappa_y y \, \sin \kappa_z z \, e^{i\omega t}, \tag{7.15}$$

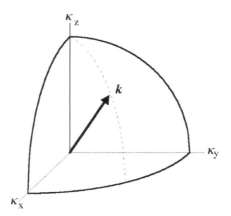

FIGURE 7.4 Reciprocal space for a three-dimensional system. The octant of the sphere of radius k encloses $N(k)$ points, each point corresponding to one stationary state; see Equation (7.16).

where, to satisfy the boundary conditions, $\kappa_x = n_x\pi/L_x$, $\kappa_y = n_y\pi/L_y$, $\kappa_z = n_z\pi/L_z$, n_x, n_y and n_z being positive non-zero integers.

As in the two-dimensional case, E cannot depend on the direction of the vector κ, so that the dispersion relation is of the form $E = E(\kappa)$, where now $\kappa^2 = \kappa_x^2 + \kappa_y^2 + \kappa_z^2$. Again we start by calculating $N(k)$, the number of states with $\kappa < k$. This is the number of states within the octant of a sphere of radius k in the three-dimensional k-space defined by the axes κ_x, κ_y, κ_z (see Figure 7.4). The volume of this octant is $\frac{1}{8}\frac{4\pi k^3}{3} = \frac{\pi k^3}{6}$. Note that this is a volume in k-space, which must be clearly distinguished from the volume of the box in real space, which is $L_xL_yL_z$.

The possible values of κ_x, κ_y, κ_z form a three-dimensional lattice in k-space. Each point in this lattice is associated with a rectangular parallelepiped with sides π/L_x, π/L_y, π/L_z so that the volume per point is $\pi^3/L_xL_yL_z$. The number of points within the octant of radius k is the volume of the octant divided by the volume per point, that is,

$$N(k) = \frac{\pi k^3}{6}\frac{L_xL_yL_z}{\pi^3}. \tag{7.16}$$

Since the volume of the box is $V = L_xL_yL_z$, the number of normal modes per unit volume of real space is:

$$G(k) = \frac{N(k)}{L_xL_yL_z} = \frac{k^3}{6\pi^2}. \tag{7.17}$$

Applying this to a non-relativistic particle with spin S, we find:

$$G(E) = \frac{(2S+1)}{6\pi^2}\left(\frac{2mE}{\hbar^2}\right)^{3/2}, \tag{7.18}$$

$$g(E) = \frac{dG}{dE} = \frac{(2S+1)}{4\pi^2}\left(\frac{2m}{\hbar^2}\right)^{3/2}E^{1/2}. \tag{7.19}$$

Particles with $S = 1/2$, such as the electron, are of particular importance, and for them:

$$g(E) = \frac{1}{2\pi^2}\left(\frac{2m}{\hbar^2}\right)^{3/2}E^{1/2}. \tag{7.20}$$

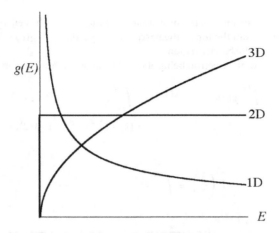

FIGURE 7.5 Density of states $g(E)$ for a non-relativistic particle in one, two, and three dimensions. The vertical and horizontal scales are arbitrary.

The one-, two-, and three-dimensional density of states functions for a non-relativistic particle, given, respectively, by Equations (7.9), (7.14), and (7.19), are shown in Figure 7.5. Note that the lower the dimensionality, the more pronounced is the singularity at $E = 0$. In three dimensions $\frac{dg}{dE}$ diverges at the origin, but $g(E)$ is continuous everywhere. In two dimensions, $g(E)$ is discontinuous at the origin, while in one dimension, $g(E)$ diverges there.

7.5 Dimensional Derivation of the Density of States

The functional form of $g(E)$ can be derived more simply by dimensional analysis. The density of states is a number per unit energy per unit volume (or area or length, according to the space dimensionality). The dimensions of energy are $M L^2 T^{-2}$, where M represents the dimension of mass, L length, and T time. Hence, if the space dimensionality is d, $g(E)$ has the dimensions $[M L^2 T^{-2}]^{-1} L^{-d}$. For a non-relativistic particle, g can depend only on the mass m, the energy E, and (since this is a quantum problem) Planck's constant $\hbar$, whose dimensions are $M L^2 T^{-1}$. Setting $g = m^\alpha E^\beta \hbar^\gamma$ and equating powers, we find $\alpha = d/2$, $\beta = d/2 - 1$, $\gamma = -d$. Thus we find that $g(E) \propto m^{1/2} \hbar^{-1} E^{-1/2}$ in one dimension, is independent of E and varies as $m \hbar^{-2}$ in two dimensions, and varies as $m^{3/2} \hbar^{-3} E^{1/2}$ in three.

For an ultra-relativistic particle (i.e., one whose energy is much greater than the rest energy mc^2, where c is the velocity of light, so that its velocity is close to c), the rest mass is irrelevant, but the velocity of light c enters as well as $\hbar$, so that $g(E)$ is a function of c, E, and $\hbar$. Setting $g = c^\alpha E^\beta \hbar^\gamma$, we find $\alpha = \gamma = -d$, $\beta = d - 1$, so that $g(E) \propto c^{-d} \hbar^{-d} E^{d-1}$. Hence, $g(E)$ is independent of E in one dimension, varies as E in two dimensions, and as E^2 in three (see Problem 7.3). This is the case for the photon.

Example 7.1: Transitions Over a Barrier

In a semiconducting device such as a diode or transistor, electrons have to overcome a potential barrier in order to contribute to the current through the device. Above the barrier is a continuum of states, whose density of states varies as $\epsilon^{1/2}$, where ϵ is the energy measured from the top of the barrier. The current is proportional to the probability of finding an electron somewhere above the barrier, whose height is ΔE. Find the temperature dependence of the current. Assume that the probability is small, so that its contribution to Z can be neglected.

From the canonical distribution, the probability that the electron has energy $\Delta E + \epsilon$, where ϵ is the energy measured from the top of the barrier, is proportional to $g(\epsilon)e^{-\beta(\epsilon+\Delta E)}$ where $g(\epsilon)$ is the density of states in the barrier region.

The total probability of the electron being above the barrier is thus proportional to:

$$\int_0^\infty g(\epsilon)e^{-\beta(\epsilon+\Delta E)}\,d\epsilon \propto \int_0^\infty \epsilon^{1/2}e^{-\beta(\epsilon+\Delta E)}\,d\epsilon$$

$$= \left(k_B T\right)^{3/2} e^{-\beta\,\Delta E}\Gamma(3/2),$$

where[1]

$$\Gamma\left(\frac{3}{2}\right) \equiv \int_0^\infty x^{1/2}e^{-x}\,dx = \frac{\pi^{1/2}}{2}.$$

Thus, the current varies as $T^{3/2}e^{-\beta\,\Delta E}$.

7.6 Effect of Intermolecular Collisions

It is reasonable to ask whether our derivation of the density of states applies to a real gas, in which the molecules collide with each other. Can one use results obtained for a freely moving particle and apply them to the real case? The answer is yes, but this is difficult to prove in full generality. If in a collision each molecule rebounded with the same energy that it started with, the plane wave single particle states that we have used in the derivation would simply be replaced by more complicated states with the same energy. The number of states of a given energy (i.e., the degeneracy) cannot change in such a replacement (in the jargon of quantum mechanics, the wave function undergoes a unitary transformation), so that such a collision would not affect the density of states. Of course, this picture is unrealistic. In a monatomic gas, collisions are indeed elastic, but only the total energy of the two colliding particles is conserved, not the individual energies. However, one can apply the same argument to the combined two-particle state, whose energy *is* conserved, and the effect of a collision is to make a unitary transformation of the entire many-particle wave function, whose energy and degeneracy is thus unaffected. In the case of a molecular gas with rotational and vibrational degrees of freedom, the problem is even more complicated and we do not attempt to deal with it here, but simply assume that we can use the same $g(E)$ for the translational states of the molecules (see Chapter 10).

7.7 Energy and Partition Function of a Free Particle in Three Dimensions

We can now obtain the partition function for a single free particle with the dispersion relation Equation (B.1) in a cavity of volume V. We substitute Equation (7.19) into Equation (7.2) to obtain:

$$Z_1 = \frac{2S+1}{4\pi^2}\left(\frac{2m}{\hbar}\right)^{3/2} V \int_0^\infty E^{1/2}e^{-\beta E}\,dE, \tag{7.21}$$

where we write Z_1 for the partition function as a reminder that Equation (7.21) applies to a *single* particle. The factor V comes from the fact that $g(E)$ is the number of states per unit energy *per unit volume*, so that the total number of states in the energy range E to $E + dE$ is $V\,g(E)\,dE$.

---

[1] Abramowitz and Stegun, p. 255.

The mean energy of the particle can be found from Equation (7.21) without actually evaluating the integral. From Equation (6.9):

$$\langle E \rangle = -Z_1^{-1} \frac{\partial Z}{\partial \beta} = \frac{\int_0^\infty E^{3/2} e^{-\beta E} \, dE}{\int_0^\infty E^{1/2} e^{-\beta E} \, dE}. \tag{7.22}$$

The numerator can be integrated by parts to give:

$$\int_0^\infty E^{3/2} e^{-\beta E} \, dE = \frac{3}{2\beta} \int_0^\infty E^{1/2} e^{-\beta E} \, dE, \tag{7.23}$$

since the integrand vanishes at both limits.

Substituting in Equation (7.22), and remembering that $\beta \equiv \frac{1}{k_B T}$, we find:

$$\langle E \rangle = \frac{3}{2} k_B T. \tag{7.24}$$

In one kmole there are N_A atoms, so that $U = N_A \langle E \rangle$, and the kmolar heat capacity of an ideal (i.e., a low density gas of non-interacting particles) monatomic gas at constant volume is:

$$C_V = \left(\frac{\partial U}{\partial T} \right)_V = \frac{3}{2} R_0, \tag{7.25}$$

since $R_0 \equiv N_A k_B$. We could have obtained this result directly from the equipartition theorem (see Section 7.9 and Appendix G). It is in agreement with experiment for monatomic gases.

To obtain Z_1, the partition function for a single particle in volume V, we substitute $x = \beta E$ in the integral in Equation (7.21):

$$\int_0^\infty E^{1/2} e^{-\beta E} \, dE = (k_B T)^{3/2} \int_0^\infty x^{1/2} e^{-x} \, dx.$$

As we saw in Example 6.1, the integral is $\pi^{1/2}/2$, so that:

$$\int_0^\infty E^{1/2} e^{-\beta E} \, dE = \frac{\pi^{1/2}}{2} (k_B T)^{3/2}. \tag{7.26}$$

Substituting Equation (7.26) in Equation (7.21) then gives:

$$Z_1 = (2S + 1) \left(\frac{m k_B T}{2\pi \hbar^2} \right)^{3/2} V. \tag{7.27}$$

Note that Z is dimensionless, and we can write Equation (7.27) more concisely as:

$$Z_1 = \frac{V}{V_q}, \tag{7.28}$$

where V_q is called the *quantum volume* and is defined as:

$$V_q \equiv (2S + 1)^{-1} \left(\frac{2\pi \hbar^2}{m k_B T} \right)^{3/2}. \tag{7.29}$$

In later chapters we shall find it more convenient to use the *quantum concentration*[2] $n_q \equiv 1/V_q$, so that:

$$\boxed{Z_1 = n_q V,} \tag{7.30}$$

where:

$$\boxed{n_q \equiv (2S+1)\left(\frac{mk_B T}{2\pi\hbar^2}\right)^{3/2}.} \tag{7.31}$$

The physical significance of V_q can be understood as follows. Consider a very small cubic box of volume V_q, containing a single particle with zero spin. The lowest energy quantum state (see Equation 7.15) in this box has $\kappa = \left(\frac{\pi}{L}, \frac{\pi}{L}, \frac{\pi}{L}\right)$, where $L^3 = V_q$, so that $\kappa = \frac{\sqrt{3}\pi}{L}$, and the energy of this state is:

$$E_1 = \frac{\hbar^2 \kappa^2}{2m} = \frac{3\pi^2\hbar^2}{2mL^2} = \frac{3\pi^2\hbar^2}{2mV_q^{2/3}} = \frac{3\pi}{4}k_B T = \frac{\pi}{2}\langle E\rangle, \tag{7.32}$$

where $\langle E\rangle$ is the thermal average energy given by Equation (7.24). Thus in a box of volume V_q the zero point energy would be approximately equal to the thermal energy, and only the lowest few quantum states would be significantly occupied. When $V \to V_q$, the continuum approximation breaks down; we must take the wave nature of matter and consequent quantization of the states into account.

An equivalent way of looking at V_q is to relate it to the de Broglie wavelength λ of a particle with energy $\langle E\rangle$ and momentum $(2m\langle E\rangle)^{1/2}$. This is given by:

$$\lambda = \frac{h}{(2m\langle E\rangle)^{1/2}} = \frac{h}{(3mk_B T)^{1/2}}.$$

Comparing this with Equation (7.29), we see that for a particle with zero spin,

$$V_q^{1/3} = \frac{h}{(2\pi m k_B T)^{1/2}} \approx 0.7\lambda. \tag{7.33}$$

Thus, V_q is the volume of a cube whose side is roughly equal to λ. In Chapter 9 we shall find that quantum effects become important when the density of particles approaches the quantum concentration n_q, where their average separation is of order λ.

Of course, what we are really interested in is the partition function for a large number of particles in a box. This problem is treated in Chapter 9.

7.8 Other Uses of the Density of States: Debye's Theory of the Heat Capacity of Solids

So far we have used the density of states only to obtain the single particle partition function Z_1 for a system with continuous energy levels, by substituting $g(E)$ in Equation (7.2). However, the density of states has other uses. Suppose, for example, we have a collection of harmonic oscillators of different frequencies ν_i. We can calculate Z for an individual oscillator and obtain the mean energy $\langle E\rangle$ of that oscillator, using

---

[2] Often misleadingly called the effective density of states.

Equation (6.9). In general, $\langle E \rangle$ is a function not only of temperature but also of the frequency ν. The energy of the entire collection of oscillators is:

$$U = \sum_i \langle E(T, \nu_i) \rangle, \tag{7.34}$$

where the sum is over all the oscillators.

If ν were the same for all the oscillators, as in the Einstein model of a solid, we could use Equation (6.10) and obtain $U = N \langle E(T, \nu) \rangle$, where N is the number of oscillators. However, in general this is not the case. If ν is a continuous variable, we can define a density of states *per unit frequency* $g(\nu)$, where the number of oscillators with frequency between ν and $\nu + d\nu$ is $g(\nu)d\nu$, and Equation (7.34) becomes:

$$U = \int_0^\infty g(\nu) \langle E(T, \nu) \rangle \, d\nu. \tag{7.35}$$

An important application of Equation (7.35) is in Debye's theory of the heat capacity of solids at low temperature.

Example 7.2: Debye's Model for the Heat Capacity of a Solid

In the previous chapter, we found that the temperature dependence of the heat capacity of a solid could be qualitatively understood in terms of the Einstein model, in which all the atoms were assumed to vibrate at a single frequency. However, Einstein's model is clearly wrong when applied to a solid, since it ignores the fact that the atomic vibrations are coupled together. In fact, the normal modes of vibration of the solid have a wide range of frequencies, and the Einstein model predicts much too low a heat capacity at low temperatures.

Debye's model starts from the opposite point of view, treating the solid as a continuum; that is, the atomic structure is ignored. A continuum has vibrational modes of arbitrarily low frequency, and at sufficiently low temperatures only these low frequency modes are excited. Low frequency implies long wavelength, the wavelength being c_s/ν, where c_s is the velocity of sound. If the wavelength is sufficiently long relative to the interatomic spacing, the continuum approximation is valid, and c_s is independent of ν. These low frequency normal modes are simply standing sound waves,[3] and the number per unit volume with $\kappa < k$ is $G(k)$, given by Equation (7.17), multiplied by 3 since a sound wave in a solid can have three polarizations (two transverse and one longitudinal). The dispersion relation $\nu(k)$ for a wave of constant phase velocity c_s is:

$$\nu = \frac{c_s k}{2\pi}, \tag{7.36}$$

so that, from Equation (7.17):

$$G(\nu) = 3 \frac{1}{6\pi^2} \left(\frac{2\pi\nu}{c_s} \right)^3$$

and

$$g(\nu) = \frac{dG}{d\nu} = \frac{12\pi\nu^2}{c_s^3}. \tag{7.37}$$

---

[3] By analogy with the quantum of radiation (the photon), the quantum of vibrational energy in a solid, whether or not the density of states is given by Equation (7.37), is called the *phonon*.

Note that Equation (7.37) only holds for frequencies sufficiently low that the continuum approximation is valid. Hence, it can only be used when the temperature is so low that only vibrational modes with wavelengths long relative to the atomic spacing are excited.

Each normal mode is a quantized harmonic oscillator, whose energy levels are given by Equation (6.13). The mean energy of each normal mode at temperature T is given by Equation (6.15a):

$$\langle E(T, v)\rangle = hv \left[\frac{1}{2} + \left(e^{\beta hv} - 1 \right)^{-1} \right], \tag{7.38}$$

where, as before, $\beta = \frac{1}{k_B T}$ and v is the frequency of the mode.

Thus, the total energy per unit volume is, from Equation (7.35):

$$U = \int_0^\infty g(v)\langle E(T, v)\rangle \, dv$$
$$= U_0 + h \int_0^\infty \frac{vg(v)}{e^{\beta hv} - 1} \, dv, \tag{7.39}$$

where:

$$U_0 = \frac{h}{2} \int_0^\infty vg(v) \, dv. \tag{7.40}$$

The U_0 term comes from the zero-point motion of the atoms. Note that the integral in Equation (7.40) is finite since $g(v)$ drops to zero above a certain frequency (the maximum frequency at which the atoms can vibrate). U_0 reduces the cohesive energy of the solid,[4] but since it does not depend on temperature it does not contribute to the heat capacity.

If only long wavelength modes are excited, we can substitute $g(v)$ from Equation (7.37) and find:

$$U = U_0 + \frac{12\pi h}{c_s^3} \int \frac{v^3}{e^{\beta hv} - 1} \, dv. \tag{7.41}$$

The lower limit of the integral in Equation (7.41) is 0, but what is the upper limit? If the temperature is low enough for the continuum approximation to be valid, the denominator is large unless v is very small, so that the integral can be taken to infinity without significant error. Substituting $x = \beta hv$, we obtain:

$$U = U_0 + \frac{12\pi k_B^4 T^4}{h^3 c_s^3} \int_0^\infty \frac{x^3}{e^x - 1} \, dx. \tag{7.42}$$

The integral is $\pi^4/15$ (see Problem 7.7). Differentiating with respect to T then gives for the heat capacity *per unit volume*:

$$C = \frac{dU}{dT} = \frac{16\pi^5 k_B^4}{5h^3 c_s^3} T^3. \tag{7.43}$$

The velocity of sound in a solid depends on direction and polarization, so that c_s^{-3} in Equation (7.43) has to be taken as the average $\langle c_s^{-3}\rangle$.

---

[4] The zero point motion in helium is sufficient to prevent solidification altogether, so that at atmospheric pressure helium remains liquid however low the temperature.

Equation (7.43) can be written more concisely in terms of the *Debye temperature* θ_D, defined in Equation (7.47b), giving for the kmolar heat capacity:

$$C_V = \frac{12\pi^4}{5} R_0 \left(\frac{T}{\theta_D}\right)^3. \tag{7.44}$$

Thus, Debye's model predicts that in the limit of sufficiently low temperature, the heat capacity due to vibrations of the atoms[5] must vary as T^3, and gives the constant of proportionality. This T^3 limiting behavior is observed (see Figure 13.3), although in most solids a temperature below 5K is required for the limit to be reached.

Since few data at such low temperature were available when Debye published his theory in 1912, he needed to interpolate between Equation (7.44), accurate at very low temperature, and $C_V = 3R_0$ Equation (6.17), which is accurate at high temperature and is predicted by equipartition (see Section 7.9). Debye's solution was to use the fact that, since each atom can vibrate in three directions, the total number of normal modes per unit volume is $3n$, where n is the number of atoms per unit volume.[6] Hence,

$$\int_0^\infty g(v)\,dv = 3n. \tag{7.45}$$

In Debye's time, the form of $g(v)$ was not known except for small v, so he arbitrarily assumed that $g(v)$ retains the value given by Equation (7.37) up to a certain value of v, called v_D, and is zero above it. He chose v_D to satisfy Equation (7.45), so that:

$$\int_0^{v_D} g(v)\,dv = 3n, \tag{7.46}$$

from which (see Problem 7.6):

$$v_D = c_s \left(\frac{3n}{4\pi}\right)^{1/3}. \tag{7.47a}$$

It is convenient to define the Debye temperature as:

$$\theta_D \equiv \frac{h v_D}{k_B} = \frac{h c_s}{k_B}\left(\frac{3n}{4\pi}\right)^{1/3}. \tag{7.47b}$$

The energy per unit volume is then, from Equation (7.41),

$$U = U_0 + \frac{9k_B T^4}{\theta_D^3} \int_0^{\frac{\theta_D}{T}} \frac{x^3}{e^x - 1}\,dx, \tag{7.48}$$

in which the integral has to be evaluated numerically. The heat capacity is found by differentiating Equation (7.48) with respect to T.

Equation (7.48) gives a considerably better account of the experimental data than the Einstein model Equation (6.16). A comparison of the predicted molar heat capacity with data on eighteen different elements and compounds is shown in Figure 7.6.[7] The fitted Debye temperatures range

---

[5] This qualifier is necessary because, as we shall see in Chapter 13, in a metal the electrons also contribute to the heat capacity.

[6] It is shown in most textbooks of mechanics that the number of normal modes of any vibrating system is equal to the number of spatial coordinates needed to describe the motion of the system. There are three such coordinates per atom.

[7] It was probably this review that Erwin Schrödinger was working on when his professor Peter Debye told him that he was not doing anything very important and that he should take a look at Louis de Broglie's thesis on the wave nature of matter. This led directly to Schrödinger's discovery of the wave equation that bears his name. The episode was described in a talk by Felix Bloch, who was Debye's student at the time, and is recorded in A. P. French and E. F. Taylor, *An Introduction to Quantum Physics* (Norton, 1978), p. 104.

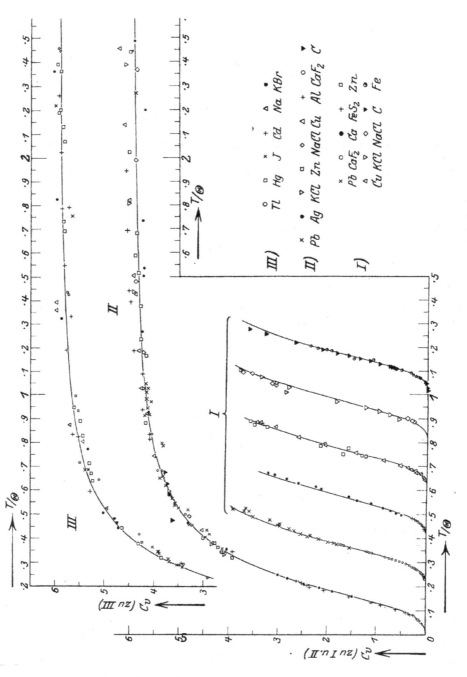

FIGURE 7.6 Comparison of Debye's curve for the molar heat capacity in cal/mole (lines) with the data points for 18 substances. The fitted Debye temperatures Θ are (in K): Pb 88, Tl 96, Hg 97, I 106, Cd 168, Na 172, KBr 177, Ag 215, Ca 226, KCl 230, Zn 235, NaCl 281, Cu 315, Al 398, Fe 453, CaF_2 474, FeS_2 645, C 1860 (many of these values have subsequently been modified to fit more recent data). Some of the curves are displaced for clarity. (From Schrödinger, E., "Specific heats," in *Handbuch der Physik*, vol. 10, ed. F. Henning, Springer, 1926.)

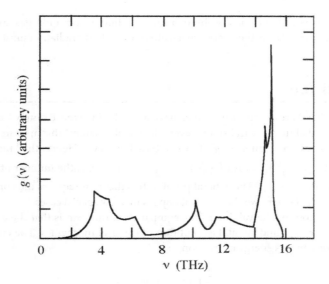

FIGURE 7.7 Density of phonon states $g(\nu)$ for silicon, calculated from a model dispersion relation whose parameters were fitted to neutron diffraction data. [From Dolling, G. and Cowley, R. A., *Proc. Phys. Soc.* (London), **88**, 463, 1966.]

over a factor of more than 20, from 88 K (lead) to 1860 K (diamond). However, the value of θ_D giving the best fit to the data is not usually the same as that calculated from Equation (7.47b); it probably helped that at the time, few data on the velocity of sound at low temperature were available, so that this comparison could not be made. It must be remembered that Equation (7.48) is based on good physics *only at the limits of low and high temperature*; in between, it is merely an interpolation formula. It cannot be expected to give a perfect fit, since the assumption regarding the density of states $g(\nu)$ is quite inaccurate, even in the simplest solids. An example of an actual $g(\nu)$ obtained by fitting neutron diffraction data is shown in Figure 7.7. The reasonable fit of the Debye curve to the wide variety of data shown in Figure 7.6 is a consequence of the fact that the heat capacity is an integral over all the phonon modes and is insensitive to their precise distribution.

Unfortunately, in the roughly forty years that intervened between the publication of Debye's theory in 1912 and the development of accurate experimental and theoretical methods of determining $g(\nu)$, the so-called Debye model in the form of Equation (7.48) acquired something of the status of holy writ, and the story of the attempts to force the data to fit it was described with some exaggeration by M. Blackman (who was one of the first to attempt to calculate $g(\nu)$ for a model solid) in an unpublished talk as "one of the most disgraceful episodes in the history of physics."

Debye's theory of heat capacity is based on the analogy between long wavelength sound waves in a solid and electromagnetic waves in free space, and the reader may justifiably wonder why we have not tackled the problem of "black body" radiation, which was solved by Planck in 1900. Planck was the first to introduce the concept of quantized energy states, and in the low temperature limit Debye's theory is virtually identical to Planck's. The reason why we have not followed Planck's treatment is that for radiation, $g(\nu)$ is given by Equation (7.37) at *all* frequencies, there being no cutoff as there is in solids. Thus, the zero point energy U_0, defined in Equation (7.40), which is finite for the vibrational modes of a solid, is infinite for radiation. Of course, Planck was unaware of the existence of zero point energy, which was only discovered when the full quantum mechanical treatment of the harmonic oscillator was carried out in the 1920s. We can avoid the difficulty[8] by treating radiation as a photon gas, and we postpone the derivation of Planck's formula for the energy spectrum of black body radiation until we treat the photon gas in Chapter 11.

7.9 Equipartition

Equations (7.4) and (7.24) are examples of results that can be obtained more simply from the equipartition theorem, which states that in a classical system every degree of freedom[9] that appears quadratically in the expression for the energy contributes $\frac{1}{2}k_B T$ to the internal energy.[10] Hence, the kmolar heat capacity at constant volume is $C_V = \left(\frac{\partial U}{\partial T}\right)_V = N_A \left(\frac{\partial \langle E \rangle}{\partial T}\right)_V = \frac{n_f}{2}R_0$, where n_f is the number of degrees of freedom per microsystem and $R_0 = N_A k_B$. The general proof of this theorem requires the concept of phase space and is given in Appendix G; here we show how it applies in a few special cases.

The essential condition for the validity of the equipartition theorem is that the energy levels be continuous; this is, of course, a fundamental assumption of classical mechanics. The energy levels of a free particle are continuous and its energy expression is:

$$E = \frac{1}{2}\left(mv_x^2 + mv_y^2 + mv_z^2\right), \tag{7.49}$$

so that three degrees of freedom, v_x, v_y, and v_z, appear in the expression for E, and they enter quadratically. We have seen [in Equations (7.24) and (7.25)] that $\langle E \rangle = \frac{3}{2}k_B T$ and $C_V = \frac{3}{2}R_0$, in agreement with the theorem.

As another example of the use of equipartition, we consider the classical harmonic oscillator, of mass m, force constant K, displacement x, and velocity v. The energy expression is, from standard classical mechanics,

$$E = \frac{1}{2}\left(mv^2 + Kx^2\right). \tag{7.50}$$

There are two degrees of freedom, corresponding to the kinetic and potential energy respectively. By the equipartition theorem, the mean value of each of these contributions to the energy is $\frac{1}{2}k_B T$, so that $\langle E \rangle = k_B T$ in agreement with Equation (7.4). For a three-dimensional oscillator, as in the Einstein model of a solid, $n_f = 6$, so that $\langle E \rangle = 3k_B T$ and the kmolar heat capacity is $3R_0$ Equation (6.17). Note that this result only applies in the high temperature limit $k_B T \gg h\nu$, since only then can the levels be treated as continuous.

For a free *point* particle, the energy expression is Equation (7.49). However, there is no physical object (with the possible exception of the electron) that is a mathematical point,[11] so that one might think that rotational degrees of freedom should be included, and the full expression for the energy should be:

$$E = \frac{1}{2}\left(mv_x^2 + mv_y^2 + mv_z^2 + I_{xx}\omega_x^2 + I_{yy}\omega_y^2 + I_{zz}\omega_z^2\right), \tag{7.51}$$

where the Is are the principal moments of inertia and $\boldsymbol{\omega} = (\omega_x, \omega_y, \omega_z)$ is the angular velocity vector. In this case, there are six degrees of freedom per particle and C_V should be $3R_0$ rather than $\frac{3}{2}R_0$. In a polyatomic molecular gas (e.g., CCl_4), C_V is indeed found to be $3R_0$ as predicted, but in a monatomic gas

---

[8] This infinite zero point energy is still present in quantum electrodynamics (QED), but since it is inherently unobservable it is swept under the rug by the process known as renormalization.

[9] Degree of freedom in statistical mechanics has a slightly broader definition than the one that is often used in classical mechanics, since it includes not only position coordinates but velocity (or momentum) as well.

[10] The theorem is more general than this (see Appendix G), but the quadratic case is the most common in practice.

[11] See Appendix G for a more general discussion of this topic.

such as He, the experimental value is $\frac{3}{2}R_0$. Diatomic gases such as O_2 and N_2 are intermediate; such gases are found to have $C_V = \frac{5}{2}R_0$, suggesting that a diatomic molecule has two rather than three rotational degrees of freedom.

Classically, equipartition is entirely independent of the magnitude of the moments of inertia; however small they are, the theorem requires that the contribution to $\langle E \rangle$ of *every* rotational degree of freedom be $\frac{1}{2}k_BT$. Clearly, this is not the case experimentally; the moment of inertia about a particular axis has to be large (on an atomic scale) if that degree of freedom is to contribute. As in the case of the harmonic oscillator, this failure of equipartition is due to energy quantization. If the moment of inertia is small, the separation of the rotational levels exceeds k_BT and the equipartition theorem does not hold. This is observed in H_2, which has a small molecular moment of inertia and hence a relatively large separation between the rotational levels; furthermore, H_2 remains gaseous down to a lower temperature than any other gas except He, which is monatomic. The kmolar heat capacity C_V of H_2 drops from $\frac{5}{2}R_0$ to $\frac{3}{2}R_0$ as the temperature is lowered, showing that the rotational degrees of freedom cease to contribute to U at low temperature (see Figure 6.7).

It was known from the middle of the nineteenth century that each element has its own characteristic optical spectrum, implying that atoms (if they exist at all) have internal structure. Classically, equipartition should apply to the internal degrees of freedom of the atom just as it applies to the translations and molecular rotations. However, it was found experimentally that these degrees of freedom make no contribution to the heat capacity. Molecular vibrations should also contribute, but in light molecules they do not, except at high temperature. The existence of such paradoxes, which required quantum theory to resolve, became clear towards the end of the nineteenth century, and encouraged many philosophically inclined scientists to reject the atomic model of matter altogether. This was particularly true in Germany, where the authority of Ernst Mach was paramount. To Mach, who despised model building, these paradoxes merely confirmed a conclusion already reached on philosophical grounds.[12]

7.10 Brownian Motion

An important application of the equipartition theorem is to Brownian motion, which is the random motion of a very small object suspended in a fluid due to the random impact of molecules on it. Since the particle has three translational degrees of freedom, it must have a mean translational energy $\frac{3}{2}k_BT$; it can never be at rest at a non-zero temperature. If the mass of the particle is M, its root mean square (thermal) velocity $v_{th} \equiv \langle v^2 \rangle^{1/2} = \left(\frac{3k_BT}{M}\right)^{1/2}$. Because the molecular velocity is changing in magnitude and direction on a very short time scale, v_{th} is not directly measurable. Instead, one measures the displacement of a particular particle as a function of time, and in the quantitative analysis of Brownian motion, viscosity must be taken into account. The theory was developed independently by Einstein and by Smoluchowski[13] in 1905, and is explained very clearly by Feynman.[14] For a particle of mass 10^{-14} kg, the thermal velocity v_{th} is about 1 mm/s at 300 K, producing a random motion that is readily visible under the microscope. It was

---

[12] Ernst Mach (1838–1916), Austrian physicist and philosopher. In 1893 he wrote in *Science of Mechanics* (transl. T. J. McCormick, Open Court Press, 1960) p. 589: "Atoms... are things of thought. *Furthermore*, atoms are invested with properties that absolutely contradict the attributes hitherto observed in bodies" (my emphasis). No amount of experimental data could convince him that atoms, with all their strange properties, do in fact exist.

[13] Marion von Smolan Smoluchowski (1872–1917), Polish physicist. Before Einstein and Smoluchowski, the French mathematical physicist Henri Poincaré (1854–1912) took an interest in Brownian motion, and apparently encouraged his student L. Bachelier to develop a theory of it. However, Bachelier preferred to apply his theory to stock market fluctuations, and his thesis, *Théorie de la Spéculation* (Gauthier-Villars, Paris, France, 1900), is an early example of the now popular practice of physicists applying their skills in the world of finance. [see M. F. M. Osborne, "Reply to comments on 'Brownian motion in the stock market'", *Oper. Res.* **7**, 807–811 (1959); S. G. Brush, *The Kind of Motion We Call Heat* (Elsevier, 1976), pp. 101, 671].

[14] R. P. Feynman, R. B. Leighton, and M. Sands, *The Feynman Lectures on Physics* (Addison Wesley, 1963), vol. 1, Section 41–4.

the experimental confirmation by Perrin[15] of Einstein's quantitative calculations of Brownian motion that convinced Ostwald[16] and other sceptics (though not Mach) of the existence of atoms. Another example of Brownian motion is given in Problem 7.9.

7.11 Electrical (Johnson) Noise

Another illuminating application of equipartition is to the "noise" due to thermal fluctuations in an electrical circuit. This section presupposes a knowledge of electrical transmission lines, in particular the concepts of characteristic impedance, matching termination, and equivalent circuit. These are explained in most textbooks of electronics.[17]

These thermal fluctuations are known as *thermal* or *Johnson noise*.[18] "Noise" in this context means random fluctuations of the voltage across a resistor, and thermal noise is the electrical analog of Brownian motion. There are other, non-thermal, sources of electrical noise which we ignore here.

Consider the transmission line shown in Figure 7.8a. Its length is L and each end is terminated by resistors whose resistance R (not to be confused with the gas constant R_0) is equal to the characteristic impedance of the line. Electromagnetic waves travel along the line with velocity c, and if the ends were short-circuited, the normal modes would be standing waves like the waves on a string discussed in Section 7.3. The resistors, on the other hand, absorb all the energy that comes to them along the line, but we assume L to be so large that the resistors at the ends have negligible effect on the density of normal modes. We can then use equipartition and the results of Section 7.3 to find the thermal energy contained in the normal modes.

From Equation (7.6), the number of modes[19] per unit length with κ less than k is $G(k) = k/\pi$. Substituting $\nu = c/2\pi k$ and differentiating with respect to ν gives $g(\nu) = 2/c$. Each mode is a harmonic oscillator, and since frequencies in electrical circuits are low relative to $\frac{k_B T}{h}$, except at extremely low temperatures, we can use the classical equipartition result Equation (7.4) for the mean energy of an oscillator,

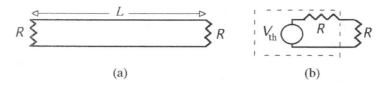

(a)   (b)

FIGURE 7.8   (a) Transmission line of length L, terminated at each end by a matching resistor. (b) Equivalent circuit of terminating resistor (within dashed lines), feeding a transmission line of characteristic impedance R.

---

[15] Jean Perrin (1870–1942), French physicist, who is generally credited with establishing the existence of atoms by his careful measurements of Brownian motion.

[16] Wilhelm Ostwald (1853–1932), German chemist, leader of the "energetics" school of chemistry, which regarded the atom as a hypothetical and unnecessary entity.

[17] See, for example J. J. Brophy, *Basic Electronics for Scientists*, 5th ed. (McGraw Hill, 1990), pp. 340–344.

[18] The existence of thermal noise in an electrical circuit was first predicted by Walter Schottky (Swiss-German physicist, 1888–1976) during the 1914–1918 War. Schottky did not believe that thermal noise would be distinguishable from the other, usually much stronger, sources of noise present in an amplifier, in particular "shot" noise which is due to the discrete nature of electric charge. However, seven years later, J. B. Johnson of Bell Telephone Laboratories demonstrated the existence of thermal noise experimentally, and in 1928, Harry Nyquist (Swedish-American physicist, 1889–1976) derived Equation (6.52), which was found to be in good agreement with Johnson's data. The early history of research on noise is recounted by Johnson in an article completed on the day of his death at the age of 83 entitled "Electronic noise: the first two decades," (*IEEE Spectrum*, February 1971, pp. 42–46.

[19] Only one polarization is possible on a transmission line.

so that the thermal energy in each mode is $k_B T$. A mode is a standing wave and is the sum of two counter-propagating waves, so that the energy in each propagating wave is $k_B T/2$. The energy per unit length on the line in frequency interval (bandwidth) $\Delta \nu$, moving (say) to the right, is thus $\frac{1}{2} k_B T g(\nu) \Delta \nu = \frac{k_B T \Delta \nu}{c}$. This energy propagates at the velocity of light c, so that the energy delivered to one of the resistors per unit time (i.e., the power) in this frequency interval, is $k_B T \Delta \nu$. Since the system is in thermal equilibrium, on average the net power transfer between the line and the resistor must be zero; hence, the resistor must deliver an equal power to the line. Figure 7.8b shows the equivalent circuit of the resistor, enclosed within the dashed lines. It consists of a noise generator (which of course does not exist, but the resistor behaves as if it did) whose open circuit voltage is V_{th}, in series with a resistance R, feeding into the transmission line whose characteristic impedance is also R. The voltage across the transmission line is $V_{th}/2$, so that the resistor delivers a power $V_{th}^2/4R$ to the line. In equilibrium, this must equal the power delivered by the line to the resistor, $\frac{k_B T \Delta \nu}{c}$, so that the Johnson noise voltage is given by:

$$V_{th} = \left(4 R k_B T \Delta \nu\right)^{1/2}. \tag{7.52}$$

If $T = 300$ K, $R = 1$ MΩ, and $\Delta \nu = 1$ MHz, $V_{th} \approx 130$ μV, a significant input voltage in a high gain amplifier.

7.12 The Classical Partition Function: Orientational Multiplicity

Boltzmann and Gibbs did not have quantum theory to help them in their calculation of the partition function of an ideal gas. They did not use the concept of a density of states, but rather that of phase space (see Appendix G), replacing $g(E)$ in the expression for Z Equation (7.2) by the corresponding volume in phase space. This approach is not only more difficult to follow than the one used here, but also only gives Z to within an unknown constant multiplicative factor with dimensions of [action]$^{-3}$ (this factor was found later to be $\hbar^{-3}$). However, there are systems which are more easily dealt with classically, and here we will discuss one class of such systems: those whose only degree of freedom is orientational. In this case, the volume of phase space is solid angle, which is dimensionless, and the constant factor in Z can be taken as 1.

Example 7.3: Dielectric Susceptibility

For specificity, we will consider one such system: a polar molecule (i.e., a molecule with a permanent electric dipole) in an electric field. The molecule is fixed in space but is free to rotate. The energy of a dipole $\boldsymbol{p}$ in an electric field $\boldsymbol{E}$ is $E = -\boldsymbol{p} \cdot \boldsymbol{E}$ (see any textbook of electromagnetism; bold symbols are vectors). We define the direction of $\boldsymbol{E}$ as the z-axis. In the absence of a field, the angle between $\boldsymbol{p}$ and the z-axis is random, so that the probability that this angle lies between θ to $\theta + d\theta$ is $\frac{d\Omega}{4\pi}$, where $d\Omega = 2\pi \sin \theta \, d\theta$ is the corresponding solid angle[20] (see Figure H.1). In the language of Appendix G, $d\Omega$ is the relevant volume of phase space.
(a) Find the partition function for a single dipole, and find a high temperature ($T \gg pE/k_B$) approximation to it, valid to order E^2.

$$E = -pE \cos \theta.$$

$$Z = \frac{1}{4\pi} \int e^{-\beta E} \, d\Omega = \frac{1}{2} \int_0^\pi e^{\beta pE \cos \theta} \sin \theta \, d\theta.$$

We substitute $x = \cos \theta$, so that $dx = -\sin \theta \, d\theta$.

---

[20] Do not confuse this Ω (solid angle) with the multiplicity defined in Chapter 4.

Then:

$$Z = \frac{1}{2} \int_{-1}^{1} e^{\beta p E x} \, dx$$
$$= \frac{\sinh(\beta p E)}{\beta p E}. \tag{7.53}$$

At high T $\beta p E \ll 1$, so that $\sinh(\beta p E)$ can be expanded Equation (A.4):

$$Z \approx \frac{\beta p E + (\beta p E)^3 / 6}{\beta p E} = 1 + \frac{1}{6}(\beta p E)^2. \tag{7.54}$$

(b) The isothermal dielectric susceptibility is $\chi^e \equiv \left(\frac{\partial P}{\partial E}\right)_T$, where P is the electric polarization. Find χ^e for N dipoles at high T.

The differential of the free energy of an incompressible dielectric is, by analogy with the corresponding magnetic expression[21] Equation (5.41),

$$dF = -S \, dT - P \, dE, \tag{7.55}$$

where P is the electric polarization.

Hence, using $F = -N k_B T \ln Z$ Equation (6.28b),

$$P = -\left(\frac{\partial F}{\partial E}\right)_T = N k_B T \left(\frac{\partial \ln Z}{\partial E}\right)_T. \tag{7.56}$$

From Equation (7.54), using $\ln(1 + x) \approx x$ Equation (A.5),

$$\ln Z \approx \frac{1}{6}(\beta p E)^2,$$

so that:

$$P = \frac{1}{3} N \beta p^2 E,$$

and:

$$\chi^e = \frac{N p^2}{3 k_B T}. \tag{7.57}$$

Note that if the elementary magnetic moment μ is substituted for the elementary electric dipole p, Equation (7.57) gives Curie's law for the magnetic susceptibility of N classical elementary magnets: compare Equation (7.57) with Equation (6.39) when $S \to \infty$.

(c) Use the exact expression for Z to find an expression for P valid at all temperatures and fields.

Z is given by Equation (7.53), so that:

$$\left(\frac{\partial \ln Z}{\partial E}\right)_T = \beta p \mathfrak{L}(\beta p E),$$

---

[21] For a careful discussion of the electric and magnetic energy, see G. Carrington, *Basic Thermodynamics* (Oxford, UK, 1994), Chapter 8.

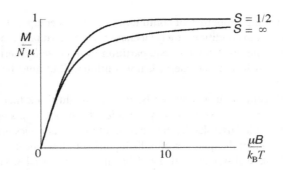

FIGURE 7.9 Comparison of the magnetization curves for $S = 1/2$ and $S = \infty$, the latter being the classical Langevin curve Equation (7.58). The field is scaled down by a factor of 3 for the $S = 1/2$ case so as to give the same initial slope.

where $\mathcal{L}(x)$ is the Langevin[22] function defined by:

$$\mathcal{L}(x) \equiv \coth(x) - \frac{1}{x}, \tag{7.58a}$$

and:

$$P = Np\mathcal{L}(\beta pE). \tag{7.58b}$$

The classical expression for the magnetization is identical, with M replacing P, B replacing E, and μ replacing p.

While the mathematical form of Equation (6.58) looks very different from the $\tanh(x)$ dependence obtained for the spin 1/2 magnetic system (see Problem 6.3a), the actual field dependences at low field are qualitatively similar if one scales the field to allow for the factor of three difference in the initial slope. However, at high field and low temperature, where $\mu B \gg k_B T$, the behavior is very different. If we start from $\frac{k_B T}{\mu B} \approx 0$ and increase T, the fractional decrease in the magnetization is initially $\frac{k_B T}{\mu B}$ in the classical case, while for spin 1/2 the fractional decrease is $2 \exp\left(-\frac{\mu B}{2k_B T}\right)$, a much smaller amount (see Problem 7.16). The two curves are compared in Figure 7.9.

7.13 Envoi

In this chapter, we have taken the statistical methods developed in the previous chapters for systems with discrete energy levels and applied them to systems with continuous levels. The essential new concept introduced is the density of states $g(E)$, which enables us to replace the sum over states in the partition function with an integral. The density of states is obtained in three steps. Starting from the wave equation, without specifying the dispersion relation for the particle, we derive the number $G(k)$ of normal modes per unit volume with wave vector less than k (equivalently, with momentum less than $\hbar k$). $G(k)$ depends only on the dimensionality of the system. We then use the dispersion relation $E(k)$ to convert $G(k)$ to $G(E)$, the number of states with energy less than E. The density of states with respect to energy, $g(E)$, is simply the derivative of $G(E)$.

---

[22] Paul Langevin (1872–1946), French physicist, of whom Einstein wrote "(Langevin's) scientific thought displayed an extraordinary clarity and vivacity combined with a quick and sure intuition for the essential point...he would have developed the special theory of relativity if that had not been done elsewhere."

One of the most interesting properties of continuous systems is equipartition (treated more fully in Appendix G): subject to certain conditions, thermal energy is shared equally amongst the degrees of freedom of the system. It was the breakdown of equipartition in the case of thermal radiation that forced Planck reluctantly to the conclusion that energy is not continuous but must be quantized. This problem is taken up in Chapter 11.

A gas is a system with continuous energy levels, and one might think that we could now obtain its partition function. However, so far we have only considered the case of a single particle, and because the particles in the gas are indistinguishable, the extension to many particles is not as straightforward as might appear. We shall find that the question of indistinguishability can be finessed if we first extend our statistical treatment to the case of systems with a variable number of particles. This is the topic of the next chapter.

7.14 Problems

7.1 What is the density of the normal modes of electromagnetic waves in a transparent medium of refractive index n_r?

7.2 Find the fractional root mean square fluctuation in energy, $\frac{\langle \Delta E^2 \rangle}{\langle E \rangle}$, of a single atom in a monatomic ideal gas

7.3 The dispersion relation for a relativistic particle of rest mass m (i.e., one whose kinetic energy E is not small relative to mc^2) is:

$$E = \left[(mc)^2 + (\hbar k)^2\right]^{1/2} c - mc^2,$$

where $\hbar k$ is the momentum (the de Broglie relation). Find $g(E)$ for the three-dimensional case and thus show that in the extreme relativistic limit, where $\hbar k \gg mc$, the density of states is:

$$g(E) = \frac{2S+1}{2\pi^2} \frac{E^2}{\hbar^3 c^3}, \tag{7.59}$$

where S is the spin of the particle. Note that Equation (7.59) is consistent with the dimensional derivation given in Section 7.5

7.4 (a) If one neglects the effect of gravity, the dispersion relation for surface (capillary) waves on a liquid is $\nu = \alpha k^b$, where α is a constant that depends on the surface energy σ and the mass density ρ. Find the exponent b by dimensional analysis.[23] This only holds if the wavelength is long relative to the interatomic separation, but short enough that the influence of gravity is negligible

   (b) Find the temperature dependence of the surface wave contribution (per unit area of surface) to the heat capacity of the liquid at a very low temperature. There is no need to evaluate the coefficient

   (c) If the interatomic spacing is a, estimate the temperature where your result ceases to be valid. Warning: Don't forget that this is a two-dimensional problem

7.5 Show that the partition function for a single particle in two dimensions can be expressed in the form $Z_1 = A/A_q$, where A is the area to which the particle is confined and A_q is the two-dimensional analog of V_q. Find A_q for a spinless particle, and the average energy $\langle E \rangle$. Compare $\langle E \rangle$ with the equipartition value

7.6 (a) Derive Equation (7.47a)

   (b) Show by expanding the exponential in Equation (7.48) that at high temperature ($T \gg \theta_D$), the Debye kmolar heat capacity agrees with the equipartition value $3R_0$

---

[23] See Cooper and West.

7.7 Expand the integrand in Equation (7.42) as a series in e^{-x}, and hence express the integral in terms of the Riemann zeta function $\zeta(n) = \sum_{k=0}^{\infty} k^{-n}$, using the fact that $\int_0^{\infty} x^n e^{-x}\, dx = n!$. Look up $\zeta(n)$ [Abramowitz and Stegun, p. 807] and thus obtain the result given in the text

7.8 Find the fractional root mean square fluctuation of the internal energy, $\frac{\langle \Delta U^2 \rangle^{1/2}}{U}$, for a solid body with N atoms, at a temperature sufficiently low that the heat capacity is given by Equation (7.44). Here U is the thermal energy and the zero point energy U_0 is not included. Estimate $\frac{\langle \Delta U^2 \rangle^{1/2}}{U}$ for 1 mm^3 of solid silicon at 0.6 mK. For silicon, $\theta_D \sim 600$ K, and its density is $n \approx 5 \times 10^{28}$ atoms/m^3. Explain in physical terms why the relative fluctuation is so large

7.9 [Thermal (Brownian) motion of a galvanometer.] The moving part of a mirror galvanometer is a light coil suspended by a thin fiber or fibers that produce a weak restoring torque $-K\theta$, where θ is the angle of twist. This is a torsional oscillator with energy given by:

$$E = \frac{1}{2}I\omega^2 + \frac{1}{2}K\theta^2,$$

where ω is the angular velocity, I the moment of inertia, and θ is measured from the equilibrium position. Find $\langle \theta^2 \rangle^{1/2}$, the root mean square fluctuation in θ, at 300 K if $K = 10^{-13}$ N m

7.10 (Noise and dissipation.) Figure 7.10 shows the equivalent circuit of a leaky capacitor. The effective bandwidth (i.e., the range of frequencies present in the noise) $\Delta\nu$ can be shown by standard circuit theory to be $(4RC)^{-1}$. The energy of a capacitor C charged to a voltage V is $CV^2/2$. Find the root mean square Johnson noise voltage across the capacitor, and show that it is equal to that obtained by applying the equipartition theorem to the capacitor, for which V is the only degree of freedom. Using the fact that power $= V^2/R$, show that as $R \to \infty$, the noise power goes to zero, so that a perfect capacitor generates no noise. This is an example of the *fluctuation-dissipation* theorem,[24] which states that if a system is lossless it is also noiseless, a fact that is used to advantage in the design of very low noise amplifiers

7.11 Consider a classical anharmonic oscillator with the energy function:

$$E = \frac{1}{2}mv^2 + Kx^4, \tag{7.60}$$

where K is a constant. What is the heat capacity of N_A such oscillators? You will need the generalized equipartition theorem; see Equation (G.9)

7.12 (Phonons in liquids.) Longitudinal (but not transverse) phonons exist in liquids. However, only helium remains liquid down to a sufficiently low temperature for the Debye model to be applicable. Find the Debye temperature and the coefficient of T^3 in the heat capacity of liquid helium at low T. The velocity of sound is 240 m/s and the density 2.2×10^{28} atoms/m^3. Compare your result with the observed kmolar heat capacity of $80T^3$ J K^{-1} kmole^{-1}. The T^3 dependence is observed up to about 0.6 K, above which the effects of deviation of the dispersion relation from the Debye model become important

FIGURE 7.10 Equivalent circuit of a leaky capacitor.

---

[24] See F. Reif, *Fundamentals of Statistical and Thermal Physics* (Wiley, 1965), p. 572.

7.13 (a) Show that the kmolar entropy of a Debye solid at low temperature ($T \ll \theta_D$) is:

$$S = \frac{4\pi^4}{5} R_0 \left(\frac{T}{\theta_D}\right)^3 \qquad (7.61)$$

(b) Show that for $T \gg \theta_D$, the entropy is:

$$S = 3R_0 \left[\ln\left(\frac{T}{\theta_D}\right) + \frac{4}{3}\right]. \qquad (7.62)$$

Note that Equation (7.62) is identical to the result for the Einstein solid Equation (6.43) if we put $\theta_D = e^{1/3}\theta_e \approx 1.4\theta_e$. Since θ_D and θ_e are merely fitting parameters, these different models give the same result at high temperature. Why is this?

7.14 (a) Given that typical molecular dipole moments are of order 1 Debye (3.3×10^{-30} C-m), estimate the temperature range over which the high T expression Equation (7.57) for the dielectric susceptibility at an electric field of 10^7 V/m is accurate to within 2%, by evaluating the next term in the expansions of Z and $\ln Z$

(b) Show that Equation (7.58) reduces to Equation (7.57) at small field or high temperature[25]

7.15 In Example 5.3, we considered a model of an unbranched polymer with N links of length ℓ_0. It was assumed that the links could only point in two directions, parallel and anti-parallel to the applied force. This model is mathematically analogous to the spin 1/2 system, with force f substituting for the field B, the length of the polymer ℓ for the bulk magnetic moment M, and ℓ_0 for the elementary moment μ. Since in a real polymer the links can be oriented in any direction, a classical model bearing the same analogy to the model of a dielectric given in Example 7.3 is more realistic.[26] Use this model to find an expression for $\ell(f)$ valid at all f

7.16 (a) The low field (Curie's Law) susceptibility of a classical paramagnet is much less than that of a spin 1/2 paramagnet [compare Equations (7.57) and (6.39)]. Explain in qualitative physical terms why this is. Hint: Think of the spin 1/2 magnet as a classical one with very restricted freedom to reorient

(b) At very high field and low temperature, the magnetization of either type of magnet converges on its maximum value of $N\mu$. Find an expression for the classical partition function at high field accurate to first order in $\frac{k_B T}{\mu B}$. If the temperature is raised from zero, show that in a classical system, the fractional reduction in magnetization is initially $\frac{k_B T}{\mu B}$, while it is $2\exp\left(-\frac{2\mu B}{k_B T}\right)$ for spin 1/2. Explain in physical terms why the magnetization decreases so much more rapidly in the classical case

7.17 In many disordered solids, such as glass, electrons are localized on individual atoms and do not form the energy bands described in Chapter 14, although at non-zero temperature they can migrate from one atom to another by jumping over an energy barrier. It is often a good approximation to assume that the energy of electron states at different points in the solid has a Gaussian distribution; that is, the density of states is given by:

$$g(E) = n_s \left(2\pi \langle \Delta E^2 \rangle\right)^{-1/2} \exp\left(-\frac{E^2}{2\langle \Delta E \rangle^2}\right),$$

where n_s is the number of possible sites per unit volume, $\langle \Delta E \rangle^2$ is a measure of the degree of disorder, and the average energy is taken as zero. Show that in equilibrium, when the probability of

---

[25] Note that since $\coth(x)$ diverges as $x \to 0$, it cannot be expanded in a Taylor series about $x = 0$. However, one can find the small x limit of Equation (7.56) by putting $\coth(x) = \frac{1}{\tanh(x)}$, expanding $\tanh(x)$ to third order in x, and then using the binomial expansion.

[26] See Section 6 of H. M. James and E. Guth, "Theory of the elastic properties of rubber," *J. Chem. Phys.* **11**, 455 (1943).

the electron occupying any one site is given by the canonical distribution, the number of occupied sites with energy E has a Gaussian distribution, and find the energy at which this number has a maximum, as a function of temperature

Mathematical hint: Complete the square in the exponent in the integrand

7.18 Graphene has been produced in single layer form that would qualify as a 2D system.[27] Describe how you would design a calorimetry experiment to discern the difference between a truly 2D single layer of graphene as opposed to a 3D sample of carbon that is simply very thin. Is there a definitive experiment that can realistically be performed?

7.19 For relativistic particles, the energy, E, of a particle with rest mass, m, and momentum $p = \hbar k$ is given by the relationship:

$$E^2 = p^2c^2 + m^2c^4.$$

An accelerator is used to bring a stream of electrons to relativistic speeds in a beam that is essentially one electron wide

(a) In the extremely relativistic limit ($p^2 >> m^2c^2$), determine the density of states and, with that, the partition function for a particle in that beam

(b) Determine the average wave vector excited in the beam, and from it, the average energy of a single electron in the beam

(c) The beam of N electrons is allowed to strike a target surface. Find an expression for the pressure as a function of temperature that the momentum transfer of the beam causes at the target surface

(d) Can the same answers in (a) (c) be easily derived by simply considering the degrees of freedom in the problem. Why or why not? If so, what is the only necessary assumption (and is that assumption satisfied by the given conditions)?

7.20 A model for predicting the height of a tree is a continuum of energy states mgh corresponding to the potential energy of the water internal to the structure. Conceptually, this sounds reasonable because the higher potential energy would seem to put a thermodynamic limit on the average height of the tree. Perform a continuum calculation of such a model and discuss the result it predicts for h. Consider the relative scales of water's potential energy and the typical Boltzmann energy $k_B T$ in explaining the issues with your result

---

[27] Z. Yan, Z. Peng, and J. M. Tour, "Chemical Vapor Deposition of Graphene Single Crystals," *Acc. Chem. Res.* **47**, 1327 (2014).

Systems with a Variable Number of Particles: The Chemical Potential

I'm on the verge of a major breakthrough, but I'm also at the point where physics ends and chemistry begins, so I'll have to drop the whole thing.

From a cartoon by Sidney Harris

Gibbs (was) the only physicist who really understood statistical thermodynamics.

Niels Bohr

8.1 Gedanken

Oxygen and nitrogen are different species. So, how does one species (at the same temperature and pressure) preserve and the other oxidize? We return to our first chapters and realize that for oxygen and nitrogen to be different, they must have a non-identical thermodynamic quantity for each atom of each species. This difference separates them chemically, but if we ignore it, each obeys the same results as one finds in partial pressures for a gas,[1] each species being well-described by the ideal gas law. However, we also know that the ideal gas law does not predict changes of phase from gas to liquid and liquid to solid. Not all gases liquify at 100 degrees Celsius and not all liquids solidify at 0 degrees Celsius. So we've reached a point where we must modify our treatment of gases to allow them the phase space to differ due to their chemistry.

8.2 A System in Contact with a Particle Reservoir: Diffusive Equilibrium

So far, we have only considered systems in which the number of particles is fixed, while energy can be exchanged with the outside world. Now we extend our analysis to the case where matter as well as energy can be exchanged. Just as a diathermal (thermally conducting) barrier allows energy to pass through in the form of heat, we can conceive of a barrier that is also permeable to matter, allowing particles to pass through. A diathermal barrier allows thermal equilibrium to be established when there is (on average) no

---

[1] Dalton's law, which is a simple result of each species being in contact with the same thermal bath of energy $k_B T$.

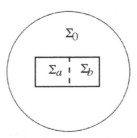

FIGURE 8.1 Two systems, Σ_a and Σ_b, separated by a permeable barrier through which they can exchange energy and matter, in thermal contact with a heat bath Σ_0.

net transfer of energy; similarly, when there is no net transfer of particles through a permeable barrier, diffusive equilibrium is established. It is not possible to have flow of particles without a flow of energy, so that a permeable barrier is also diathermal, and we assume that when there is diffusive equilibrium, there is also thermal equilibrium.

Consider two systems Σ_a and Σ_b, of fixed volume, immersed in a heat bath Σ_0 at temperature T. They contain N_a and N_b particles, respectively, and have energies U_a and U_b. They are separated by a permeable and diathermal barrier, so that they can exchange particles and energy with each other, but can exchange only energy with the heat bath. This situation is illustrated schematically in Figure 8.1, in which the continuous line represents a diathermal, but impermeable barrier, while the dashed line represents a barrier permeable to matter as well as to energy in the form of heat.

We know from Chapter 5 that if Σ_a and Σ_b were separated by a diathermal wall impermeable to matter, and otherwise isolated, equilibrium would be reached when the total entropy $S_a + S_b$ is maximized, subject to the condition that the total energy $U_a + U_b$ is fixed. This was shown to imply that $T_a \equiv \left(\frac{\partial U_a}{\partial S_a}\right)_V = \left(\frac{\partial U_b}{\partial S_b}\right)_V \equiv T_b$. If the two systems start in arbitrary initial states, energy is exchanged until the temperatures are equal. This is the condition in which the total free energy $F = F_a + F_b$ is minimized (see Section 5.8).

With the actual situation shown in Figure 8.1, equilibrium still occurs when the total entropy $S_{\text{tot}} = S_0 + S_a + S_b$ is maximized, but this is now subject to *two* conditions: the total energy $U_{\text{tot}} = U_0 + U_a + U_b$ and the total number of particles $N_a + N_b$ are both fixed. Equilibrium in which transfer of matter is possible, but the net transfer of matter is zero is called *diffusive equilibrium*.[2] The name refers to the fact that particles diffuse down a density gradient to equalize the density.

8.3 The Chemical Potential

The change in S_{tot} when dN particles and energy dU are transferred from Σ_a to Σ_b is:

$$dS_{\text{tot}} = \left[\left(\frac{\partial S_b}{\partial N_b}\right)_U - \left(\frac{\partial S_a}{\partial N_a}\right)_U\right] dN + \left[\left(\frac{\partial S_b}{\partial U_b}\right)_N - \left(\frac{\partial S_a}{\partial U_a}\right)_N\right] dU, \tag{8.1}$$

where the derivatives are taken with volume, magnetic field, and any other parameter that affects the energy levels, kept constant. Energy may be transferred between the systems (in fact, transferring a particle usually transfers energy at the same time) and also between either system and the heat bath Σ_0. Since the heat bath is, by definition, much larger than Σ_a or Σ_b, in thermal equilibrium, both Σ_a and Σ_b are at the

---

[2] Sometimes rather misleadingly called chemical equilibrium. The concept of diffusive equilibrium includes chemical equilibrium, as usually understood, as a special case.

temperature T of the heat bath and have the same $\left(\frac{\partial S}{\partial U}\right)_N$; hence, the coefficient of dU in Equation (8.1) is zero. Then in diffusive equilibrium, for which S_{tot} is a maximum so that $dS_{\text{tot}} = 0$ for any small change, we must have:

$$\left(\frac{\partial S_b}{\partial N_b}\right)_U = \left(\frac{\partial S_a}{\partial N_a}\right)_U. \tag{8.2}$$

We initially define the *chemical potential* ζ as:

$$\zeta \equiv -T\left(\frac{\partial S}{\partial N}\right)_{U,V}. \tag{8.3}$$

As before, all parameters, such as volume, that affect the energy levels are kept constant. Two systems at the same temperature that can exchange particles as well as energy have the same chemical potential when in diffusive equilibrium. We shall soon see why the factor $-T$ is useful. If there are different species of particle present, each has its own chemical potential ζ_i.

A *reservoir* is a heat bath that can exchange particles as well as heat with the system or systems of interest, and, like a heat bath, is much larger than these systems. The chemical potential plays the same role for diffusive equilibrium as temperature plays for thermal equilibrium. Just as systems in thermal equilibrium with each other or with a heat bath all have the same temperature, systems that can exchange particles with each other or with a reservoir all have the same chemical potential when in diffusive equilibrium. Hence ζ, like T, is a property of the reservoir.

For a single system, the entropy can be expressed as a function of U, V, and N, so that:

$$dS = \left(\frac{\partial S}{\partial U}\right)_{N,V} dU + \left(\frac{\partial S}{\partial V}\right)_{U,N} dV + \left(\frac{\partial S}{\partial N}\right)_{U,V} dN. \tag{8.4}$$

From the triple product rule (Equation D.2):

$$\left(\frac{\partial S}{\partial V}\right)_{U,N} = -\left(\frac{\partial U}{\partial V}\right)_{S,N}\left(\frac{\partial S}{\partial U}\right)_{N,V}.$$

Using Equations (5.8), (5.12), and (8.3), we can make the substitutions:

$$\left(\frac{\partial S}{\partial U}\right)_{N,V} = \frac{1}{T}, \quad \left(\frac{\partial U}{\partial V}\right)_{S,N} = -p, \quad \text{and} \quad \left(\frac{\partial S}{\partial N}\right)_{U,V} = -\frac{\zeta}{T}.$$

Equation (8.4) then becomes:

$$dS = \frac{dU}{T} + p\frac{dV}{T} - \zeta\frac{dN}{T}$$

or

$$\boxed{dU = T\,dS - p\,dV + \zeta\,dN.} \tag{8.5}$$

This is the first law of thermodynamics (Equation 5.13), generalized to include the transfer of matter as well as of energy. In a reversible process, the first term is the heat transfer, the second is the work done, and the third is the energy change due to particle transfer at *constant volume and entropy*:

$$\zeta = \left(\frac{\partial U}{\partial N}\right)_{S,V}. \tag{8.6}$$

It is very difficult to add a particle to a system without changing the entropy, so that Equation (8.6) is not a very useful definition in practice. On the other hand, it is easy to keep the temperature constant, and we saw in Chapter 5 that in isothermal processes, the free energy $F = U - TS$ is a more useful quantity than the internal energy U. From the definition of F,

$$dF = dU - T\,dS - S\,dT.$$

Substituting for dU from Equation (8.5) gives:

$$dF = -S\,dT - p\,dV + \zeta\,dN, \tag{8.7}$$

so that:

$$\boxed{\zeta = \left(\frac{\partial F}{\partial N}\right)_{T,V}.} \tag{8.8}$$

This is the standard definition of chemical potential and is equivalent to Equation (8.3). We see that ζ is the *change in free energy per particle added at constant temperature*. In general, such a transfer changes the entropy as well as the internal energy, and ζ has contributions from both.

Now suppose that the two systems Σ_a and Σ_b are not in diffusive equilibrium with each other, although at the same temperature, and that $\zeta_b > \zeta_a$. The total free energy is $F = F_a + F_b$. If dN particles are transferred from Σ_b to Σ_a then $dN_a = dN$, $dN_b = -dN$, and the change in F is:

$$dF = \zeta_a\,dN - \zeta_b\,dN = (\zeta_a - \zeta_b)\,dN.$$

Since F tends to a minimum as a system tends to equilibrium, $dF < 0$, and $(\zeta_a - \zeta_b) < 0$, so that $dN > 0$. Thus, particles diffuse from Σ_b to Σ_a; that is, from high chemical potential to low, just as heat flows from high to low temperature.

8.4 Free Enthalpy of a Multicomponent System

As pointed out in Section 2.6, constant pressure processes are more common, and safer, than constant volume ones. In such processes, the free enthalpy G is a more useful quantity than the free energy F, the two quantities being related by $G = F + pV$. It can easily be shown [see Problem 8.1(a)] that:

$$\zeta = \left(\frac{\partial G}{\partial N}\right)_{T,p}, \tag{8.9}$$

and that at constant temperature and pressure, G is minimized in equilibrium.

Furthermore, G is a function of p, T, and N, of which only N is an extensive quantity. It follows that since G is an extensive quantity, it must be proportional to N, so that $\left(\frac{\partial G}{\partial N}\right)_{T,p}$ is independent of N and:

$$G = N\zeta. \tag{8.10}$$

Note that Equation (8.10) is only true of the *bulk* contribution to G; surfaces and interfaces, whose contribution is not proportional to N, can also contribute to G in a small system. The surface of a solid or liquid has a certain energy per unit area, called the surface energy or surface tension (see Problem 2.5). This is due to the fact that atoms near the surface have fewer neighbors than those in the bulk, so that the interatomic interaction energy (which is negative) is reduced in magnitude. Interfaces between different substances also have energy and contribute to G; for example, the large interfacial energy of the oil-water boundary is responsible for the fact that oil and water do not mix even though their entropy would be increased by mixing. A detergent reduces the interfacial energy, thus lowering the free energy of the mixed state and allowing mixing to occur. Further examples of the importance of surface energy in systems with at least one small dimension are given in Problem 2.5. Surface energy is of great importance in first order phase transitions, making possible the existence of metastable supercooled and superheated phases; see Chapter 15.

Even when surface effects can be neglected, so that Equation (8.10) holds, it would be a mistake to conclude that G/N and ζ are synonymous. The chemical potential ζ is a property of the reservoir, like temperature, while G and N are properties of the system.

Equation (8.7) can be generalized to include the contribution of different species of particle (*components*):

$$dF = -S\,dT - p\,dV + \sum_i \zeta_i\,dN_i, \tag{8.11}$$

where N_i is the number of particles of species i in the system and:

$$\zeta_i = \left(\frac{\partial F}{\partial N_i}\right)_V = \left(\frac{\partial G}{\partial N_i}\right)_p.$$

The derivatives are taken keeping T constant and also all the particle numbers other than N_i.

Similarly, Equation (8.10) can be generalized (with the same restriction to bulk properties) to:

$$G = \sum_i N_i\zeta_i. \tag{8.12}$$

This can be shown as follows. G is an extensive quantity, as are the N_i. This means that if the volume of the system is changed by a factor λ, the intensive quantities pressure, temperature, and densities all remaining fixed, all extensive quantities change by this factor. Thus, if we write $G = G(p, T, N_1, N_2, \dots)$, we have $\lambda G = G(T, p, \lambda N_1, \lambda N_2, \dots)$. Differentiating both sides of this equation with respect to λ, keeping T, p, and all the N_i constant and using the chain rule, and finally setting $\lambda = 1$, we obtain Equation (8.12). This is an example of Euler's theorem of homogeneous functions.[3]

For a multicomponent system, Equation (8.5) becomes:

$$dU = T\,dS - p\,dV + \sum_i \zeta_i\,dN_i. \tag{8.13}$$

It follows from Equation (8.12) [see Problem 8.1(c)] that:

$$\sum_i N_i d\zeta_i + S\,dT - V\,dp = 0. \tag{8.14}$$

Equation (8.14) is called the *Gibbs-Duhem* relation. It implies that the dependence of the chemical potential of one component on the density of the other components present is not completely arbitrary. At constant T and p, not only must $\sum_i N_i d\zeta_i = 0$, but also:

$$\sum_i N_i \left(\frac{\partial \zeta_i}{\partial N_j}\right)_{T,p} = 0, \text{ all } j; \tag{8.15}$$

that is, the weighted sum of the derivatives of the chemical potentials [labelled i in Equation (8.15)] with respect to the quantity of any one component, (labelled j) must vanish. Equation (8.15) follows directly from Equation (8.12), by differentiating with respect to N_j and using Equation (8.9).

---

[3] See, for example, M. L. Boas, *Mathematical Methods in the Physical Sciences*, 2nd ed. (Wiley, 1983), p. 197. Note that if we do the same for $\lambda F = F(T, \lambda V, \lambda N_1, \lambda N_2, \dots)$, we obtain $F = V(\partial F/\partial V)_T + \sum_i N_i\zeta_i$. From Equation (5.23), $(\partial F/\partial V)_T = -p$, so this result is consistent with the original definition of $G \equiv F + pV$. See Problem 8.6 for an alternative route to the same result.

Equation (8.14) also implies that [see Problem 8.1(d)]:

$$\left(\frac{\partial \zeta_i}{\partial T}\right)_p = \frac{S}{N_i}, \tag{8.16}$$

and:

$$\left(\frac{\partial \zeta_i}{\partial p}\right)_T = -\frac{V}{N_i} = -n_i^{-1}, \tag{8.17}$$

where n_i is the particle density of the i'th component.

8.5 Electrochemical Potential

Suppose that there is a conservative force (i.e., one that can be derived from a potential) on the particles (e.g., an electric or gravitational field). The *electrochemical potential* is defined as the sum of the chemical and all other potentials, and, in isothermal equilibrium, it must be the same everywhere (this will be shown by consideration of some specific examples below). This is the basis of the operation of a battery. Chemical reactions in the battery generate a difference in the chemical potential for electrons at the two electrodes. Charge then flows until the electrostatic potential exactly balances this difference, so that the electrochemical potential is uniform. This electrostatic potential is the open circuit voltage that one measures between the electrodes. However, it must be emphasized that ζ only acts as a potential *in a constant temperature situation*; see, for example, Problem 10.6(d).

Example 8.1: Work Function and Contact Potential

The difference between the chemical potential of an electron at rest outside a conducting solid and its chemical potential inside the solid[4] is called the work function ϕ_w. If two solids A and B with different work functions ϕ_{wA} and ϕ_{wB} touch, a *contact voltage* V_c develops between them. Why? Find V_c, defined as positive when the voltage is higher on solid B than on solid A, in terms of ϕ_{wA}, ϕ_{wB}, and the charge on the electron $-q$.

Since ϕ_w is the difference between a fixed reference chemical potential (the value of ζ outside either solid) and ζ in the solid, the larger ϕ_w is, the lower is ζ. Hence, $\zeta_B - \zeta_A = \phi_{wA} - \phi_{wB}$. For definiteness, let us take $\phi_{wA} - \phi_{wB}$ as positive; that is, we define the solid with the higher ϕ_w, and hence the lower ζ, as A. When the two solids are first brought into contact, electrons flow from B, which has the higher chemical potential, to A. Thus A becomes electrically negative, and B positive (remember that the charge on the electron is negative). Equilibrium is reached when the electrochemical potentials are equal; that is, when:

$$-qV_A + \zeta_A = -qV_B + \zeta_B.$$

Hence,

$$V_c = V_B - V_A = \frac{\zeta_B - \zeta_A}{q} = \frac{\phi_{wA} - \phi_{wB}}{q}.$$

Energy is often expressed in electron volts (eV), in which case q should be taken as 1 in these equations.

---

[4] Since the electron inside the solid is held there by the electrostatic attraction of the positive ions, one might think that its chemical potential should be called the electrochemical potential. However, to do so would be to obscure an important distinction, since *all* chemical forces are ultimately electrical in origin. We distinguish between the contribution of external electrostatic potentials, which can be measured with a voltmeter, and internal ones, such as the attraction of the electron to the nucleus, and include the latter in the chemical potential.

Example 8.2: Diffusive Equilibrium of an Ideal Gas in a Gravitational Field

Consider a molecule in the atmosphere, which we take for the purposes of this example as isothermal and in equilibrium (in fact, this is rarely the case). We can treat this system in two ways. One way is to look at it as a one-particle system, in which the probability of the molecule being at a certain height (and therefore having a certain energy) is given by the Boltzmann factor, as in Section 6.3. This gives us (within a constant factor) the density as a function of height. The other is to look at it as a system in diffusive equilibrium, in which the electrochemical potential (we might better call it the gravichemical potential in this problem) is independent of height in equilibrium. Since this gravichemical potential is the sum of the chemical potential ζ and the gravitational potential, and we know the latter, we can determine ζ as a function of height. Comparing the two results gives us ζ as a function of density, within an as yet undetermined constant, which will be calculated in Chapter 9.

We first consider the one-particle problem, and ask what the probability is of a molecule being at a certain height z above the ground. The molecule experiences a gravitational potential mgz, where m is the mass of the molecule and g is the acceleration due to gravity (taken to be independent of height). The density of states per unit volume does not vary with height; nor does the average kinetic energy of a molecule, since in an ideal gas this depends only on temperature, which we have assumed to be constant. Only the potential energy varies, so that the canonical distribution in the form Equation (7.1) tells us that in equilibrium, the probability of a gas molecule being a height between z and $z + dz$ is:

$$P(z)\, dz = Z^{-1} e^{-\beta mgz}\, dz, \tag{8.18}$$

where Z is a constant that we need not calculate.

The particle density $n(z)$ is $NP(z)$, where N is the total number of gas molecules in the atmosphere per unit area of the earth's surface, so that:

$$n(z) = n_0 e^{-\beta mgz}, \tag{8.19}$$

where n_0 is the density at ground level.

Now let us look at the problem as one in which the "system" is a layer of gas at a particular height, which is in diffusive equilibrium with the layers above and below it. Since the gas is assumed also to be in thermal equilibrium, the total gravichemical potential must be the same everywhere; that is,

$$\zeta(z) + mgz = \zeta_0,$$

where ζ_0 is the chemical potential of the gas at density n_0. Solving Equation (8.19) for mgz as a function of density, we find:

$$mgz = -k_B T \ln\left(\frac{n}{n_0}\right).$$

Since the chemical potential ζ is a property of the gas, and T is constant, ζ can only depend on z through the density $n(z)$, so that:

$$\zeta(n) = \zeta_0 - mgz = \zeta_0 + k_B T \ln\left(\frac{n}{n_0}\right); \tag{8.20}$$

that is, at a given temperature the chemical potential increases logarithmically with the particle density. The magnitude of ζ_0, and its dependence on T, will be found when we treat the ideal gas in Chapter 9. The exponential distribution of density can be regarded as the result of competition between the gravitational potential gradient, which tries to concentrate all the molecules at ground level, and the chemical potential gradient, which is only zero when the density is uniform. We can now see why such equilibrium is called diffusive equilibrium, since the downwards motion due to gravity is balanced by the upwards diffusion due to the density gradient. If there were no gravity, and the density were initially non-uniform, the resulting chemical potential gradient would drive diffusion until the density was the same everywhere. In Chapter 16, we shall consider the transport processes, such as diffusion, that lead to equilibrium.

8.6 Probability of Occupancy: The Grand Partition Function

In Chapter 6 we considered a system with a fixed number of particles in thermal equilibrium with a heat bath (i.e., a thermal reservoir) at temperature T. We found that the probability $P(E_i)$ of finding the system in a particular state of energy E_i is given by the canonical distribution $P_i = Z^{-1}e^{-\beta E_i}$, where, as always, $\beta = \frac{1}{k_B T}$ and Z is the partition function defined in Equation (6.6). We now generalize this result to the case where particles as well as energy can be exchanged with the reservoir.

Suppose that we have a system Σ that can exchange energy and particles with a large reservoir Σ_0. We assume that both Σ and Σ_0 have fixed volume and are isolated from the surroundings, so that the total energy and particle number of the system plus reservoir is fixed. Σ and Σ_0 are initially at different temperatures and chemical potentials. If they are brought into thermal and diffusive contact, they come into equilibrium by transferring a certain amount of energy in the form of heat, and a certain number of particles. However, the effect of this transfer on the state of Σ_0 is very small compared with its effect on the small system Σ, so that when equilibrium is reached, Σ is at the temperature and chemical potential of Σ_0.

Now we suppose that Σ is a microsystem with only one state, which is non-degenerate and whose energy is E. This state may or may not be occupied by a particle. When the state is occupied, Σ has an energy E, while when the state is unoccupied, the energy of Σ is zero. An example is a positively charged atom (positive ion), to which an electron can bind and so neutralize its charge (in this case, E is negative). We ask what happens to the total entropy S_{tot} of the combined system (Σ plus the reservoir Σ_0) when one particle is transferred from Σ_0 to Σ. The total energy is U_{tot}, and the total number of particles is N_{tot}. When Σ is unoccupied, its energy is zero, so that the total energy is the energy of the reservoir $U_0 = U_{\text{tot}}$, and the number of particles in the reservoir is $N_0 = N_{\text{tot}}$. When Σ is occupied, $U_0 = U_{\text{tot}} - E$, while the number of particles in the reservoir is $N_0 = N_{\text{tot}} - 1$. Since Σ is in a non-degenerate state whether or not it contains a particle, its entropy change is zero. The entropy S_{tot} is equal to the entropy of the reservoir, S_0, and there are two contributions to its change, ΔS_{tot}: one contribution is due to its decrease in energy and the other to the removal of one particle. Hence, the change in total entropy is:

$$\Delta S_{\text{tot}} = -\left(\frac{\partial S_0}{\partial U_0}\right)_N E - \left(\frac{\partial S_0}{\partial N_0}\right)_U$$
$$= \frac{\zeta - E}{T}, \tag{8.21}$$

where we have used $\left(\frac{\partial S}{\partial U}\right)_N = \frac{1}{T}$ (Equation 5.8) and $\left(\frac{\partial S}{\partial N}\right)_U = -\frac{\zeta}{T}$ (Equation 8.3).

We can extend this argument to the case where the state in Σ can be occupied by n particles. Do not confuse the occupancy n, which is a pure number, with the particle density n, which has dimensions [volume]$^{-1}$. We still consider only a single non-degenerate state of Σ, so that its entropy remains zero. The energy of this state is in general a function of n, and the transfer of n particles from the reservoir to the system changes the total entropy by:

$$\Delta S_{\text{tot}} = \frac{n\zeta - E(n)}{T}. \tag{8.22}$$

Note that ζ, like T, is held constant by the reservoir.

From the original definition of entropy (Equation 5.4), the ratio of the multiplicity of the occupied state to that of the unoccupied state is:

$$\frac{\Omega_{occ}}{\Omega_{unocc}} = e^{\Delta S/k_B} = e^{\beta[n\zeta - E(n)]}. \tag{8.23}$$

In general, for any given occupancy there are many states, usually of different energy. For example, the neutral hydrogen atom has electron occupancy $n = 1$, and an infinite number of states, any one of which can be occupied. If no states are occupied (i.e., $n = 0$), we have the H^+ ion.

As in the derivation of the canonical distribution (Chapter 6), we can apply the argument that led to Equation (8.23) to each occupancy, and to each state of a given occupancy of the system, in turn, and obtain the probability that the system contains n particles and has energy E_i;

$$P(n, E_i) = \Xi^{-1} e^{\beta(n\zeta - E_i)}, \tag{8.24}$$

where $e^{\beta(n\zeta - E_i)}$ is called the *Gibbs factor* (analogous to the Boltzmann factor), and Ξ is called the *grand partition function*.[5] Ξ is the generalization of the partition function Z to the case where the particle number can vary. Since the total probability, summed over all possible occupancies and all states for each occupancy, adds up to 1, the grand partition function is the sum of all possible Gibbs factors:

$$\Xi = \sum_{i,n} e^{\beta(n\zeta - E_i)}. \tag{8.25}$$

Note that in general, the E_i depend on the occupancy n.

It is often convenient to express the chemical potential in terms of the *absolute activity*[6] α, where:

$$\alpha \equiv e^{\beta\zeta}. \tag{8.26}$$

For brevity, in this book we call α the *activity*. Note that the logarithmic dependence of ζ on density n for an ideal gas (Equation 8.20) implies that in this case $\alpha \propto n$.

Since $e^{n\beta\zeta} = \alpha^n$ and the many-particle partition function for a fixed number n of particles is $Z_n = \sum_i e^{-\beta E_i}$, Equation (8.25) can be written:

$$\Xi = \sum_n \left[\alpha^n \sum_i e^{-\beta E_i} \right] = \sum_n \alpha^n Z_n, \tag{8.27}$$

where the sum is over all possible occupancies. It is important to distinguish between the many-particle partition function Z_n (Z_N in the case of a macrosystem) and the grand partition function Ξ. Z_N was defined in Section 6.5 as the sum over all possible states of a macrosystem with a *fixed* number of particles N, in thermal equilibrium with a heat bath. The grand partition function Ξ, on the other hand, is a sum over *all possible numbers of particles* as well as over all states, for a system that is in thermal and diffusive equilibrium with a reservoir. Ξ for a macrosystem is expressed in terms of Z_N by Equation (8.27), with N substituted for n.

---

[5] Sometimes called the Gibbs sum.

[6] The name "activity" is due to the American chemist G. N. Lewis (1875–1946). It refers to the fact that a chemical reaction evolves to a state of minimum α. The modifier "absolute" is necessary since in chemistry it is convenient to express activities relative to a standard reference state.

The probability that a system contains n particles, of whatever energy, is:

$$P(n) = \Xi^{-1} \sum_i e^{\beta(n\zeta - E_i)}, \tag{8.28}$$

the sum being over all energy states of occupancy n.

Since $e^{\beta\zeta} = \alpha$ and $\sum_i e^{-\beta E_i} = Z_n$, Equation (8.28) can also be written:

$$P(n) = \Xi^{-1} \alpha^n Z_n. \tag{8.29}$$

The mean occupancy; that is, the average number of particles in the system, is:

$$\langle n \rangle = \sum_n n P(n) = \Xi^{-1} \sum_{n,i} n e^{\beta(n\zeta - E_i)}.$$

Hence,

$$\boxed{\langle n \rangle = k_B T \Xi^{-1} \left(\frac{\partial \Xi}{\partial \zeta} \right)_T = \left(\frac{\partial \ln \Xi}{\partial \ln \alpha} \right)_T = \frac{\alpha}{\Xi} \left(\frac{\partial \Xi}{\partial \alpha} \right)_T.} \tag{8.30}$$

In a system where $\langle n \rangle$ is known, one can use Equation (8.30) to find ζ or (equivalently) α.

Example 8.3: Ionizable Atom Immersed in an Electron Gas

Every atom has a certain ionization energy I; that is, it takes an energy I to remove an electron from it, leaving a positively charged ion which we denote A^+. This process can be represented as a chemical reaction:

$$A^0 \rightleftharpoons A^+ + e^-,$$

where we write A^0 for the neutral atom, and e^- for the electron released. For an atom at ordinary temperatures $I \gg k_B T$; nevertheless, as we shall see, it is quite possible for most of the atoms to be ionized. For simplicity we neglect the electron spin, and assume that both the neutral state A^0 and the ionized state A^+ are non-degenerate. We also ignore the excited states of the neutral atom. Suppose that the atom is in the presence of an electron gas whose chemical potential is ζ; for example, an atom in the ionosphere or a stellar atmosphere, or in the plasma of a fusion reactor. This is also a good model of an impurity atom in a semiconductor, such as phosphorus in silicon (see Chapter 14), although in that case the ionization energy is usually only a few meV and one has to go to low temperatures to achieve $I \gg k_B T$. We want to find the mean occupancy $\langle n \rangle$, which is the fraction of atoms that are neutral.

If we choose the energy of the dissociated atom, $A^+ + e^-$, to be zero, the energy of A^0 is $-I$. If n is the number of electrons attached to the atom, $n = 0$ for A^+ and $n = 1$ for A^0. These are the only possibilities, so the grand partition function is:

$$\Xi = \sum_n \alpha^n e^{-\beta E(n)} = 1 + \alpha e^{\beta I}, \tag{8.31}$$

where the first term comes from A^+ ($n = 0$) and the second from A^0 ($n = 1$). Thus, the mean occupancy is:

$$\langle n \rangle = \frac{\alpha}{\Xi} \frac{\partial \Xi}{\partial \alpha} = \frac{\alpha e^{\beta I}}{1 + \alpha e^{\beta I}} = \frac{1}{\alpha^{-1} e^{-\beta I} + 1}$$
$$= \frac{1}{e^{-\beta(\zeta + I)} + 1}. \tag{8.32}$$

The chemical potential ζ is a function of the density n of the electron gas, and its determination will be discussed in subsequent chapters. We shall find in Chapter 9 that for an ideal gas α is less than 1 and is proportional to the density n. Hence, ζ ($= k_B T \ln \alpha$) is negative and increases (i.e., becomes less negative) logarithmically with n. If $\zeta(n)$ is such that $\beta(\zeta + I) = 0$; that is, when the chemical potential is equal to the energy $-I$ of the bound state, we see from Equation (8.32) that the mean occupancy $\langle n \rangle = 1/2$; that is, half of the atoms are neutral and half ionized. If the ambient electron density n is high, so that $\zeta + I \gg k_B T$, $\langle n \rangle \rightarrow 1$, and most of the atoms are neutral. On the other hand, at low density $\zeta + I < 0$; if $-(\zeta + I) \gg k_B T$, the exponent in Equation (8.32) is large and positive, so that the denominator is large, the mean occupancy $\langle n \rangle$ is very small, and most of the atoms are ionized. This result makes physical sense, since if the density is very low, a positive ion has little chance of meeting an electron and becoming neutralized. Note that at sufficiently low density the atom is ionized even when $k_B T \ll |I|$. The situation is quite different from the case of an excited state at energy $|I|$, which is only significantly occupied when $k_B T \sim |I|$ or greater (see Example 6.1). This result justifies our neglect of excited states in the calculation of $\langle n \rangle$ [see Problem 8.7(b) for a specific example].

As a numerical example, let us consider the sun's photosphere (the surface region from which most of the visible radiation comes), as we did in Example 6.1. Then we asked what fraction of the neutral Li atoms are in the excited state; now we ask what fraction of the Li atoms are neutral. For comparison, we also ask what fraction of H atoms are neutral. To simplify the problem, we neglect the electron spin (so that the ground state of each neutral atom is taken to be non-degenerate) and the occupancy of excited states. The temperature is 6400 K ($k_B T = 0.55$ eV), and the electron density is 6×10^{19} m^{-3}. The ionization energy of Li is 5.4 eV and that of H is 13.6 eV.

We shall see in the next chapter that the activity of an ideal monatomic gas of density n is $\alpha(n) = \frac{n}{n_q}$, where n_q is the quantum concentration given by Equation (7.31), so long as $n \ll n_q$. For electrons at 6400 K, neglecting spin,

$$n_q = \left[\frac{9.1 \times 10^{-31} \times 1.38 \times 10^{-23} \times 6400}{2\pi (1.05 \times 10^{-34})^2} \right]^{3/2} = 1.25 \times 10^{27} \mathrm{m}^{-3}.$$

Hence $\alpha = 4.8 \times 10^{-8}$, $\zeta = k_B T \ln \alpha = 0.55 \times \ln(4.8 \times 10^{-8}) = -9.3$ eV. For Li, the mean occupancy is:

$$\langle n \rangle = \frac{1}{e^{(9.3-5.4)/0.55} + 1} = 8 \times 10^{-4}.$$

Thus, even though $k_B T \ll I$, almost all the Li atoms are ionized. For H, on the other hand,

$$\langle n \rangle = \frac{1}{e^{(9.3-13.6)/0.55} + 1} \approx 1 - e^{-7.8} = 0.9996,$$

so that only a fraction 4×10^{-4} of the H atoms are ionized. This is close to what is observed[7]; however our calculation has neglected the fact that neutral hydrogen can capture a second electron, becoming H$^-$ (see Problem 8.8). This negative ion has a small ionization energy and is responsible for the opacity of stellar atmospheres.

---

[7] The numbers assumed here for the solar photosphere are from D. Mihalas, *Stellar Atmospheres* (Freeman, 1978), Table 7-1, "The Harvard-Smithsonian reference atmosphere." We take $h = 0$, where h is the height measured from the point at which the opacity at 500 nm is 1 (i.e., the point at which light of wavelength 500 nm, entering the solar atmosphere vertically from outside, would be attenuated by a factor e^{-1}).

8.7 Chemical Potential and Free Energy

We shall see in the next chapter that if the particles interact only weakly, it is often relatively easy to calculate the chemical potential $\zeta(n)$ as a function of the number n of particles in the system. Since ζ is defined as the change in free energy F when one particle is added to the system at constant volume, we can obtain F for a system containing N particles by summing over these changes:

$$F = \sum_{n=1}^{N} \zeta(n). \tag{8.33}$$

Knowing the free energy, we can find all other quantities of interest, such as the entropy and pressure, by the methods of Sections 5.8 and 5.9. We shall find Equation (8.33) useful in the following chapters.

8.8 Fluctuations in Occupancy

Just as we used the partition function to find the fluctuations in the energy of a system with a fixed number of particles (see Equation 6.22), held at a constant temperature, so we can use the grand partition function to find the fluctuations in particle number of a system held at a constant chemical potential.

The mean square fluctuation in occupancy is (see Equation D.10):

$$\langle \Delta n^2 \rangle = \langle n^2 \rangle - \langle n \rangle^2. \tag{8.34}$$

$$\langle n^2 \rangle = \sum_{n=0}^{\infty} n^2 P(n) = \Xi^{-1} \sum_i n^2 e^{\beta(n\zeta - E_i)},$$

where E_i is the energy of the i'th state of occupancy n.

Since $\Xi = \sum e^{\beta(n\zeta - E_i)}$ (Equation 8.25), $\left(\frac{\partial^2 \Xi}{\partial \zeta^2} \right)_{V,T} = \beta^2 \sum n^2 e^{\beta(n\zeta - E_i)}$, so that:

$$\langle n^2 \rangle = (k_B T)^2 \Xi^{-1} \left(\frac{\partial^2 \Xi}{\partial \zeta^2} \right)_{V,T}. \tag{8.35}$$

From Equation (8.30),

$$\langle n \rangle = k_B T \Xi^{-1} \left(\frac{\partial \Xi}{\partial \zeta} \right)_{V,T},$$

so that:

$$\langle \Delta n^2 \rangle = \langle n^2 \rangle - \langle n \rangle^2 = (k_B T)^2 \left[\Xi^{-1} \frac{\partial^2 \Xi}{\partial \zeta^2} - \Xi^{-2} \left(\frac{\partial \Xi}{\partial \zeta} \right)^2 \right]$$

$$= (k_B T)^2 \frac{\partial}{\partial \zeta} \left(\Xi^{-1} \frac{\partial \Xi}{\partial \zeta} \right)$$

$$= k_B T \left(\frac{\partial \langle n \rangle}{\partial \zeta} \right)_{V,T}. \tag{8.36}$$

Since we usually know $\langle n \rangle$ and obtain ζ from it, Equation (8.36) can be more usefully written as:

$$\langle \Delta n^2 \rangle = k_B T \left(\frac{\partial \zeta}{\partial \langle n \rangle} \right)^{-1}. \tag{8.37}$$

Thus, if the chemical potential is insensitive to $\langle n \rangle$, large fluctuations in occupancy can be expected.

8.9 Envoi

As its name implies, the concept of chemical potential was introduced by Gibbs to deal with problems involving chemical reactions, since a reaction changes the number of particles of any particular molecular species.[8] Except for a few special cases, chemical reactions are not within the scope of this book. However, as we shall see in the next and later chapters, the chemical potential is an extremely useful concept in any multiparticle system; in particular, it is the key parameter in determining the thermodynamic properties of gases and plays an important role in metals and semiconductors.

8.10 Problems

8.1 (a) Prove Equation (8.9)
   (b) Obtain an expression for dG from Equation (8.11) and hence show that for any process at constant temperature and pressure, diffusive equilibrium is reached when G is minimized
   (c) Derive the Gibbs-Duhem relation (Equation 7.14)
   (d) Derive Equations (8.15) through (8.17)
8.2 (a) The equation of state of an ideal gas is $p = nk_BT$, where n is the particle density. Starting from Equation (8.17), show that the chemical potential is $\zeta = k_BT \ln(An)$, where A can depend only on temperature. It will be shown in the next chapter that $A = V_q$, where V_q is the quantum volume defined in Equation (7.29)
   (b) In a mixture of gases (not necessarily ideal), show that the pressure is:

$$p = \sum_i p_i, \tag{8.38}$$

   where $p_i \equiv \int_0^{n_i} n_i \, d\zeta_i$, n_i being the density of the i'th species of molecule, and the integral is evaluated at constant T. p_i is called the partial pressure of the i'th gas. You will need to make the reasonable assumption that $p = 0$ when all the n_i are zero
   (c) Hence, show that if the gases are ideal, as in (a),

$$p = k_BT \sum_i n_i, \tag{8.39}$$

   so that for a mixture of ideal gases, the partial pressures are the actual pressures, $n_i k_BT$, that the gases would exert had they been present alone.
8.3 A tree stands in soil saturated with water. The leaves at the top of the tree are surrounded by air with relative humidity R_h (*relative humidity* is defined as the density of H_2O in the air expressed as a fraction of the saturated density at that temperature). By considering the equilibrium of a water molecule in a gravitational and chemical potential gradient, find the maximum height the tree can have if sap is to reach the top. Assume that at this height the sap ceases to rise, that the water inside the leaves is in equilibrium with that outside, that the water in the roots is in equilibrium with the water in the soil, and that the temperature is uniform. Note that at ground level, water vapor is in equilibrium with the liquid and has its saturated density (i.e., $R_h = 1$), but away from the ground it is *not* in equilibrium; the situation is different from Example 8.2. Treat water vapor as an ideal gas, so that $\alpha \propto R_h$ at a given temperature, and treat the sap as pure water; that is, ignore the effect of salts dissolved in it. Find the maximum height for $R_h = 0.8$, $T = 300$ K. You will probably be astonished by the result.[9]

---

[8] See, for example, McGlashan.
[9] The question of why the column of sap in a tall tree is mechanically stable has only recently been settled. It turns out that the hydrostatic pressure at the top of the tree is in fact negative, being maintained by the tensile strength of water, so that $G < F$. See E. Steudle, "Trees under tension," *Nature* **378**, 663 (1995).

8.4   (a) Starting from Equation (2.15) and the fact that magnetic field is an intensive quantity, show that in an incompressible magnetic system $F = N\zeta$. It follows that in this case:

$$G = F + \mu_0 HM \neq N\zeta$$

(b) Starting from Equation (6.38) for the free energy of a spin 1/2 magnet at high temperature, find $\zeta = \left(\frac{\partial F}{\partial N}\right)_{T,H}$, and hence confirm by direct calculation that $F = N\zeta$, $G \neq N\zeta$, and that $\left(\frac{\partial G}{\partial N}\right)_{T,M} = \zeta$

8.5   (a) Show that for a macrosystem, Equation (8.37) can be written:

$$\langle \Delta N^2 \rangle = \alpha \left(\frac{\partial \alpha}{\partial \langle N \rangle} \right)^{-1},$$

where $\langle N \rangle = nV$.

(b) Use the result in (a), and the fact that $\alpha \propto n$ for an ideal gas, to show that the mean square fractional fluctuation in the density of a fixed volume V of an ideal gas in diffusive equilibrium with a reservoir is:

$$\frac{\langle \Delta n^2 \rangle}{n^2} = \langle N \rangle^{-1}. \tag{8.40}$$

This result was obtained in Problem 6.13 by another method

8.6   We have usually written the free energy F as a function of N, V, and T. In some textbooks, it is said that the reason why $F \neq N\zeta$ in a one component system is that F is a function of two extensive variables, N and V, whereas G is a function of only one. It is not obvious that this argument is valid, since F can equally well be written as a function of density n (which is an intensive variable) rather than volume, and G is often a function of many N_i. Treating $F = F(N, n, T)$ in the same way that $G = G(N, p, T)$ was treated in the derivation of Equation (8.12), and without using $G = N\zeta$, show that:

$$F = N \left(\frac{\partial F}{\partial N} \right)_{T,n} = N\zeta - pV.$$

Note that $\left(\frac{\partial F}{\partial N}\right)_{T,n} \neq \left(\frac{\partial F}{\partial N}\right)_{T,V}$, since when N changes, V must change if the density is to remain constant

8.7   (a) Repeat Example 8.3, giving the degeneracy of A^0 its correct value of 2 due to electron spin. How much difference does including spin make to the fractions of Li and H which are neutral in the sun's outer atmosphere? Caution: Use the correct expression, including spin, for n_q

(b) Now suppose that the electron bound in the neutral atom has a second doubly degenerate level, ϵ above its ground level (which is at $-I$ and is also doubly degenerate). Find the mean number of neutral atoms $\langle n \rangle$ as a function of temperature. Explain your result physically by considering the limiting cases: $k_B T \ll \epsilon$ and $k_B T \gg \epsilon$. Note that $\epsilon < I$, so that in both levels the electron is bound and the atom is neutral.

8.8   (Electron transfer between atoms.) Consider the reaction:

$$2H^0 \rightleftharpoons H^+ + H^-, \tag{8.41}$$

in which an electron is removed from one neutral hydrogen atom H^0, leaving a positively charged atom (positive ion) H^+, and becomes attached to another neutral atom, making it a negative ion H^-. On the left-hand side of Equation (8.41), we have $n = 1$ for each atom, while on the right we have $n = 0$ for one atom and $n = 2$ for the other. The H^0 level is doubly degenerate, corresponding to the two orientations of the electron spin. The ionized states are non-degenerate; H^+ because it has no electron, H^- because the spins are paired

The ionization energy of H^0 is Δ; that is, the $n = 1$ level is at an energy $E(1) = -\Delta$ relative to the energy of the $n = 0$ level (which we take as zero). The ionization energy of H^- is ϵ; this is the energy required to remove one of the electrons in H^-, leaving the other behind, so the energy of the $n = 2$ level is $E(2) = -(\Delta + \epsilon)$. Because of the Coulomb repulsion between the electrons, ϵ is less than Δ

(a) Draw the level scheme for the three possible occupancies and find the grand partition function for one atom

(b) Use the condition $\langle n \rangle = 1$ to find the activity α and the chemical potential ζ. This is a valid approximation in the case of the solar photosphere since the free electron concentration is less than the hydrogen concentration by a factor greater than 10^3 (see the reference in Footnote 6 in this Chapter)

(c) Hence, find the fractions of H^+ and H^- ions. Show that at low temperature, this fraction is $e^{-\beta(\Delta-\epsilon)/2}$. The factor 2 in the activation energy comes from the fact that *two* atoms are involved in the reaction described by Equation (8.41)

(d) For hydrogen, $\Delta = 13.6$ eV, $\epsilon = 0.75$ eV. Find the fraction of H^- ions in the atmosphere of the sun ($T = 6400$ K). The process described in this example is important in understanding the opacity to visible radiation of stars such as the sun, which consist predominantly of hydrogen. The outer atmosphere of the sun is not hot enough for its thermal radiation to be absorbed appreciably by H^0, but it is nevertheless opaque in the visible because of the presence of H^- [see H. A. Bethe and E. E. Salpeter, *Quantum Mechanics of One and Two-Electron Atoms*, 2nd ed. (Plenum, 1977)]

8.9 In Problem 8.8, suppose that $\epsilon > \Delta$

(a) What are the occupancies of the various charge states at $T = 0$?

(b) What is the entropy at $T = 0$? Hint: This is a case where it is convenient to use Equation (6.31) for S

(c) What is $\langle \Delta n^2 \rangle$ at $T = 0$?

Explain your results in physical terms. The phenomenon is called *disproportionation*. It occurs for some impurities, called negative U centers (here $U \equiv \Delta - \epsilon$ and is not the internal energy), in solids. The Coulomb repulsion between the two electrons is more than canceled out by the energy gained by local distortion of the solid which becomes possible when the impurity is negatively charged. An example is the oxygen impurity in gallium phosphide, which can bind zero, one, or two electrons, producing $+$, 0, and $-$ charge states, respectively. This phenomenon is an exception to the third law of thermodynamics in the form, "any system in equilibrium has zero entropy at absolute zero" (see Section 5.13)

8.10 At the end of Section 8.3, two systems a and b were treated to interact at the same temperature to show that particles diffuse from b to a, that is, high chemical potential to low chemical potential as the system approaches diffusive equilibrium. Beginning from Equation 8.4 and remaining in the entropy representation, demonstrate that this diffusive equilibrium, with the same relative values of the chemical potential for b and a, is approached as a process that maximizes entropy

III

The Thermodynamics
of Gases

9

Perfect Gases

All science is full of statements where you put your best face on your ignorance, where you say: ...we know awfully little about this, but more or less irrespective of the stuff we don't know about, we can make certain useful deductions.

Hermann Bondi

Q: How many physicists does it take to milk a cow? A: First, we assume a spherical cow...

Anonymous

9.1 Gedanken

We've been relying on our trustworthy ideal gas for quite some time now, and we've been able to accomplish quite a lot of interesting physics. Ultimately, we have to acknowledge that our ideal gas will be insufficient to describe where we're ultimately heading, and that is to describe a phase transition from a gas to a liquid, liquid to solid, or a gas to a solid. The ideal gas law predicts only a gas all the way down to absolute zero. So now we imagine taking our ideal gas and making it less than ideal, by compressing the gas into a smaller and smaller volume so that the particle-particle interactions which the ideal gas assumes are negligible are now greatly important to the behavior of the gas. As we compress the gas, one of three things can (should) happen. The particles resist interaction requiring a quantitatively larger force to continue to compress the gas, the particles will demonstrate an affinity for interaction requiring a quantitatively smaller force to continue to compress the gas, or the particles could demonstrate no interaction with each other whatsoever. This last behavior would be a perfect gas.

9.2 Distribution Functions

We are now in a position to approach the problem that we have been postponing for so long: how to deal with a many-particle system in which the individual particles cannot be distinguished from each other. The simplest and most useful example of such a system is a "perfect" gas, of which the "ideal" gas is a special case. Note the distinction, the need for which will become clear later in this chapter.[1] A *perfect gas* is a gas in which the energy of any one molecule is independent of the presence of the others; that is, interactions between the molecules have a negligible effect on their energy states. An *ideal gas* is a perfect gas whose density is sufficiently low that quantum effects can be neglected. Because this distinction

---

[1] Both terms must be distinguished from the terms "noble" or "rare" gas, which are synonymous generic terms for the elements in group 8 of the periodic table: He, Ne, Ar, Kr, Xe, and Rn. These terms refer to a specific group of gases identified by their electronic structure and chemical inertness, while "ideal" and "perfect" refer to the physical properties of the gas, regardless of chemical composition. At room temperature, all the noble gases are (to a very good approximation) ideal gases, but so are many other gases; for example, hydrogen, nitrogen, and oxygen.

is not always observed in the literature, some authors prefer to call a perfect gas a *non-interacting gas*. However, this terminology is somewhat misleading. Except at extremely low density, gas molecules collide frequently (see Chapter 16) and hence interact; if they did not the gas could not reach internal equilibrium. Thus even in a perfect gas there are intermolecular interactions, but they are not strong enough to affect the energy states.

Our purpose is to find the *distribution function* of a perfect gas. In the present context, this term has a specialized meaning and is defined as follows. Suppose that we know all the possible states of a single particle in the system (e.g., the states of a particle in a box, calculated in Chapter 7). We call these single particle states, and here we define occupancy n as the number of particles in a given single particle state. In a perfect gas, the energy of any one particle is not affected by the presence of the other particles, so that the single particle states are independent of the occupancies. It follows that a macrostate of a perfect gas with a given particle density is completely defined if we know the mean occupancy $\langle n(E) \rangle$ for all E. The mean occupancy is called the distribution function and is written $f(E)$:

$$f(E) = \langle n(E) \rangle. \tag{9.1}$$

While $f(E)$ is often less than unity (much less in the case of an ideal gas), it is *not* a probability. As we shall see, it can exceed unity in a Bose gas; furthermore, the sum of the occupancies over all possible states is $\sum f(E) = N$, the number of particles in the system, whereas probability P is normalized so that $\sum P = 1$. In general, $f(E)$ depends parametrically on temperature and chemical potential.

There are three different forms of $f(E)$, depending on the class of particle we are considering and on the density. We shall find that in the low density (i.e., the ideal) limit, the distribution function, which is called the Maxwell-Boltzmann or Boltzmann distribution, has a strong mathematical resemblance to the canonical distribution (Equation 6.4), but it is important to realize that these are two quite different quantities and must not be confused. The canonical distribution is a probability distribution; it depends parametrically only on the temperature and applies to *any* system with a fixed number of particles. The Maxwell-Boltzmann distribution, on the other hand, is a distribution function; it depends parametrically on the chemical potential as well as on the temperature, and applies *only* to an ideal gas.

We know from quantum mechanics that there are two general classes of particles: fermions and bosons. A fermion has odd half-integer spin,[2] its many-particle wave functions are anti-symmetric, and because of this anti-symmetry, it obeys the Pauli principle that only one particle can occupy a given single particle state; that is, $n = 0$ or 1. Examples of fermions are the electron, neutrino, muon, proton, neutron, quark, and all nuclei with odd mass numbers(e.g., ^{3}He). A boson has zero or integer spin and has symmetric many-particle wave functions, so that any number of bosons can occupy the same single particle state; that is, n can have any non-negative value. Examples are the photon (γ), pion (π^+), gluon (g), and any nucleus with even mass number (e.g., the deuteron ^{2}H and the alpha particle ^{4}He). We must deal with fermions and bosons separately, since their statistics are very different.

9.3 Fermions: The Fermi-Dirac Distribution

We start by considering a system that consists of just one single fermion state of energy E. The system is in contact with a reservoir whose chemical potential is ζ. The state may or may not be occupied. We choose the zero of energy as the energy of the system when the state is unoccupied. For fermions, there are only two possibilities: the occupation number n is either 0 or 1. The grand partition function for this system with one single particle state is then:

$$\Xi = 1 + e^{\beta(\zeta - E)}, \tag{9.2}$$

---

[2] The relation between spin and statistics is a necessary consequence of relativistic quantum theory. See F. Mandl and G. Shaw, *Quantum Field Theory*, 2nd ed. (Wiley, 1993), pp. 72–73.

where the first term corresponds to the empty state ($n = 0$), and the second term corresponds to the occupied state ($n = 1$). As always, $\beta = 1/(k_B T)$. The mean occupancy of the state is, from Equation (8.30):

$$f(E) = \langle n(E) \rangle = k_B T \Xi^{-1} \left(\frac{\partial \Xi}{\partial \zeta} \right)_T = \frac{e^{\beta(\zeta - E)}}{1 + e^{\beta(\zeta - E)}},$$

so that:

$$f(E) = \frac{1}{e^{\beta(E - \zeta)} + 1}. \tag{9.3}$$

This is called the Fermi-Dirac distribution. It applies to any perfect gas of fermions.

The chemical potential ζ is determined by the temperature and the particle density $n \equiv N/V$. In volume V, the number of states with energy between E and $E + dE$ is $Vg(E)\,dE$, where $g(E)$ is the density of states defined in Section 7.2. The mean occupancy of each of these states is $f(E)$, so that the number of particles with energy between E and $E + dE$ is $Vg(E)f(E)\,dE$. Hence, the total number of particles in volume V is:

$$N = \int_0^\infty Vg(E)f(E)\,dE, \tag{9.4}$$

and the particle density is:

$$n = \int_0^\infty g(E)f(E)\,dE \tag{9.5}$$

$$= \int_0^\infty \frac{g(E)}{e^{\beta(E - \zeta)} + 1}\,dE. \tag{9.6}$$

Equation (9.6) enables us to obtain n if ζ is known. More commonly, we know n and need to find ζ.

This is easy to do in the limit $T \to 0$; that is, $\beta \to \infty$. The value of ζ at absolute zero ($\beta = \infty$) is called the Fermi energy E_F. Putting $\beta = \infty$ in Equation (9.3), we see that if $E > E_F, f(E) = 0$, while if $E < E_F$, $f(E) = 1$. Thus, $f(E)$ has the form shown in Figure 9.1, dropping abruptly from 1 to 0 at $E = E_F$. This distribution has a simple physical interpretation. Suppose that we start with all the states empty and add particles one by one, filling the lowest energy states first. Because each state can only accommodate one particle, states of increasing energy are successively filled until all the particles are used up. The highest occupied level for a given number of particles has energy E_F. Since the levels are continuous, the energy required to add one more particle is E_F. Since $F = U - TS$ (Equation 5.18), $F = U$ at $T = 0$, so that the chemical potential, which is the change in F produced by the addition of one particle, is $\zeta = E_F$.

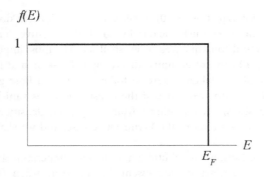

FIGURE 9.1 Fermi-Dirac distribution function at absolute zero.

Since n is positive, E_F must also be positive, and Equation (9.6) becomes:

$$n = \int_0^{E_F} g(E) \, dE. \tag{9.7}$$

If we know $g(E)$ and n, we can find E_F from Equation (9.7). A perfect gas of fermions at a temperature so low that $\zeta \approx E_F$ is called a *degenerate Fermi gas*, usually abbreviated to *Fermi gas*. This use of the word "degenerate" should not be confused with the use of the same word in connection with energy levels; it simply means that the density is so high that the quantum nature of the particles cannot be ignored. An important example of a degenerate Fermi gas is the electron gas in a metal, which makes it a conductor. We will discuss this gas more fully in Chapter 13.

9.4 Bosons: The Bose-Einstein Distribution

Any number of bosons can occupy a single state. The occupation number n is no longer restricted to 0 or 1, but can be any non-negative integer. The addition of each particle adds an energy E, where E is the energy of the single particle state, so that there is now a ladder of energy states, with energies $E_n = nE$ (remember that the gas is assumed to be perfect, so that the energy of a state is not affected by its occupancy). The grand partition function for a Bose gas is:

$$\Xi = \sum_{n=0}^{\infty} e^{\beta(n\zeta - E_n)} = \sum_{n=0}^{\infty} e^{\beta(n\zeta - nE)} = \sum_{n=0}^{\infty} \left(e^{\beta(\zeta - E)} \right)^n. \tag{9.8}$$

This is a geometric series, which only converges if $E > \zeta$. If this is the case, the series can be summed (as in the case of the harmonic oscillator; see Example 6.3) to give:

$$\Xi = \frac{1}{1 - e^{\beta(\zeta - E)}}. \tag{9.9}$$

From Equation (8.30),

$$f(E) = \langle n(E) \rangle = k_B T \Xi^{-1} \left(\frac{\partial \Xi}{\partial \zeta} \right)_T = \frac{e^{\beta(\zeta - E)}}{1 - e^{\beta(\zeta - E)}},$$

so that:

$$\boxed{f(E) = \frac{1}{e^{\beta(E - \zeta)} - 1} = \frac{1}{\alpha^{-1} e^{\beta E} - 1},} \tag{9.10}$$

where $\alpha = e^{\beta\zeta}$ is the activity. Equation (9.10) is called the Bose-Einstein distribution. While the only mathematical difference from the Fermi-Dirac distribution is the negative sign in the denominator, this difference has dramatic physical consequences, as we shall see. The most important consequence is that in the Bose-Einstein case, $f(E)$ can exceed unity, diverging as $\zeta \to E$, and is nonexistent for $\zeta > E$. If the lowest allowed value of E is taken as zero, it follows that for a Bose gas the chemical potential must always be negative ($\zeta = 0$ is permitted if the density of states vanishes at $E = 0$, as is usually the case in three dimensions). Determining ζ from the particle density n, using Equation (9.5), is mathematically more complicated than in the Fermi-Dirac case, and we shall return to Bose gases in Chapters 11 and 12.

The Fermi-Dirac and Bose-Einstein distributions at non-zero temperature are compared in Figure 9.2. Note that the two distribution functions are essentially the same when $f(E)$ is small. This makes physical sense, since if the mean occupancy is much less than one, the difference between fermions and

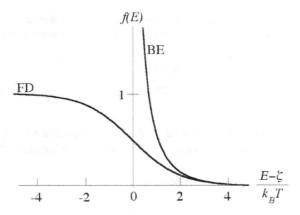

FIGURE 9.2 Fermi-Dirac (FD) and Bose-Einstein (BE) distribution functions in the vicinity of $E = \zeta$, at non-zero temperature.

bosons is irrelevant. On the other hand, the distributions diverge rapidly as $E \to \zeta$ from above; a gas of fermions has $f(\zeta) = 1/2$, while for a gas of bosons $f(E) \to \infty$ as $E \to \zeta$. As noted above, a gas for which $f(E)$ is close to or greater than unity at some E is said to be degenerate.

9.5 The Ideal Monatomic Gas and the Maxwell-Boltzmann Distribution

We now consider the "dilute" or low-density limit in which $f(E) \ll 1$ at all E; that is, the mean occupancy is low for all states. We rearrange Equations (9.3) and (9.10) to give:

$$e^{\beta(E-\zeta)} = \frac{1}{f(E)} \pm 1, \tag{9.11}$$

where the upper sign applies to bosons and the lower to fermions. If $f(E) \ll 1$, 1 can be neglected relative to $1/f(E)$ and both distributions reduce to:

$$\boxed{f(E) = e^{\beta(\zeta - E)} = \alpha e^{-\beta E}.} \tag{9.12}$$

Equation (9.12) is called the Maxwell-Boltzmann (or Boltzmann) distribution, and a perfect gas with this distribution is called an ideal gas. As pointed out above, while Equation (9.12) looks very like the canonical distribution (Equation 6.4) for the probability of occupation of a state, $P(E)$, $f(E)$ is a different quantity from $P(E)$ and Equation (9.12) applies *only* to an ideal gas.

The condition for the validity of the low density approximation can also be expressed in terms of ζ. The condition $f(E) \ll 1$ requires $\beta(\zeta - E)$ to be large and negative for all possible values of E. Since E is a kinetic energy, its lowest value is zero. Hence, for $f(E)$ to be small, ζ must be negative, $|\zeta| \gg k_B T$, and $\alpha \ll 1$. Note, however, that ζ is strongly temperature dependent, and the condition $|\zeta| \gg k_B T$ *does not* imply that the temperature must be low; on the contrary, as we shall see, Equation (9.12) holds in the high temperature limit but breaks down at low temperature [see the discussion of Equation (9.15) below].

As we saw, the quantum mechanical distinction between fermions and bosons disappears in the low density limit, and in fact Equation (9.12) can be derived by purely classical arguments. However, the derivation here is not only simpler, but makes clear that the condition for its validity is that the mean occupancy of any state must be small. This dilute (or low density) limit is often called the classical limit, since in this limit, quantum effects become almost irrelevant, although not quite, as we shall see in Chapter 10.

In this low density limit, the chemical potential can easily be obtained from the particle density n. The number of particles in volume V is nV, and we sum over all states in this volume to obtain:

$$nV = \sum f(E) \tag{9.13}$$

$$= \alpha \sum e^{-\beta E} = \alpha Z_1, \tag{9.14}$$

where $Z_1 = \sum e^{-\beta E}$ is the partition function for a *single* particle (Equation 6.7). Z_1 for a free particle with spin S but no internal structure, confined to a volume V, was found in Equation (7.30) to be:

$$Z_1 = n_q V,$$

where, from Equation (7.31),

$$n_q = (2S + 1) \left(\frac{m k_B T}{2\pi \hbar^2} \right)^{3/2}.$$

Hence, the activity and chemical potential are:

$$\boxed{\alpha = \frac{n}{n_q}, \quad \zeta = -k_B T \ln \left(\frac{n_q}{n} \right).} \tag{9.15}$$

The condition for the validity of the ideal gas approximation, $\alpha \ll 1$, is thus equivalent to $n \ll n_q$. Note that n_q varies as $T^{3/2}$, so that if the temperature is reduced at constant density, α increases towards 1. The condition $n \ll n_q$ requires that $T \gg \frac{2\pi \hbar^2}{m k_B} \left(\frac{n}{2S+1} \right)^{2/3}$, a condition that is satisfied by all gases except helium, since they liquefy at temperatures well above this.

The physical reason why ζ must be negative in an ideal gas, while it is positive in a degenerate Fermi gas, can be seen as follows. The multiplicity Ω, which is the number of microstates in a macrostate, is a rapidly increasing function of the number of particles, since increasing the number of particles increases the number of ways that the energy can be divided between them. Hence, adding a particle to an ideal gas increases the entropy as well as the energy, and the entropy term in $F = U - TS$ increases by more than the energy term [this is shown in the next chapter; compare Equations (10.14) and (10.16)], so that the free energy decreases. In a Fermi gas at $T = 0$, on the other hand, the entropy is zero and remains so when a particle is added. The particle must come with energy E_F if it is to find an unoccupied state, so that ζ is positive. In Chapter 13, this argument is extended to non-zero temperature.

We can now quantify the remark in Chapter 7 that quantum effects become important when the density approaches the quantum concentration n_q; that is, when the interparticle separation approaches the de Broglie wavelength. If $n \ll n_q$, $\alpha \ll 1$, and from Equation (9.12), $f(E) \ll 1$ for all E, so that the condition for classical behavior is satisfied. If $n \to n_q$, $\alpha \to 1$; the condition $f(E) \ll 1$ is not satisfied, and we have to take into account the distinction between fermions and bosons. This is the subject of Chapters 11 through 13.

Example 9.1: Chemical Potential of a Monatomic Ideal Gas

Helium is a monatomic gas with atomic weight 4 and spin zero. Assuming it to be an ideal gas with the equation of state $p = n k_B T$, find α and ζ at atmospheric pressure (10^5 Pa) at $T = 300$ K and at $T = 3$ K. Express ζ in eV (electron volts). Is the ideal gas approximation valid at 3 K?

To find α and ζ, we use Equation (9.15) with $n = p/(k_B T)$ and $(2S + 1) = 1$, giving:

$$\alpha = \left(\frac{2\pi\hbar^2}{m} \right)^{3/2} \frac{p}{(k_B T)^{5/2}}.$$

For He, $m = 4M_H = 6.7 \times 10^{-27}$ kg.

At 300 K and 10^5 Pa, $\alpha = 3 \times 10^{-6}$, $\zeta = k_B T \ln \alpha = 0.025 \ln(3 \times 10^{-6}) = -0.32$ eV. At 3 K and 10^5 Pa, $\alpha = 0.3$; $\zeta = 2.5 \times 10^{-4} \ln(0.3) = -3 \times 10^{-4}$ eV.

The ideal gas approximation is not good at 3 K for two reasons:

1. At 3 K and 10^5 Pa, helium is in fact a liquid, showing that interatomic interactions cannot be neglected, as they are in an ideal gas

2. Even if these interactions were not present, α is too near to 1 for the ideal gas approximation to hold. The fact that the He atom is a boson must be taken into account. In fact, liquid He undergoes a transition resembling Bose-Einstein condensation at 2.2 K; see Chapter 12.

Since helium is the lightest gas, except for hydrogen and the electron gas, and since n_q increases with increasing mass, Example 9.1 shows that ordinary gases at room temperature have very small α, typically $< 10^{-6}$. As the temperature is reduced, α increases, but all gases except helium and the electron gas liquefy while α is still much less than 1.

9.6 Envoi

In this brief chapter, we have introduced the concept of distribution function $f(E)$, which specifies the average occupancy of the levels of a perfect gas as a function of energy and contains the chemical potential ζ and the temperature T as parameters. We derived three forms of $f(E)$. The Fermi-Dirac distribution applies to particles whose many-particle wave functions are anti-symmetric under exchange, so that they obey the Pauli exclusion principle. The Bose-Einstein distribution applies to particles with symmetric many-particle wave functions; these do not obey the Pauli principle. We found that the low density limit of both distributions is the same, the Maxwell-Boltzmann distribution. In the next four chapters, we use these results to obtain the thermodynamic properties of different types of perfect gases.

9.7 Problems

9.1 (a) Show that if the Fermi-Dirac distribution function (Equation 9.3) is expressed as a function $f(\epsilon)$ where $\epsilon \equiv E - \zeta$, it has the following symmetry:

$$f(-\epsilon) = 1 - f(\epsilon). \tag{9.16}$$

In other words, $f(\epsilon) - 1/2$ is an odd function of ϵ; that is, $f(\epsilon) - 1/2 = -[f(-\epsilon) - 1/2]$

It follows from Equation (9.16) that if the density of states $g(E)$ does not vary appreciably over the relevant range of energy, for every state of energy $\zeta + \epsilon$ ($\epsilon > 0$) occupied by an electron, there is a vacant state, or *hole*, with energy $\zeta - \epsilon$

(b) Show that if we are only interested in the "tails" of the distribution where $|\epsilon| \gg k_B T$, the Maxwell-Boltzmann distribution is a good approximation, both for the electron distribution $f(\epsilon)$ and for the hole distribution $1 - f(\epsilon)$ ($\epsilon < 0$). This is often the situation in a semiconductor,

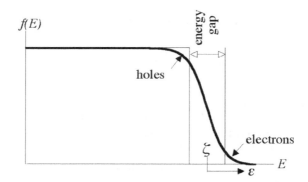

FIGURE 9.3   Fermi-Dirac distribution in a semiconductor, which has no states (i.e., $g(E) = 0$) within the energy gap.

where the chemical potential usually lies in an energy range where there are no states (the "energy gap"); see Figure 9.3. Semiconductors are discussed in more detail in Chapter 14

9.2   Find the chemical potential at $T = 0$ (i.e., the Fermi energy) for a two-dimensional electron gas of areal density σ electrons per unit area (don't forget the electron spin)

9.3   (a) Starting from Equation (8.35), show that the mean square fluctuation in the occupancy n of a single particle state is:

$$\langle \Delta n^2 \rangle = \langle n \rangle (1 \pm \langle n \rangle), \qquad (9.17)$$

where the plus sign applies to bosons and the minus sign to fermions, and $\langle n \rangle$ is the mean occupancy. Explain what these results mean in physical terms

   (b) Show that the fractional mean square fluctuation in the particle density $\frac{\langle \Delta n^2 \rangle}{n^2}$ of a given volume V of a perfect degenerate Fermi gas; that is, one in which $\zeta \gg k_B T$ so that $\zeta = E_F$, is less than that of an ideal gas by a factor $\sim \frac{k_B T}{E_F}$. Explain this factor in physical terms. Find the numerical factor for the three-dimensional case

9.4   (a) Electrons trapped on the surface of liquid helium by an electric field perpendicular to the surface can be treated as a perfect monatomic two-dimensional Fermi gas. Without doing any algebra, find the average kinetic energy per electron at zero temperature in terms of the Fermi energy E_F (see Chapter 7 for the density of states in two dimensions)

   (b) The field is removed and the electron gas expands irreversibly into a vacuum, which is, of course, three-dimensional. The final volume is so large that the electron gas is ideal at the end of the expansion. No heat flows in or out during this operation, and no work is done. What is the final temperature of the electrons in terms of E_F, the Fermi energy in (a)?

   (c) Show that the assumption that the electron gas is ideal in its final state is valid if $V/A \gg (54/\sigma)^{1/2}$, where σ is the initial areal density of the electron gas, and V/A is the ratio of the final volume to the initial area

9.5   (a) Find the activity α and chemical potential ζ at $T = 300$ K of an electron gas of density 10^{24} m^{-3}, assuming that it is an ideal gas

   (b) Repeat the calculation in (a) for $T = 3$ K. Show that the ideal gas assumption is no longer justified. Why would it be a better approximation to treat the gas as a degenerate Fermi gas? Find ζ using this approximation

9.6 Show from Equation (8.37) that in a mixture of perfect degenerate Fermi gases at 0 K, the partial pressures are the actual pressures that the gases would exert had they been present alone, as in an ideal gas. In fact, this result is true for any mixture of perfect gases, as one might expect, since the molecules in a perfect gas act independently of each other

9.7 Using Figures 9.1 and 9.3 as a guide, sketch a corresponding guess as to what the Bose occupation states would be at $T = 0$ and just above $T = 0$. Comment on the relationship between your sketches and the form of the curves in Figure 9.2

10

Ideal Gases and Solutions

One of the most beautiful hypotheses ever propounded in physics is...the Dynamical Theory of Gases.

Simon Newcomb, 1861

It appears, therefore, that gases are distinguished from other forms of matter, not only by their power of indefinite expansion so as to fill any vessel ...but by the uniformity and simplicity of the laws which regulate [them].

James Clerk Maxwell, 1871

10.1 Gedanken

We now return to one of our prior gedanken experiments, albeit in a slightly different form: Imagine two volumes separated by a divider. In one volume is a gas of particles that can be distinguished from those on the other side of the divider. When this gedanken is usually introduced in a class the instructor asks the students to imagine that particles on one side of the divider are painted red, and particles on the other side of the divider are painted blue. The two gases are both at the same temperature T but when we remove the divider and then replace it later, the two species have mixed. Entropy (disorder) has increased. Now we re-imagine the experiment, except this time, we can't tell the difference between the particles on the right and the particles on the left. Imagine now they are all the same color. Remove the divider, and some time later replace it. Was there an increase in entropy? If you answer no, realize that particles from the left and right sides have mixed. If you answer yes, how do you measure the increase in entropy?

10.2 Ideal Monatomic Gases: Introduction

In the last chapter, we showed that the distinction between bosons and fermions vanishes in the low density (ideal) limit, and that the mean occupancy of the energy levels of an ideal gas of either particle is given by the Maxwell-Boltzmann distribution (Equation 9.12):

$$f(E) = e^{\beta(\zeta - E)} = \alpha e^{-\beta E},$$

where ζ is the chemical potential and α ($\equiv e^{\beta \zeta}$) is the activity. In deriving Equation (9.12), we assumed that the gas consists of particles with no internal structure other than spin; that is, we assumed a monatomic gas such as helium. Later in this chapter, we extend our treatment to ideal polyatomic gases.

The activity α is very simply related to the particle density n. For a three-dimensional monatomic gas, Equation (9.15) gives:

$$\alpha = \frac{n}{n_q},$$

where n_q is the quantum concentration given by Equation (7.31),

$$n_q = (2S + 1)\left(\frac{mk_B T}{2\pi\hbar^2}\right)^{3/2}.$$

In this chapter, we explore the experimental consequences of these results, and extend them to polyatomic ideal gases. It should always be remembered that the results only apply when $\alpha \ll 1$.

10.3 The Maxwell Velocity Distribution in an Ideal Gas

One result that we can obtain right away is the distribution of molecular energies and velocities. The probability that a molecule has an energy between E and $E + dE$ is, from Equation (7.1),

$$P(E)\, dE = Z_1^{-1} g(E) e^{-\beta E}\, dE,$$

where we write Z_1 rather than Z as a reminder that this is the single particle partition function. For unit volume, we put $V = 1$ in Equation (7.30) so that $Z_1 = n_q$. From Equation (7.19),

$$g(E) = \frac{2S + 1}{4\pi^2}\left(\frac{2m}{\hbar^2}\right)^{3/2} E^{1/2} = 2\pi^{-1/2} E^{1/2} \beta^{3/2} n_q.$$

Thus, the probability distribution is:

$$P(E) = 2\pi^{-1/2}\beta^{3/2} E^{1/2} e^{-\beta E}. \tag{10.1}$$

Note that $P(E) \to 0$ as $E \to 0$, because the density of states goes to zero.

The distribution can also be expressed as a function of velocity v rather than of energy, the two being related by:

$$E = \frac{1}{2}mv^2. \tag{10.2}$$

The probability $P(v)\, dv$ that a particle has a velocity between v and $v + dv$ is the same as that of finding it between E and $E + dE$, where $dE = mv\, dv$ from Equation (10.2):

$$P(v)\, dv = P(E)\, dE = P(E)mv\, dv.$$

Substituting from Equation (10.1) gives:

$$P(v) = \left(\frac{2}{\pi}\right)^{1/2} (m\beta)^{3/2} v^2 \exp\left(-\frac{1}{2}\beta mv^2\right). \tag{10.3}$$

This is called the Maxwell velocity distribution. Note that Planck's constant has vanished from the equation, showing that it is a purely classical result. Maxwell derived it by an ingenious but shaky

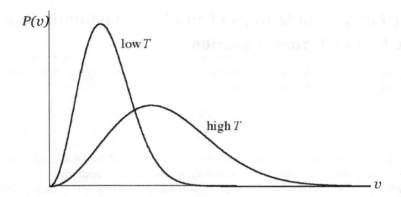

FIGURE 10.1 Maxwell velocity distribution in an ideal gas, for two temperatures differing by a factor of 4.

argument[1] before the development of statistical mechanics. The Maxwell distribution is shown in Figure 10.1 for two different temperatures. Note that because the distribution is not symmetric in v, the root mean square velocity, $\left(\frac{3k_BT}{m}\right)^{1/2}$, is not equal to the most probable velocity; that is, the velocity where $P(v)$ is a maximum, which is $\left(\frac{2k_BT}{m}\right)^{1/2}$.

The Maxwell distribution provides an intuitively appealing interpretation of Arrhenius' law (Equation 6.44) for the rate $R_c(T)$ of a chemical reaction, at least as it applies to reactions in the gas phase. In the absence of a catalyst, such reactions require that two atoms or molecules collide with a sufficient kinetic energy to overcome some potential barrier. If the activation energy (i.e., the minimum energy required for the reaction to occur) is E_a, the rate of the reaction (i.e., the probability per unit time of the reaction occurring) is proportional to the probability that the combined kinetic energy of the colliding molecules exceeds E_a. If we suppose for simplicity that one molecule is stationary so that all the required energy is carried by the other, we have for the rate:

$$R_c(T) = A \int_{E_a}^{\infty} E^{1/2} e^{-\beta E} \, dE, \tag{10.4}$$

where A is a constant. The integral can be simplified by making the substitution $x = \beta E$, giving:

$$A\left(k_B T\right)^{3/2} \Gamma\left(\frac{3}{2}, \beta E_a\right), \tag{10.5}$$

where $\Gamma(a, b)$ is the incomplete gamma function $\int_b^{\infty} x^a e^{-x} \, dx$.

Usually E_a is of the order of electron volts and is thus much greater than $k_B T$, so that $E_a \gg 1$. When $b \gg 1$, $\Gamma(a, b) \approx b^{a-1} e^{-b}$ (see Abramowitz and Stegun, p. 263), so that:

$$R_c(T) \approx A k_B T E_a^{1/2} e^{-\beta E_a}. \tag{10.6}$$

Apart from the slowly varying prefactor, Equation (10.6) is the same as Equation (6.44), and can be regarded simply as a directly observable expression of the canonical distribution.

---

[1] See M. Born, *The Natural Philosophy of Cause and Chance* (Oxford, UK, 1949), p. 50.

10.4 Free Energy and Entropy of an Ideal Monatomic Gas: The Sackur-Tetrode Equation

We saw in Equation (8.33) that we can obtain the free energy F by summing the chemical potentials as we add particles to the system, starting from zero:

$$F = \sum_{r=1}^{N} \zeta(r),$$

where r is the (varying) number of particles in volume V, and $N = nV$ the total number. Knowing F, we can obtain the many-particle partition function and all thermodynamic quantities.

We can apply Equation (8.33) to the ideal gas by substituting r/V for n in Equation (9.15), obtaining:

$$\zeta(r) = k_B T \ln\left(\frac{r}{n_q V}\right) = k_B T \left[\ln(r) - \ln(n_q V)\right].$$

$$F = k_B T \sum_{r=1}^{N} \left[\ln(r) - \ln(n_q V)\right]$$

$$= k_B T \left[\ln(N!) - N \ln(n_q V)\right]. \tag{10.7}$$

In Chapter 6, we saw that $F = -k_B T \ln Z_N$ (Equation 6.28a), where Z_N is the many-particle partition function for the N-particle system, so that for an ideal (i.e., low density) gas:

$$Z_N = e^{-\beta F} = \frac{(n_q V)^N}{N!} = \frac{Z_1^N}{N!}, \tag{10.8}$$

since $Z_1 = n_q V$ (Equation 7.30).

N is a very large number, so that $\ln N \ll N$, and we can use Stirling's formula in the form Equation (D.6) to replace $\ln N!$ in Equation (10.7) by $N(\ln N - 1)$. The free energy of an ideal monatomic gas is then:

$$F = Nk_B T \left[\ln\left(\frac{N}{n_q V}\right) - 1\right]$$

$$= -Nk_B T \left[\ln\left(\frac{n_q}{n}\right) + 1\right]. \tag{10.9}$$

We could have obtained Equation (10.9) directly from Equation (8.33) by treating r as a continuous variable and integrating at constant volume.

Comparing Equation (10.8) with the expression for Z_N for distinguishable systems, $Z_N = Z_1^N$ (Equation 6.12), we see that there is an extra factor $1/N!$. This factor is a consequence of the fact that the particles in the gas are indistinguishable; indistinguishability was implicit in our derivation of the distribution function and hence of ζ. Note that even though Equation (10.8) applies only in the low density limit, which is "classical" in the sense that the distinction between fermions and bosons disappears, the strictly quantum concept of indistinguishability still plays an essential role. If the factor $N!$ were omitted, non-sensical results would be obtained. For example, if we were to calculate the free energy from the classical many-particle partition function Equation (6.12), we would obtain:

$$F = -k_B T \ln Z_N = -Nk_B T \ln Z_1 = -Nk_B T \ln(n_q V).$$

Substituting $V = N/n$, we find:

$$F = -Nk_B T \left[\ln\left(\frac{n_q}{n}\right) + \ln N\right], \tag{10.10}$$

which differs from the correct expression (Equation 10.9) by the $\ln N$ term inside the bracket.

Suppose that we increase N at constant temperature, increasing the volume so as to keep $n = N/V$ constant. Then F, being an extensive quantity like U, should be proportional to N. However, this is not the case for Equation (10.10), because of the term $-Nk_BT \ln N$, which is not present in Equation (10.9). This anomaly, which demonstrates the internal inconsistency of classical statistical mechanics, is connected with the so-called Gibbs' paradox, which is discussed in Section 10.5.

We can derive other thermodynamic quantities from Equation (10.9), in particular the pressure, internal energy, and the entropy. The equation of state relating the pressure to the density and temperature is, from Equation (5.22):

$$p = -\left(\frac{\partial F}{\partial V}\right)_{T,N} = \frac{Nk_BT}{V} = nk_BT. \tag{10.11}$$

For ν kmoles, $N = \nu N_A$, and Equation (9.11) takes the familiar form:

$$pV = \nu R_0 T, \tag{10.12}$$

where $R_0 \equiv N_A k_B$. Note that Equations (10.11) and (10.12) only apply to an *ideal* gas. Deviations can occur for two reasons: the gas may not be at a sufficiently low concentration for the distinction between bosons and fermions to be negligible, so that the gas, while perfect, is not ideal; or the interactions between molecules may have a significant effect on their energy levels (i.e., the gas is imperfect). These complications are discussed in subsequent chapters.

We can obtain the entropy from the free energy (Equation 10.9) in the same way as we obtained the pressure, using Equation (5.24) and remembering that n_q is the only temperature dependent term inside the square brackets in Equation (10.9):

$$S = -\left(\frac{\partial F}{\partial T}\right)_{V,N} = Nk_B\left[1 + \ln\left(\frac{n_q}{n}\right) + T\frac{d\ln(n_q)}{dT}\right]. \tag{10.13}$$

Since $\frac{dT}{T} = d\ln T$, and $n_q \propto T^{3/2}$ from Equation (7.31), the third term in the bracket is $\frac{d\ln(n_q)}{d\ln T} = \frac{3}{2}$. Hence, the entropy of an ideal monatomic gas is:

$$S = Nk_B\left[\frac{5}{2} + \ln\left(\frac{n_q}{n}\right)\right], \tag{10.14}$$

which is necessarily positive since $n < n_q$. Equation (10.14) is called the Sackur-Tetrode equation[2] for a monatomic gas.

Even though we are in the ideal limit, the Sackur-Tetrode equation for S, unlike the equation for U (Equation 9.16) below, contains Planck's constant through n_q and could not have been obtained by classical theory. It is in excellent agreement with the experimental value of S obtained from[3]:

$$S = \int_0^T \frac{dH}{T}. \tag{10.15}$$

---

[2] For a careful discussion of the Sackur-Tetrode equation and its implications, see P. Ehrenfest and V. Trkal in *Collected Scientific Papers of Paul Ehrenfest*, ed. M. J. Klein (North-Holland, 1959), p. 414.
[3] See Kittel and Kroemer, p. 167.

The integral is over H rather than U, since it is experimentally convenient to make the measurements at constant pressure, so that $dH = đQ$ [see Equations (2.18) and (5.14)]. The third law of thermodynamics in the form $S = 0$ at $T = 0$ (see Chapter 5) is assumed to hold. Although Equation (10.14) only applies to an ideal gas, the integral in Equation (10.15) is over the liquid and solid as well as the gaseous state; since S is a state function, its value does not depend on how the state is reached. Except at phase transitions, $dH = C_p(T)\,dT$. At a phase transition (e.g., from the liquid to the gaseous state) the heat capacity diverges, but its integral (the latent heat L) remains finite, so that each transition contributes an amount $\Delta S = L/T_t$ to the integral, where T_t is the temperature at which the transition takes place (see Chapter 15).

We can also obtain the internal energy from F, using Equation (5.18):

$$U = F + TS$$
$$= Nk_BT\left[\ln\left(\frac{n}{n_q}\right) - 1 + \frac{5}{2} + \ln\left(\frac{n_q}{n}\right)\right] = \frac{3}{2}nk_BT. \tag{10.16}$$

10.5 Entropy of Mixing of Ideal Gases: Gibbs' Paradox

Consider the process illustrated by Figure 2.2, where an ideal gas, initially contained in a volume V_1, is allowed (by the removal of a partition) to expand irreversibly to a volume $V_2 = V_1 + V_0$. No work is done and no heat flows, so that U, and therefore T and n_q, remain constant. For simplicity, we take $V_2 = 2V_1$. The number of gas molecules N is fixed, so that the density is halved, and Equation (10.14) gives for the increase in entropy:

$$\Delta S = Nk_B\ln 2. \tag{10.17}$$

Now suppose that instead of a vacuum, we start with a different gas in the right-hand volume, at the same pressure, also with N molecules. When the partition is removed, this gas also expands to twice its volume, and its entropy increases by the same amount, so long as the gases do not interact with each other. Thus, the total entropy of the system has increased by $2Nk_B\ln 2$. This extra entropy is an example of entropy of mixing (see Section 5.7); it arises from the fact that the two gases have many more microstates available to them after mixing than they had before. While we have used Equation (10.14), which was derived from statistical mechanics, to obtain ΔS, the same result can be obtained by pure thermodynamics, using the first law (Equation 5.13) and the equation of state [see Equation (5.17) and Problem 5.5c]. Note that this increase in entropy involves no heat flow; as we have said, mixing is an irreversible process so that the relation $dS = \frac{đQ}{T}$ (Equation 5.14) does not apply.

While the pressure of each gas (its partial pressure) has halved in this mixing process, the total pressure, which is the sum of the partial pressures, is unchanged. What if both gases are the same? Then nothing is changed by the removal of the partition; the entropy before and after its removal must be the same. In classical theory this result is a paradox, since one could imagine the two gases being made more and more alike, until it became impossible to distinguish them by their properties; yet the increase in entropy due to mixing is independent of how different the two gases are. Quantum theory, which draws a clear distinction between distinguishable and indistinguishable particles, resolves the paradox. If the two gases are different, a molecule of one is distinguishable from a molecule of the other, and entropy is increased by mixing. If the gases are truly identical, the molecules are indistinguishable, and there is no entropy increase. For example, suppose that each gas is isotopically pure, but the two gases consist of different isotopes of the same element. In this case the molecules of one gas are distinguishable from those of the other (though not by chemical means), and there is an increase in entropy on mixing. However, if we start with the *same* isotopic mixture on either side of the partition, the entropy of mixing is already present before the partition is removed and does not increase further. Since the entropy of mixing exists even in pure thermodynamics, Gibbs' paradox illustrates the fact that even this classical theory requires the concept of indistinguishability, which derives from quantum theory, to make it self-consistent.[4]

Entropy of mixing is not confined to ideal gases. For example, in Chapter 5 we discussed the binary alloy AB, which is formed by mixing N atoms of element A with N of element B, and found that if the alloy is disordered, there is an increase of entropy $2Nk_B \ln 2$ relative to the separate elements. Thus, the free energy of a disordered alloy is likely to be lower than that of the separate elements, at least at high temperature where the entropy term in the free energy dominates.

10.6 Particles with Internal Structure: Ideal Polyatomic Gases

An ideal gas need not be monatomic. In general, it is composed of molecules that have internal degrees of freedom, the most important of which for our purposes are those corresponding to rotation and vibration.[5] These degrees of freedom give rise to excited states with energies E_i, which are independent of the translational state. For example, a diatomic molecule has a single mode of vibration, with a ladder of vibrational states spaced by $h\nu$, where ν is the frequency of vibration (see Appendix B). The molecule also has rotational states, which we will consider later. We call such rotational and vibrational states *internal states* of the molecule. Do not confuse this use of the word "internal," which refers to states of an individual molecule, with its use in the term "internal energy," (i.e., U) where the word refers to the fact that the energy is internal to a macroscopic body. We now discuss how these internal molecular degrees of freedom affect the partition function and the thermodynamic properties of the gas.

The present discussion is quite general and does not depend on the nature of the internal degrees of freedom, but only on the well-justified assumption that the internal energy levels are not affected by the translational motion of the molecule. Think of a molecule at rest. The probability of occupation of the i'th internal state, whose energy is E_i, is:

$$P_i = Z_{\text{int}}^{-1} e^{-\beta E_i}, \tag{10.18}$$

where $Z_{\text{int}} = \sum_i e^{-\beta E_i}$ is the internal partition function, the sum being over internal states only. Now consider one particular translational state of the molecule, with kinetic energy E_t (this was called E in the earlier part of this chapter, where we were considering molecules with no internal degrees of freedom). Since the energies of the internal states are independent of the translational state, the energy of the molecule in the i'th internal state and with kinetic energy E_t is $E_i + E_t$. Each internal state contributes

---

[4] It is sometimes argued that the concept of distinguishability is a subjective one that should have no place in a physical science, since a physical fact (such as a change in entropy) cannot be deduced from a psychological one. This is a misunderstanding. Quantum mechanics requires that the many-particle wave function of a collection of identical particles be either symmetric or anti-symmetric under particle exchange; that is, the wave function is either unchanged (for bosons) or changed only in sign (for fermions) when the labels on particles are interchanged. This requirement limits the range of possible wave functions and therefore reduces the multiplicity. There is no such symmetry requirement for non-identical particles. The apparent subjectivity is due merely to the form of words; the word "indistinguishability" refers not to our senses but to the symmetry of the physical system, just as we can describe the symmetry of a cube by saying that the different faces are indistinguishable.

[5] As remarked in Chapter 7 in connection with equipartition, atoms and molecules have internal degrees of freedom due to their electronic structure. However, the excited atomic energy levels (and the excited electronic, as opposed to vibrational or rotational, levels of molecules) are usually several eV above the ground state, while at normally accessible terrestrial temperatures $k_B T < 0.4$ eV. As a result, the atom or molecule remains in its ground electronic state and the excited levels are usually irrelevant.

a term $e^{\beta(\zeta - E_t - E_i)}$ to the grand partition function Ξ for this translational state, so that for fermions, Equation (9.2) is replaced by:

$$\Xi = 1 + e^{\beta\zeta} \sum_i e^{\beta(E_t + E_i)} = 1 + e^{(\beta\zeta - E_t)} \sum_i e^{-\beta E_i}$$

$$= 1 + \alpha Z_{int} e^{-\beta E_t}. \qquad (10.19)$$

Thus, α in Equation (9.2) is replaced by αZ_{int}. It is easy to show this is also true for bosons, and the mean occupancy of this translational state is:

$$f(E_t) = \left(\frac{\partial \ln \Xi}{\partial \ln \alpha} \right)_T = \frac{\alpha Z_{int} e^{-\beta E_t}}{1 \pm \alpha Z_{int} e^{-\beta E_t}},$$

$$\approx \alpha Z_{int} e^{-\beta E_t} \qquad (10.20)$$

in the low density (ideal) limit where α is small. As in the monatomic case, in this limit, it makes no difference whether the molecules are fermions or bosons. The total number of particles is still $N = \sum f(E_t)$, where the sum is over translational states only, so that:

$$N = \alpha Z_{int} \sum e^{-\beta E_t},$$

since Z_{int} is the same for all the translational states.

As before, $\sum e^{-\beta E_t} = Z_1 = n_q V$, the single particle partition function for the translational states, so that the density is:

$$n = \alpha n_q Z_{int}. \qquad (10.21)$$

Hence,

$$\alpha = \frac{n}{n_q Z_{int}},$$

and:

$$\zeta = -k_B T \ln \left(\frac{n_q}{n} \right) - k_B T \ln Z_{int}. \qquad (10.22)$$

The second term in Equation (10.22) is simply the free energy per molecule associated with the internal degrees of freedom (see Equation 6.28b). This makes sense in terms of the definition of ζ as $\left(\frac{\partial F}{\partial N} \right)_{T,V}$; when a particle is added to the system, it brings with it not only the free energy associated with translation, which is $-k_B T \ln \left(\frac{n_q}{n} \right)$, but also an amount $-k_B T \ln Z_{int}$ due to its internal motion.

The derivation of the many-particle partition function Z_N for the polyatomic gas follows that for the monatomic case. Equation (10.22) for r molecules can be written as:

$$\zeta(r) = k_B T \left[\ln r - \ln \left(Z_{int} n_q V \right) \right], \qquad (10.23)$$

where we have used $n = r/V$.

Summing the chemical potential from 1 to N as before, we find the free energy:

$$F = \sum_{r=1}^{N} \zeta(r) = k_B T \left[\ln(N!) - N \ln(n_q V) - N \ln Z_{int} \right]$$

$$= F_t + F_{int}, \qquad (10.24)$$

where F_t is the free energy due to translational motion (Equation 10.9) and $F_{int} = -k_B T \ln Z_{int}$ is the free energy due to the internal degrees of freedom. Since F_{int} is independent of V, the equation of state (Equation 10.12) is the same as for a monatomic gas.

From Equation (6.28a), $F = -k_B T \ln Z_N$, so that:

$$Z_N = \frac{(Z_1 Z_{int})^N}{N!}.$$ (10.25)

Apart from the factor $N!$, which is due to indistinguishability and is the same as in the monatomic case (Equation 10.8), we could have obtained Equation (10.25) directly by the argument that led to Equation (6.12): for every translational state in Z_1, there is a complete set of internal states represented by Z_{int}.

The internal degrees of freedom contribute to the energy and entropy of the gas; their contribution to U is $U_{int} = N \langle E_{int} \rangle$, so that:

$$U = N \left(\frac{3}{2} k_B T + \langle E_{int} \rangle \right)$$

where:

$$\langle E_{int} \rangle = -\frac{\partial \ln (Z_{int})}{\partial \beta},$$ (10.26)

exactly as if there were no translational motion at all. Since the free energies are additive (see Equation 10.24), so are the entropies:

$$S = \frac{\partial F}{\partial T} = \frac{\partial F_t}{\partial T} + \frac{\partial F_{int}}{\partial T}.$$

If $k_B T$ is much larger than the energy level separations, as is usually the case for rotational levels (see below), the equipartition theorem applies (see Section 7.9 and Appendix G). Then each internal degree of freedom contributes $k_B T / 2$ to $\langle E_{int} \rangle$.

If $k_B T$ is not large, Z_{int} must be calculated and Equation (10.26) used to obtain $\langle E_{int} \rangle$, as we did in Section 6.6 for the harmonic oscillator and in Problem 6.5 for the rotating diatomic molecule. For degrees of freedom whose energy levels are separated by much more than $k_B T$ (e.g., the vibrational modes of light molecules at room temperature) $\langle E_{int} \rangle$ is much less than the equipartition value, so that these modes do not contribute to U or to the heat capacity.

Example 10.1: Diatomic Molecules

The vibrational frequency ν_{vib} of the N_2 molecule is 7×10^{13} Hz, giving an energy separation $h\nu_{vib} = 0.26$ eV. At about what temperature does one have to worry about the contribution of the vibrations to the heat capacity?

The vibrational contribution to the heat capacity of the diatomic gas has the same temperature dependence as the heat capacity of the Einstein solid shown in Figure 6.2; for the diatomic molecule the ordinate is C_V / R_0, since there is only one vibrational mode per molecule, and the abscissa is $k_B T / h\nu_{vib}$. This contribution to C_V is about 10% of the total when $T \sim 0.2 h\nu_{vib} / k_B$, which is about 750 K. Thus, only at rather high temperatures, such as those achieved in an internal combustion engine (see Chapter 3), do the vibrational modes of a light molecule like N_2 contribute significantly to the heat capacity.

The rotational states of a molecule are a different matter. The rotational energy levels of a diatomic molecule are given by (see Appendix B):

$$E_j = j(j+1)E_0,$$ (10.27)

where $j = 0, 1, 2, \ldots$ and $E_0 = \frac{\hbar^2}{2I_{xx}}$, I_{xx} being the moment of inertia of the molecule about an axis perpendicular to the line joining the atoms. The characteristic energy E_0 is about 2.5×10^{-4} eV for the N_2 molecule, so that levels with large j are occupied at room temperature ($k_B T \approx 0.025$ eV). The separation of these levels is $2jE_0 \sim 0.2k_B T$, so that the equipartition theorem holds to a good approximation. There are two rotational degrees of freedom; the third degree of freedom expected classically, rotation about the molecular axis, has a very much smaller moment of inertia and hence large E_0, since this rotation corresponds to electronic rather than to nuclear motion. Hence, equipartition predicts a rotational contribution $k_B T$, so that $\langle E \rangle = \frac{5}{2}k_B T$. The internal energy per kmole is $U = \frac{5}{2}R_0 T$ and $C_V = \frac{5}{2}R_0$. This is found to hold in such gases as oxygen and nitrogen at moderate temperatures.

The high temperature limit of Z_{int} in a diatomic molecule is $\frac{k_B T}{E_0}$ (see Problem 6.5c), so that:

$$\zeta = -k_B T \ln \left(\frac{n_q}{n} \right) - k_B T \ln \left(\frac{k_B T}{E_0} \right),$$

and:

$$\alpha = \frac{n}{n_q} \frac{E_0}{k_B T}. \tag{10.28}$$

Thus for a diatomic gas of given density, at a temperature high enough for equipartition to hold in the rotational levels, $\alpha \propto T^{-5/2}$. Equation (10.28) is generalized to an arbitrary number of internal degrees of freedom in Problem 10.5.

The case of the H_2 molecule is interesting since its small moment of inertia produces a value of E_0 of 7.5×10^{-3} eV, considerably larger than in any other gas. Furthermore, hydrogen remains gaseous down to a very low temperature (21 K at atmospheric pressure). At 50 K, $k_B T \sim \frac{1}{2}E_0$ and C_V drops significantly below $\frac{5}{2}R_0$. The rotational heat capacity of a heteronuclear molecule such as hydrogen deuterium (HD) is calculated in Problem 6.5 and shown as a function of temperature in Figure 6.7. Quantum mechanical effects arising from symmetry requirements on the nuclear wave function complicate the case of a homonuclear molecule,[6] but the data on H_2 are in qualitative agreement with that calculation.

All the principal moments of inertia of a non-linear[7] polyatomic molecule are of the same order of magnitude, so that equipartition applies to all three rotational degrees of freedom. The different moments of inertia about the three rotation axes make calculation of the energy levels very complicated.[6] However, at moderate temperatures the equipartition theorem still applies. Since there are three rotational degrees of freedom, the rotational contribution to $\langle E \rangle$ is $\frac{3}{2}k_B T$, giving $U = 3R_0 T$ per kmole and $C_V = 3R_0$, $C_p = 4R_0$.

Note that whether or not U has the value given by equipartition, in an ideal gas with given molecular parameters it is a function only of temperature. This follows from our initial assumption that the molecules in the gas do not interact, so that U does not depend on the intermolecular separation and is therefore independent of the volume of the gas. Furthermore, since the internal contribution to the free energy [F_{int} in Equation (10.24)] is independent of volume, it does not contribute to the pressure, $-\left(\frac{\partial F}{\partial V} \right)_{T,N}$, so that the equation of state is Equation (10.11), as for a monatomic gas.

10.7 Heat Capacity of an Ideal Gas at Constant Pressure

In Chapter 2, we defined the internal energy U and also the enthalpy $H = U + pV$. We saw that when a gas is heated at constant pressure, the energy supplied goes not only into the change of internal energy ΔU, but also into the work done in expanding the gas against the pressure p. The heat capacity at constant

---

[6] See, for example, G. Herzberg, *Infrared and Raman Spectra of Polyatomic Molecules* (van Nostrand, 1945), p. 15.

[7] A linear polyatomic molecule such as CO_2 has only two rotational degrees of freedom, like a diatomic molecule.

pressure is $C_p = \left(\frac{\partial H}{\partial T}\right)_p$ (see Example 2.5), while the heat capacity at constant volume is $C_V = \left(\frac{\partial U}{\partial T}\right)_V$. We have already seen (Equation 2.18) that the kmolar heat capacities of an ideal gas are related by:

$$C_p - C_V = R_0.$$

It was shown in Chapter 7 that when equipartition holds, $C_V = \frac{1}{2}n_f R_0$, where n_f is the number of degrees of freedom per molecule. Substituting in Equation (2.18), we obtain for the ratio γ of the heat capacities of an ideal gas, at a temperature where equipartition applies:

$$\gamma \equiv \frac{C_p}{C_V} = 1 + \frac{2}{n_f}. \tag{10.29}$$

Thus for a monatomic gas, which has $n_f = 3$, $\gamma = 1.67$. For N_2 and O_2, which have three translational and two rotational degrees of freedom, $n_f = 5$ and $\gamma = 1.4$ at normal temperatures, but n_f increases to 7, corresponding to $\gamma = 1.29$, when the temperature is high enough to excite the vibrational degrees of freedom (see Example 10.1). We have seen previously that γ is easier to measure than either heat capacity, and that it plays a crucial role in the performance of heat engines, such as the internal combustion engine, which use an approximately ideal gas as the working fluid (see Chapter 3).

Example 10.2: Ideal Gas: Particles in a Quantum Mechanical Box

An ideal monatomic gas is enclosed in a cube of side L. The cube expands, retaining its cubic shape. Find how the energy $E(n_x, n_y, n_z)$, of a state with a given set of quantum numbers n_x, n_y, n_z, defined in Appendix F, varies with volume V. Hence, show that the occupancy $f(n_x, n_y, n_z)$ of a given state is unchanged during an isentropic expansion, even though both E and T vary.

For a non-relativistic particle, $E = \frac{\hbar^2}{2m}\kappa^2$ where $\kappa^2 = \frac{\pi^2}{L^2}(n_x^2 + n_x^2 + n_z^2)$, so that:

$$E(n_x, n_y, n_z) \propto L^{-2} \propto V^{-2/3}.$$

For a monatomic gas $\gamma = \frac{5}{3}$, so that from Equation (3.17),

$$T \propto n^{\gamma-1} \propto V^{1-\gamma} = V^{-2/3}.$$

Hence, $\beta E \propto T^{-1}V^{-2/3} = $ constant.

From Equation (9.13), $\alpha = \frac{n}{n_q} \propto V^{-1}T^{-3/2} = $ constant.

It was shown in Chapter 9 that in a perfect gas, the occupancy $f(n_x, n_y, n_z)$ is a function only of α and βE. Since both are constant, f is constant. The fact that the occupancy of each state does not change in an isentropic process shows that such a process is adiabatic in the quantum mechanical sense for a system with continuous energy levels, as was shown for the discrete case in Section 6.8.

Another route to the same result is to use the theorem of classical mechanics that $\frac{E}{\nu}$ is an *adiabatic invariant*. What this means is that if the frequency of an oscillator is changed slowly (for instance, in the gradual stretching of a vibrating string), the energy in each vibrational mode varies in proportion to its frequency.[8] By Equation (6.13) the ratio of energy to frequency depends only on the occupancy, which is thus unchanged.

---

[8] See, for example, Landau and Lifshitz, *Mechanics*, 3rd ed. transl. J. B. Sykes and J. S. Bell (Pergamon, 1976), Eq. 49.12.

10.8 The Ideal Gas as a Spring: The Velocity of Sound

The bulk modulus or "springiness" of a gas (that is, its resistance to a small change in volume), is defined as $B \equiv n\frac{dp}{dn}$. Consider, for example, the apparatus illustrated in Figure 10.2, which is used in a standard undergraduate experiment to measure γ. If the ball in the neck of the flask is displaced downwards by a small distance x from its equilibrium position, the change in volume is $\Delta V = -Ax$, where A is the cross-sectional area of the neck. The fractional increase in density is:

$$\frac{\Delta n}{n} = -\frac{\Delta V}{V} = \frac{Ax}{V}.$$

This produces an increase Δp in the pressure inside the vessel, and a restoring force on the ball $f = -A\,\Delta p$. If $\frac{|\Delta V|}{V} \ll 1$, $\Delta p = B\frac{\Delta n}{n}$, and $f = -B\frac{A^2 x}{V}$, where the minus sign indicates that the force opposes the displacement. Thus, the gas behaves like a spring with force constant $K = \frac{BA^2}{V}$. If the mass of the ball is M, the frequency of small oscillations is $\nu = \frac{1}{2\pi}\left(\frac{K}{M}\right)^{1/2}$ (see any elementary textbook of mechanics), so that:

$$\nu = \frac{1}{2\pi}\left(\frac{BA^2}{VM}\right)^{1/2}. \tag{10.30}$$

The value of B depends on whether the change in volume is isothermal or isentropic. If isothermal, $p \propto n$. Differentiating this gives:

$$\frac{dn}{n} = \frac{dp}{p}. \tag{10.31}$$

Hence, the isothermal bulk modulus is:

$$B_T \equiv n\left(\frac{dp}{dn}\right)_T = p. \tag{10.32}$$

On the other hand, the isentropic bulk modulus is, from Equation (3.13),

$$B_S \equiv n\left(\frac{dp}{dn}\right)_S = \gamma p. \tag{10.33}$$

If the ball were moved very slowly, allowing time for heat to flow in or out to maintain a constant temperature,[9] the displacement as a function of pressure would give B_T. However, if the ball oscillates freely,

FIGURE 10.2 Apparatus to measure γ.

---

[9] In practice this is impossible, since the air would have time to leak past the ball.

and the oscillation frequency is high enough (a fraction of 1 Hz is sufficient), there is no time for heat to flow and the bulk modulus obtained from the vibrational frequency by Equation (10.30) is B_S. Since p is known, this gives γ.

An important application of Equation (10.33) is to the velocity of sound. It is shown in textbooks[10] of acoustics that the velocity of sound in a gas is:

$$c_s = \left(\frac{B}{\rho}\right)^{1/2}, \tag{10.34}$$

where $\rho = mn$ is the mass density of the gas, m being the mass of one molecule. At normal acoustic frequencies, the relevant bulk modulus is B_S rather than B_T. This can be seen as follows. In a sound wave, the gas is instantaneously compressed at one point in the wave while it is expanded a half wavelength away. If the temperature is to remain constant, heat must flow a distance of the order of a wavelength in a time of the order of a period of oscillation. We shall see when we consider transport processes in Chapter 16 that at atmospheric pressure, this is only possible at very short wavelengths (less than $\sim 1\mu m$). Hence, at acoustic wavelengths (3–10 m) the process is isentropic and B in Equation (10.34) must be taken as B_S. Thus $c_s = \left(\frac{\gamma p}{mn}\right)^{1/2}$. Substituting $p = nk_BT$, we find:

$$c_s = \left(\gamma \frac{k_BT}{m}\right)^{1/2}. \tag{10.35}$$

Note that the root mean square thermal velocity $v_{th} = \left(3\frac{k_BT}{m}\right)^{1/2}$, so that $c_s \approx v_{th}$. The effect of temperature on the velocity of sound is well known to wind instrument players, whose instruments tend to go out of tune on a cold day.[11]

10.9 Ideal Solutions and Osmotic Pressure

When you drop an ionic solid such as potassium chloride into water, it dissolves; that is, the positive and negative ions that make up the solid separate and move freely in the liquid. If you measure the temperature of the water carefully while the solid is dissolving you will find that it drops; this is because energy (the heat of solution) is needed to separate the ions.[12] However, although the energy of the ions is higher in solution, their entropy is much greater, so that their free enthalpy is lower and they dissolve.

An *ideal solution* is one in which the dissolved molecules or ions do not interact with each other, so that the only contributions to the free enthalpy difference between the solid and solution are the heat of solution and the entropy of mixing, which is from Equation (5.16):

$$S = -Nk_B\left[x \ln x + (1-x)\ln(1-x)\right],$$

---

[10] See, for example, A. P. French, *Vibrations and Waves* (MIT, 1971), p. 211.

[11] Newton was the first to make a serious attempt to calculate the velocity of sound [*Principia* II, Sec. 8]. At that time, the concept of isentropic change was unknown, but Boyle had established his law that density is proportional to pressure in an isothermal change. Naturally, Newton used the isothermal bulk modulus B_T, obtaining a velocity about 20% below the observed value. The story of how he tried to force his calculation and the data into agreement shows that even such a transcendent genius as Newton was not above "fudging" his results when unable to account for a discrepancy [R. S. Westfall, "Newton and the fudge factor," *Science* **179**, 751–758 (1973); *Never at Rest, a Biography of Isaac Newton* (Cambridge, 1980), pp. 734–735]. The correct calculation was made more than a century later and is generally credited to Laplace [see, however, C. Truesdell, *The Tragicomical History of Thermodynamics 1822–1854* (Springer-Verlag, 1980), pp. 31–38].

[12] The heat of solution of KCl is about 0.18 eV per $K^+ - Cl^-$ pair; this energy is less than the cohesive energy of the crystal because the dissolved ions attract the highly polarizable H_2O molecules.

where x is the concentration of the solute. We label the solute x and solvent s, writing $N_x = Nx$ for the number of dissolved particles and $N_s = N(1 - x)$ for the number of solvent particles. The free enthalpy difference between the solution and the pure materials is then:

$$\Delta G = N_x \epsilon + k_B T \left[N_x \ln x + N_s \ln(1 - x) \right], \tag{10.36}$$

where ϵ is the heat of solution per solute atom and the quantity in square brackets is negative since $x < 1$. Note that however large ϵ is, ΔG is negative for sufficiently small x, since $\ln x \to -\infty$ as $x \to 0$. It follows that a small amount of material will always dissolve. In this limit of large ϵ, the solubility (defined as the value of x for which $\Delta G = 0$) increases with temperature according to Arrhenius' law, with activation energy ϵ.

We will now examine the remarkable phenomenon of osmosis. If a solution is separated from the pure solvent by a membrane permeable to the solvent but not to the solute, the solvent will diffuse through the membrane until a sufficient pressure difference, called the osmotic pressure Δp, has built up in the solution to prevent any further diffusion. This happens because the chemical potential of the *solvent*, ζ_s, is decreased by the presence of the solute, but increases with pressure. The solvent will diffuse until the pressure difference equalizes ζ_s on both sides of the membrane.

To make this quantitative, we assume that the solution is dilute, that is $x \ll 1$, so that $\ln(1 - x) \approx -x$. Then:

$$\Delta G \approx N_x \epsilon + k_B T \left[N_x \ln x - N_s x \right]. \tag{10.37}$$

The presence of the solute thus reduces the chemical potential of the solvent by:

$$\Delta \zeta_s = \left(\frac{\partial \Delta G}{\partial N_s} \right)_{T,p,N_x} = -k_B T x. \tag{10.38}$$

There is a Maxwell relation (see Problem 10.12a):

$$\left(\frac{\partial \zeta_i}{\partial p} \right)_{T,N_i} = \left(\frac{\partial V}{\partial N_i} \right)_{T,p} = n_i^{-1}, \tag{10.39}$$

so that an increase in pressure Δp increases ζ_s by $\frac{\Delta p}{n}$, where n is the particle density of the solvent.

Hence, equilibrium is achieved when:

$$\Delta \zeta_s = -k_B T x + \frac{\Delta p}{n} = 0.$$

Hence,

$$\Delta p = n_x k_B T, \tag{10.40}$$

where $n_x = nx$ is the particle density of the solute and we have neglected x relative to 1.

Equation (10.40) is known as van't Hoff's[13] law, and is identical to the ideal gas law (Equation 10.11), with the osmotic pressure substituted for the actual pressure of the gas. This is not a coincidence, since the density dependence of the chemical potential of the solute has the same form as that of an ideal gas (see Problem 10.12b). Since kinetic theory cannot possibly apply to a solution, where motion of the solute molecules is severely restricted by the solvent, this result shows that it is a mistake to think of the ideal gas law as merely a consequence of kinetic theory.

---

[13] Jacobus Henricus van't Hoff, Dutch physical chemist (1855–1911), winner of the first (1901) Nobel Prize for Chemistry.

Example 10.3: Lowering of the Freezing Point by a Solute

Solubilities are usually much lower in solids than in liquids, and for the purpose of this example, we will assume that the solute is not soluble in the solid solvent. Then, while the chemical potential of the liquid solvent containing solute is lower than that of pure liquid by $\Delta \zeta_s$, given by Equation (10.38), that of the solid is unaffected. If we start at a temperature above the freezing point and reduce it, freezing becomes possible when the chemical potential of the solid drops below that of the liquid. Since the presence of the solute reduces the chemical potential of the liquid, the liquid remains at the stable phase to a lower temperature and the freezing point is depressed. Find the change ΔT, in temperature, to first order in x.

Besides the reduction $\Delta \zeta_s$ due to the solute, we also need to know how the chemical potential difference between the pure liquid and the solid varies with temperature. If this difference is $\Delta \zeta_0(T)$, freezing will occur when:

$$\Delta \zeta_s + \Delta \zeta_0 = 0. \tag{10.41}$$

$\Delta \zeta_0$ can be calculated from the Gibbs-Helmholtz relation for G (Equation 5.45), which can be written:

$$\frac{\partial}{\partial T}\left(\frac{\zeta}{T}\right) = -\frac{H}{N_A T^2},$$

where H is the kmolar enthalpy, the derivative is at constant pressure, and we have used $G = N\zeta$ (Equation 8.10).

The change in temperature ΔT is small, and $\Delta \zeta_0 = 0$ when $\Delta T = 0$, so that:

$$\Delta \zeta_0 \approx T \frac{\partial}{\partial T}\left(\frac{\Delta \zeta_0}{T}\right)\Delta T = -\frac{L \Delta T}{N_A T}, \tag{10.42}$$

where L is the latent heat (defined as the difference in kmolar enthalpy between the pure liquid solvent and the solid).

Substituting Equations (10.38) and (10.42) into Equation (10.41) we find:

$$\Delta T = -\frac{R_0 T^2}{L}x. \tag{10.43}$$

Note that, so long as the assumption of ideality holds, $\frac{\Delta T}{x}$ is the same for all solutes, depending only on the latent heat of the pure solvent. An analogous calculation shows that the presence of a non-volatile solute raises the boiling point (see Problem 10.13b).

10.10 Envoi

Ideal gas theory has a wide range of application, since not only do most common gases approximate closely to ideality, but the model is applicable to many systems which are physically very different (e.g., the ideal solution discussed in Section 10.9 and the electron gas in lightly doped semiconductors, to be described in Chapter 14). The ideal gas was for many years the "fruit fly" of statistical mechanics, in the sense that its simplicity and ubiquity made it the touchstone of any new method of calculation: if the technique when applied to the ideal gas doesn't give $pV_m = R_0 T$, it must be wrong. From the time of Clausius and Maxwell,[14] an important field of statistical mechanics research has been the effort to improve

---

[14] In 1875, J. Clerk Maxwell summarized what was then understood about the molecular nature of imperfect gases in a masterfully clear and concise paper entitled "On the dynamical evidence of the molecular constitution of bodies," *Nature* **11**, 357.

on the ideal gas model by taking account of intermolecular forces using progressively more powerful calculational techniques.[15] However, to go more deeply into this program here would divert us from the main purpose of this book, which is to show how much of Nature can be understood in terms of very simple models.

10.11 Problems

10.1 (a) Show that the entropy of an ideal diatomic gas at a temperature where equipartition holds is $S = Nk_B \left[\frac{7}{2} + \ln \left(\frac{n_q k_B T}{n E_0} \right) \right]$, where E_0 is defined as in Section B.5

(b) Hence, show that the free energy and chemical potential of such a gas are necessarily negative

10.2 (a) Starting from Equation (8.27), show that the grand partition function of a volume V of an ideal gas containing (on average) $\langle N \rangle = nV$ molecules is $\Xi = e^{\langle N \rangle}$

(b) Hence show, using Equation (8.29), that the probability of finding N particles in this volume is given by the Poisson distribution:

$$P(N) = e^{-\langle N \rangle} \frac{\langle N \rangle^N}{N!} \tag{10.44}$$

(c) Confirm by direct calculation from Equation (10.44) that $\sum P(N) = 1$, $\sum NP(N) = \langle N \rangle$, and $\langle \Delta N^2 \rangle = \langle N \rangle$, where $\Delta N \equiv N - \langle N \rangle$

(d) Show that for large $\langle N \rangle$, Equation (10.44) is a Gaussian in ΔN for all values of N for which $P(N)$ is significantly different from zero. Hint: Take the natural logarithm of $P(N)$, use Stirling's approximation to $\ln N!$ in its crude form, and put $N = \langle N \rangle + \Delta N$, where $\Delta N \ll \langle N \rangle$

10.3 Starting from the first law in the form $dH = T\,dS + V\,dp$, show (without using Equations 3.16 or 3.17) that for an isentropic change of an ideal gas,

$$T \propto p^{(1-1/\gamma)} \tag{10.45}$$

10.4 In a BB gun, air (a diatomic gas which can be assumed to be ideal) is compressed to high pressure p_0 by a hand pump. When the gun is fired, this air is released into the barrel to propel the pellet. Consider a gun whose barrel has cross-section 1.5×10^{-5} m^2 and length 0.6 m. Before firing, the air is contained in a chamber of volume $V_0 = 2 \times 10^{-7}$ m^3. The pellet weighs 4×10^{-3} kg and leaves the gun with a velocity 150 m/s.

(a) Assuming the expansion to be isentropic, and neglecting friction, find the initial pressure p_0 in the chamber. Caution: Don't forget the work done by the expanding gas against the atmosphere

(b) If the air is initially at 300 K, what is its temperature at the moment the pellet leaves the barrel? Note that the expansion ceases to be isentropic at this point, and that the subsequent return to atmospheric pressure is irreversible

(c) Before compression, the air is at 300 K and at atmospheric pressure (1 bar). Pumping by hand is a relatively slow process, so it is a reasonable approximation to treat the compression as isothermal. Neglecting friction, find the work done compressing the air for each firing. Find an expression for the overall efficiency (kinetic energy out divided by work put in) in terms of γ, p_0, and r, the ratio of volume at the end of the isentropic expansion to that at the beginning, and evaluate it for the numbers given. Draw up an energy balance for the entire process, including all work done and heat transferred: Where has the lost energy gone?

---

[15] See, for example, D. ter Haar, *Elements of Statistical Mechanics*, 3rd ed. (Butterworth-Heinemann, 1995), and H. Gould and J. Tobochnik, *Theories of Gases and Liquids*, which is Chapter 8 of a book that can be accessed on the Web at http://stp.clarku.edu/notes.

(d) Now suppose that one pumps so quickly that the compression is isentropic, and assume the air does not cool significantly before firing. If p_0 is the same as in (a), what is the efficiency? Why is it less than 100%?

10.5 (a) Apply the Gibbs-Helmholtz Equation (5.25) to the equipartition value of U to find the contribution of the internal motion of a single molecule with n_f internal degrees of freedom to the free energy F. Note that since this contribution does not depend on V in an ideal gas, this is an ordinary differential equation. Hence, show that if equipartition holds, in a polyatomic gas

$Z_{\text{int}} = \left(\frac{k_B T}{E_0}\right)^{n_f/2}$, where E_0 is an undetermined constant with the dimensions of energy.

(b) Show that under these conditions (i.e., if equipartition holds) $\alpha = \frac{n}{n_q}\left(\frac{E_0}{k_B T}\right)^{n_f/2}$

(c) Hence, show that in an isentropic expansion of any ideal gas, α is constant

10.6 (Convective stability of the atmosphere.) In Example 8.2, we assumed for simplicity that the atmosphere is isothermal. This is not, of course, the case. A better approximation assumes that the pressure is related to density by the isentropic relation (Equation 3.16), rather than by Boyle's law. If this relation holds, a rising column of air, as it expands isentropically, is always at the same temperature and density as the surrounding air, so that it remains in hydrostatic equilibrium. This situation is called *convective equilibrium* and usually holds in steady dry weather

(a) By considering the hydrostatic equilibrium of a thin horizontal layer of air, show that $\frac{dp}{dz} = -g\rho$, where p is the pressure, z is the height, g is the acceleration due to gravity, and ρ is the mass density ($\rho = mn$, where m is the mass of a molecule)[16]

(b) Take the logarithmic derivative of Equation (3.17) to find $\frac{dp}{dT}$ in an isentropic change of an ideal gas, and hence convert the differential equation for p in (a) into an equation for $T(z)$. Show that the temperature falls linearly with height and find $\frac{dT}{dz}$ for air, which has an average molecular weight of 29. $|\frac{dT}{dz}|$ is called the *dry adiabatic lapse rate* ("dry" because we have ignored the effect of condensation of water vapor as the air cools)[17]

(c) Assuming convective equilibrium and a dry atmosphere, what is the temperature at the top of Mount Everest ($h \approx 8700$ m) if it is 35 degrees Celsius at sea level?

(d) Use the result in Problem 10.5b to find the total gravichemical potential (chemical + gravitational) as a function of height when the atmosphere is in convective equilibrium. Explain why this potential is *not* constant

10.7 The *escape velocity* of an object in the vicinity of a gravitating body, such as the earth, is the velocity that it must have if it is to escape that body's gravitational field. Calculate the escape velocity at the surface of the earth (radius = 6400 km, $g = 9.8$ m s^{-2} at the surface). Compare your result with the root mean square thermal velocity v_{th} at 300 K of (a) a nitrogen molecule N_2 and (b) a hydrogen molecule H_2. Explain why the earth's atmosphere contains nitrogen but not hydrogen. Repeat your calculation for the moon (radius = 1700 km, $g = 1.6$ m s^{-2} at the surface) and explain why the moon has no atmosphere at all. For simplicity, assume an isothermal atmosphere

10.8 A possible way to prevent accidents due to failure of an elevator cable is to make the lowest part of the elevator shaft airtight, so that if the elevator falls, it is cushioned by compression of the air in the shaft. Given that the initial temperature is T_0, the airtight section is of height h_0 and cross section A, and that the elevator of mass M falls a total height h before being brought to rest, find an equation for the final temperature T_f and pressure p_f of the air (initially at atmospheric

---

[16] This is a purely mechanical result and applies to any fluid; for example, it gives the dependence of hydrostatic pressure on depth in water, for which ρ is nearly constant (so that the equation can be integrated to give $p = p_0 + \rho g|z|$ where $-z$ is the depth and p_0 is the atmospheric pressure).

[17] On many airlines, one of the in-flight entertainment channels gives continuously updated information on the airplane's altitude and on the ambient air temperature. Next time that you fly in fair, dry weather, watch these figures while the airplane is climbing or descending, and see how valid the assumptions of this problem are.

temperature T_0 and pressure p_0). Assume that the compression is isentropic, that air is an ideal diatomic gas, and don't forget to allow for the pressure of the atmosphere on the top of the elevator If $h_0 = 10$ m, $h = 50$ m, $A = 2.5$ m^2, $M = 500$ kg, $T_0 = 300$ K, $p_0 = 1$ bar $= 10^5$ Pa, estimate T_f, using the approximation that $T_f - T_0 \ll T_0$. What is the vertical acceleration of the elevator at the moment it comes to rest? How could the design be modified in order to reduce this acceleration to a safe level and to prevent the elevator from bouncing?

10.9   One liter of helium (a monatomic gas) at standard temperature and pressure (STP) (0 degrees Celsius, 1 bar) is heated at constant volume to 100 degrees Celsius. It is then allowed to mix in an insulated container with one liter of methane which is initially at STP, the final volume being two liters. The methane molecule has three rotational degrees of freedom. What is the final temperature? What is the increase in entropy? Compare your result with the entropy change that would occur if the gases were allowed to come into thermal equilibrium with each other, but not to mix

10.10 Consider a cylinder of volume V containing a mixture of N_a particles of an ideal gas A and N_b of an ideal gas B. At one end of the cylinder there is a membrane permeable to A and not to B, while at the other there is a membrane permeable to B and not to A. The membranes are moved slowly to the center of the cylinder, concentrating A in one half of the cylinder, B in the other

   (a) If the cylinder is in good thermal contact with a heat bath at temperature T, find the change in entropy of the contents of the cylinder, the net change in entropy of the cylinder plus heat bath, the heat transferred to the bath, and the work done in moving the membranes. Neglect friction

   (b) If the cylinder is insulated so that no heat can flow in or out, and A and B are diatomic, what is the change in temperature of the gases, and what is the work done?

10.11 Consider a monatomic ideal gas subject to a long range external potential $\phi(r) = Ar^{-1}$, where r is the distance from some given point. This is a rough model of what happens in the early stages of collapse of a gas cloud to form a star. Use the generalized equipartition theorem[18] to show that the thermal average value of the potential energy of one particle is $\langle \phi \rangle = -3k_B T$. Hence, find $U = N(\langle \phi \rangle + \langle K_e \rangle)$, where K_e is the kinetic energy of one particle and N is the number of particles. Show that as the gas cloud shrinks the temperature rises in proportion to the mean inverse radius $\langle r^{-1} \rangle$

10.12 (a) Prove Equation (10.41)

   (b) Differentiate Equation (10.37) to obtain the chemical potential of the solute, and compare its density dependence with that of the chemical potential of an ideal gas

   (c) In the spring, water is sucked into the roots of a tree by osmosis. Assuming that the sap in a tree is an ideal solution, and that the roots are immersed in pure water at 285 K, what molar concentration of solute is needed to raise the sap to the top of a 15 meter tree by osmotic pressure? The molecular weight of H_2O is 18

10.13 (a) Find the depression of the freezing point of water produced by 1% w/w NaCl ("w/w" is short for "weight for weight," so that 1% w/w means that 1 g of salt is dissolved in every 100 g of water). The molecular weight of H_2O is 18, of NaCl 58.5; the latent heat of melting is 6×10^6 J/kmole, and the freezing point of pure water is 273 K. Don't forget that when one molecule of NaCl dissolves, two ions are produced. Compare your result with the observed value of 0.6 K

   (b) Why is the boiling point of the solvent *raised* by the presence of the solute, while the freezing point is lowered? How much is the boiling point of water at atmospheric pressure (373 K) raised by 1% w/w NaCl ? The latent heat of vaporization of water is 4×10^7 J/kmole

---

[18] See Appendix G. Don't forget that r is *not* a canonical coordinate, but is related to the canonical coordinates x, y, z by $r^2 = x^2 + y^2 + z^2$.

10.14 In the superheated Rankine cycle described in Example 3.3, use the Clausius-Clapeyron equation and the ideal gas law for isentropic expansion to find p_2 (in bar), T_1 and η for $T_2 = 310$ K, and $p_1 = 50$ bar. Assume that the water molecule has six degrees of freedom and that equipartition holds. Take $L = 40$ MJ/kmole

10.15 Examine the latent heat as N_A bonds breaking per mole in a solid, and use this to estimate the binding energy for single molecule. Comment on your estimate

11

Black Body Radiation and the Photon Gas

The spectral density of black body radiation... represents something absolute, and since the search for absolutes has always appeared to me to be the highest form of research, I applied myself vigorously to its solution.

Max Planck

Were the succession of stars endless, then the background of the sky would present us an uniform luminosity ... since there could be absolutely no point, in all that background, at which would not exist a star.

from *Eureka* (1848) by Edgar Allan Poe

11.1 Gedanken

We imagine one of our spin systems, at $T = 0$, where all the spins point in the same direction. Any amount of heat (energy) added to the system has any number of configurations to increase entropy, so it is not surprising that the heat capacity grows exponentially at low temperature. However, as we continue to add energy to the system (any real, finite system), we know that ultimately the number of spins that can be further flipped is limited. The quantity of $k_B T$ sets the scale by which the Boltzmann factor tells us what is more likely to happen and what is less likely for some energy, E, so one should not expect to be able to continue to add energy to any system without bound. Conversely, since T is the slope of $(\partial U/\partial S)_{V,N}$, any finite temperature tells us there is an upper bound on energy that can be added to a system, decreasing the number of conformations from that large number of possibilities from whence we proceeded at $T = 0$.

11.2 Field-Carrying Bosons: The Photon

There are two types of bosons. The first type is a composite particle which contains an even number of fermions; for example, the deuteron and the ^{4}He nucleus. So long as we restrict ourselves to a range of energy well below the dissociation energy of such a composite particle ($\sim$ MeV in the case of the nucleus), these particles are conserved. The second type is a particle associated with a field, of which the most important example is the photon, which is associated with the electromagnetic field. Such particles are not conserved; each particle carries a certain amount of energy, and if the total energy of the system changes, particles appear and disappear as required. In this chapter, we consider the second type, confining our attention to the photon. We shall see that the chemical potential of such particles is zero in equilibrium, regardless of density. This is quite different from the case of conserved particles, whose chemical potential depends on the particle density. These particles are discussed in the next chapter.

11.3 Thermodynamics of Radiation: Cavity Radiation

We first establish the classical background. The problem of *black body radiation* (that is, the electromagnetic radiation emitted by a hot surface) has been under active consideration since the middle of the nineteenth century. It was the inability of classical statistical mechanics to account for the black body spectrum (the emitted power at a given temperature as a function of frequency) that led Planck[1] in 1900 to introduce the concept of the quantum. On the other hand, in 1884 Boltzmann used thermodynamic reasoning to correctly predict the temperature dependence of the total radiation from a hot body, and we begin by discussing the thermodynamics of radiation, leaving the statistical treatment till later.

Radiation, whether from a hot body, from a fluorescent gas, or from a laser, is not under normal circumstances in equilibrium; energy is flowing outwards, and must be replenished from some source if it is to continue. However, if we are to apply thermodynamics to the radiation from a hot body, we must first find some situation where radiation is being emitted and absorbed at the same average rate, so that the body is in equilibrium with the radiation. This first step in understanding was made by Kirchhoff,[2] who considered an enclosure (cavity) inside an opaque, but thermally conducting solid body, which can be regarded as a heat bath at a constant temperature T (see Figure 11.1a).

The walls of the cavity emit and absorb electromagnetic waves. The cavity is filled with radiation that is in equilibrium with the heat bath surrounding it, so that it can be assigned a temperature T. The energy is spread over a range of frequencies, and we define $u_s(\nu, T)\,d\nu$ as the energy density (energy per unit volume) of the radiation with frequency between ν and $\nu + d\nu$ when the temperature is T. We call $u_s(\nu, T)$ the *spectral energy density*. The integral of $u_s(\nu, T)$ over all frequencies is the internal energy density of the photon gas, and is written $u(T)$. Thus,

$$u(T) = \int_0^\infty u_s(\nu, T)\,d\nu. \tag{11.1}$$

For brevity we will call u_s the spectral density, and u the integrated density. In equilibrium, u_s must be the same everywhere in the cavity, and is a function of frequency and temperature only. It is independent of the cavity volume and shape and of the material from which the walls are made. This can be shown as follows. Suppose that we have two cavities, A and B, of different volumes, whose walls are at the same temperature, but differ in shape and composition. Imagine that the cavities, shown in Figure 11.1b, are connected by a small hole covered by a filter that transmits only radiation at frequency ν. The size of the hole is assumed to be sufficiently small, relative to the size of the cavity, that it does not perturb the radiation within the cavity. Then radiation in each cavity escapes through the hole at a rate proportional

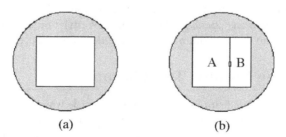

      (a)   (b)

FIGURE 11.1   (a) Cavity containing radiation, enclosed in an opaque heat bath. (b) Two cavities connected by a hole that is covered by a filter.

---

[1] Max Planck (1858–1947), German theoretical physicist. Planck was the reluctant founder of quantum theory, for which he won the Nobel Prize in 1918.
[2] Gustav Kirchhoff (1824–1887), German physicist, who in 1859 first applied the new thermodynamic reasoning of Clausius and Kelvin to radiation.

to the spectral density inside that cavity,[3] so that if the spectral densities were to differ, there would be a net flow of energy from one cavity to the other. As a result, one cavity would heat up and the other cool down. However, since they are initially at the same temperature, this temperature change would require that energy be transferred from the cooler cavity to the hotter. Since no work is done on the system, such transfer is contrary to the second law. Thus $u_s(\nu, T)_A = u_s(\nu, T)_B$; that is, u_s is independent of the nature of the cavity walls and is a function of frequency and temperature alone. Integration over frequency shows that the integrated energy density u is a function only of temperature. An analogous argument shows that $u_s(\nu, T)$ and $u(T)$ must be spatially uniform within the cavity if the walls are at a uniform temperature and the photon gas is in equilibrium.

Now consider what happens if we increase the volume V of a cavity, keeping T constant. Since u depends only on temperature and is independent of the volume, the internal energy $U = uV$ increases, this energy being supplied by the heat bath. It is very important to recognize the essential difference between the photon gas (that is, radiation) and the ideal gas considered in the last chapter. In an ideal gas, the number of particles is conserved, and the energy per particle is a function only of temperature, so that the internal energy is unchanged in an isothermal expansion. On the other hand, for a photon gas it is the energy *density*, $u = U/V$, which is unchanged, as was shown in the previous paragraph. The number of photons is not conserved, but is proportional to volume in an isothermal change.

11.4 Kirchhoff's Radiation Law and the Brightness Theorem

Some important results can be obtained from considering the thermal equilibrium between the radiation and the cavity walls. A real surface absorbs only a fraction of the radiation falling on it, the remainder being either transmitted or reflected. We call this fraction the *absorptivity* α_r. The absorptivity is a function of frequency and temperature, but it is a property of the surface, not of the amount of radiation falling on it. A surface for which $\alpha_r(\nu) = 1$ at all relevant frequencies is called a *black body*. Any hot surface emits radiation, and we define the *emissivity* $e_r(\nu)$ of a surface as the ratio of its emission to that of a black body of the same surface area at the same temperature. Like the absorptivity, the emissivity is a property of the surface. Now suppose that the surface forms part of the wall of a cavity containing radiation, as described in the previous section. This surface is continually emitting and absorbing radiation. The average amount of radiation at frequency ν which is absorbed by the surface is $\alpha_r(\nu)$ times the amount that would have been absorbed by a black body. Similarly, the amount emitted is $e_r(\nu)$ times the amount that would have been emitted by a black body. In equilibrium, the average amount of radiation absorbed by the surface at any frequency must equal the average amount emitted at that frequency. Hence, the emissivity $e_r(\nu)$ must be equal to the absorptivity $\alpha_r(\nu)$. This is known as Kirchhoff's radiation law. Note that absorptivity and emissivity depend *only* on the frequency, the surface temperature, and the material from which the surface is made; they do not depend on the actual radiation density. It follows that, even though the derivation of this law is based on an equilibrium situation, at a given frequency and surface temperature T, the emissivity $e_r(\nu, T)$ is equal to the absorptivity $\alpha_r(\nu, T)$ under all conditions, equilibrium or not.

It follows from Kirchhoff's radiation law that if a body does not absorb at a particular frequency (that is, $\alpha_r = 0$ at that frequency), it does not emit either. This can be demonstrated by heating a silica disc in a hot flame. Pure silica is transparent to visible light. If the edges are slightly dirty so that they can absorb light, they become visibly hot, while the clean surface of the silica remains invisible, even though it is at the same or higher temperature. Glassblowers know that if they are to heat silica to its melting point in a normal flame, it must be scrupulously clean; otherwise, the loss of energy by radiation will prevent the high melting temperature from being reached. Other applications of Kirchhoff's radiation law are mentioned at the end of this chapter.

---

[3] See Appendix H for the proof that this is so.

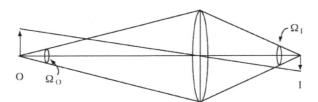

FIGURE 11.2 Brightness theorem in a simple lens system. The lens subtends a solid angle Ω_O at the object O, whose area is A_O, and subtends Ω_I at the image, area A_I. The radiative energy falling on the image cannot exceed that emitted in Ω_O, so that if B_O and B_I are the respective brightnesses, $B_O \Omega_O A_O \geq B_I \Omega_I A_I$. Simple geometrical optics show that $\Omega_O A_O = \Omega_I A_I$, so $B_O \geq B_I$.

The spatial uniformity of $u_s(\nu, T)$ means that the *brightness* of a surface, defined as the energy emitted per unit solid angle per unit area, is a function only of frequency, temperature, and emissivity. This has some immediate experimental consequences. One consequence is that an object in equilibrium in the cavity is invisible against the background, since the amount of radiation it emits is exactly equal to the amount that it absorbs.

Another important consequence of Kirchhoff's radiation law, known as the brightness theorem, is that if an optical image is formed of a self-luminous object (as, for example, in Figure 11.2), the brightness of the image cannot exceed that of the object. If it did, one could heat a body placed at the image above the temperature of the originating body, contrary to the second law. If there are no losses, so that the system is reversible, an object at I is imaged at O, and the brightness of the image must equal that of the source.[4]

The brightness theorem is exploited in the *disappearing filament pyrometer*, in which a wire (the filament) that can be heated electrically is placed at the image plane of a telescope focused on the object whose temperature is to be measured. If the temperature of the filament is less than that of the (extended) object, it appears dark against the background image, while if its temperature is higher, it appears bright. When the temperatures are equal, the filament disappears into the background. With very hot objects it is customary to attenuate the radiation intensity by a known factor, so that the filament need not be heated to the temperature being measured, which may be above its melting point.

11.5 The Stefan-Boltzmann Law

We have seen that in equilibrium, the energy emitted by a surface is equal to that absorbed, and that the energy absorbed at any frequency depends only on the absorptivity of the surface and on the spectral density $u_s(\nu, T)$. The absorptivity α_r is a property of the material, and we have defined a black body as one for which $\alpha_r = 1$ at all relevant frequencies. Our task now is to calculate $u_s(\nu, T)$ and its integral $u(T)$ for a cavity; we shall see later how we can relate them to the actual emission of a hot body.

To calculate $u_s(\nu, T)$, we shall need statistical mechanics and quantum theory. However, the temperature dependence of the integrated density $u(T)$ can be obtained by a purely thermodynamic argument, making no assumptions about the microscopic nature of the radiation, so we present this argument first. Electromagnetic waves falling on a surface exert a pressure p on it. It can be shown by standard electromagnetic theory[5] or by kinetic theory (see Equation H.8) that if the waves (photons) arrive at the surface from random directions, the pressure is related to the integrated energy density by:

$$p = \frac{u}{3}. \tag{11.2}$$

---

[4] For a more general treatment, see M. Born and E. Wolf, *Principles of Optics*, 6th ed. (Pergamon Press, Oxford, UK, 1980), pp. 188–190.

[5] See, for example, B. I. Bleaney and B. Bleaney, *Electricity and Magnetism* (Oxford University Press, Oxford, UK, 1976), p. 248.

We use this relation to derive $u(T)$ as follows. By the same reasoning that showed that the integrated energy density is a function of T alone, the free energy density $f = F/V$ is also.[6] Do not confuse the free energy density f with the distribution function $f(E, T)$, defined in Chapter 9. From Equation (5.23),

$$p = -\left(\frac{\partial F}{\partial V}\right)_T = -f,$$

so that:

$$f = -p = -\frac{u}{3}. \tag{11.3}$$

The Gibbs-Helmholtz relation [Equation (5.25a)] can be divided through by V and rearranged to give:

$$f = u + T\frac{df}{dT}. \tag{11.4}$$

Combining Equations (11.3) and (11.4) gives:

$$-\frac{u}{3} = u - \frac{T}{3}\frac{du}{dT}.$$

Hence,

$$\frac{du}{u} = 4\frac{dT}{T}. \tag{11.5}$$

Integrating Equation (11.5) gives:

$$\boxed{u(T) \propto T^4,} \tag{11.6}$$

where the constant of proportionality cannot be obtained by thermodynamic arguments alone.

As pointed out above, what is actually measured in an experiment is not the energy density in a cavity, but the radiation from a hot surface. We consider a perfectly black surface (that is, one with $\alpha_r = e_r = 1$), so that it absorbs all the radiation falling on it. If the surface were in equilibrium with the radiation, it would emit whatever it absorbed. It follows that the emitted energy at temperature T must equal the amount of energy that would fall on the surface if it were in a cavity at that temperature. The energy crossing unit area in unit time is called the *energy flux* J_u. It is shown in Appendix H that in a perfect gas (such as the photon gas), the particle flux J_n is proportional to the particle density n and their mean velocity $\langle v \rangle$. For photons $\langle v \rangle = c$, so that from Equation (H.4),

$$J_n = \frac{1}{4}nc.$$

Each photon carries an average energy $\langle E \rangle = \frac{u}{n}$, so that the energy flux falling on the surface is:

$$J_u = J_n\langle E \rangle = \frac{1}{4}uc = \sigma_B T^4, \tag{11.7}$$

---

[6] This can be seen most easily by applying Kirchhoff's argument [which proved that $u_s = u_s(T, v)$] to the entropy density in a cavity. Consider two cavities at the same temperature and connected by a window, as in Figure 11.1b. A photon passing through the window carries some entropy with it. In equilibrium there can be no net transfer of entropy, so the entropy flux in each direction must be the same. Since the flux is proportional to the density, the entropy density must be the same in both cavities, and hence must be a function of T alone. Since $F = U - TS$, the same is true of the free energy density.

where σ_B is called Stefan's constant. By Kirchhoff's radiation law, this is also the energy flux emitted by a black (that is, perfectly absorbing) surface[7] at temperature T. Equation (11.7) was established experimentally by Stefan[8] in 1879, while Equation (11.6) was derived thermodynamically five years later by Boltzmann; Equations (11.6) and (11.7) together are known as the Stefan-Boltzmann law.

The experimental value of σ_B is 5.67×10^{-8} W K^{-4} m^{-2}. This may look like a small number, but consider the case of a human body at 310 K, at which temperature $\sigma_B T^4 \approx 500$ W/m^2. While the emissivity of skin is considerably less than 1, it emits sufficient infrared radiation to be easily detectable by modern techniques.[9] At 2800 K, the filament temperature of a typical incandescent lamp, $\sigma_B T^4 \approx 3.5$ MW/m^2.

11.6 Isentropic Expansion: The Cosmic Microwave Background

From Equations (11.7) and (11.3), the free energy of a photon gas of volume V and temperature T is:

$$F = -\frac{4}{3c}\sigma_B V T^4.$$

The entropy of the photon gas can be obtained from this using Equation (5.24):

$$S = -\left(\frac{\partial F}{\partial T}\right)_V = \frac{16}{3c}\sigma_B V T^3. \tag{11.8}$$

Thus in an isentropic expansion, VT^3 = constant. The most important application of this result is to the cosmic background radiation. According to currently accepted cosmology, the universe began with the "Big Bang," which left it filled with radiation at a very high temperature. As the universe expands, the photon wavelengths stretch in proportion. This is an adiabatic process, so that the occupancies do not change and the radiation expands isentropically[10] (compare Example 1.2). Since $V \propto R^3$, where R is the time-dependent "radius of the Universe," it follows from Equation (11.8) that $T \propto R^{-1}$. The temperature at the present time is 2.735 K (see Figure 11.4). This radiation is detected as a "cosmic microwave background" coming from all directions and is quite distinct from the radiation from stars and galaxies. Its existence is generally considered to be the most direct evidence for the Big Bang theory of the origin of the universe.

---

[7] While there is no such thing as a perfectly black body in nature, one can construct a very good approximation to one by ensuring that incident radiation is reflected or scattered many times by the surface of the body before escaping. If on average the radiation strikes the surface i times, the effective reflectivity is $(1 - \alpha_r)^i$, which rapidly becomes very small as i increases. This can be achieved by making the surface irregular, or better by making a narrow, but deep hole in an opaque body.

[8] Josef Stefan (1835–1893), Austrian physicist.

[9] The applications are obvious, especially military ones. See, for example, G. C. Holst and S. W. McHugh, Review of thermal imaging performance, and other papers in *Infrared Imaging Systems: Design, Analysis, Modeling and Testing III*, ed. G. C. Holst (International Society for Optical Engineering, Bellingham, WA, 1991).

[10] This is only true if the radiation is decoupled from the matter in the Universe. In the standard Big Bang model of the origin of the universe, this decoupling occurred about 300,000 years after the Big Bang, when the temperature dropped to the point where neutral atoms form, so that there were no longer free charged particles to absorb and emit radiation. Since that time, the radiation has expanded independently of the matter. Careful measurements of the dependence of the radiation intensity on direction have shown that at the time that decoupling occurred, the density of the universe only deviated from perfect uniformity by a few parts per million, very different from the "clumped" matter that we observe today. These results (in particular, the apparent scale of the angular variation) are of profound cosmological significance (see, e.g., the link to "The Parameters of Cosmology" at the Website of the Microwave Anisotropy Probe mission at http://map.gsfc.nasa.gov/m_mm/sg-parameters1.html).

11.7 Radiative Transfer

As remarked in Chapter 2, one can insulate an object thermally from its environment by surrounding it with a vacuum jacket. However, while this prevents heat being conducted to or from the object, energy can still be transferred by radiation. One way to reduce this radiative transfer is to reduce the absorptivity of the surfaces; by Kirchhoff's radiation law this reduces the emissivity in the same proportion. Making the surfaces transparent is not going to help, so they must be reflective. Note that what matters is the reflectivity at the predominant frequencies emitted at the temperatures involved. For example, copper is not a good reflector in the visible (it absorbs in the blue, which is why it looks red), but it is an excellent reflector ($\sim$98%) in the infrared and so makes a suitable surface material at temperatures not too far above room temperature. A still higher degree of insulation can be achieved by inserting one or more heat shields (see Problem 11.6).

Example 11.1: The Dewar Flask

A vacuum or Dewar flask[11] (sometimes called a Thermos flask) consists of a cylindrical container surrounded by a thin evacuated jacket as in Figure 11.3. The entire surface of the evacuated region is coated with a material such as silver, which is opaque and has a high reflectivity. The reflectivity is r, and the temperatures inside and outside the flask are T_a and T_b, respectively. What is the energy flux (the rate of radiative energy transfer per unit area)?

Since the separation between the walls is much less than the diameter of the cylinder, they can be treated as plane parallel surfaces.

By the Kirchhoff and Stefan-Boltzmann laws, the inner surface radiates at a rate $(1 - r)\sigma_B T_a^4$ watts per square meter, and the outer one at $(1 - r)\sigma_B T_b^4$. Each surface reflects a fraction r of the radiation falling on it. Let the total ingoing flux be J and the total outgoing flux be J'. Then:

$$J = (1 - r)\sigma_B T_a^4 + rJ', \quad J' = (1 - r)\sigma_B T_b^4 + rJ.$$

The net ingoing flux is thus:

$$J - J' = \frac{1 - r}{1 + r}\sigma_B \left(T_a^4 - T_b^4\right).$$

If $r = 0.98$, the net flux is reduced to about 1% of the value it would have if the surfaces were black bodies.

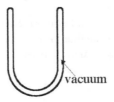

FIGURE 11.3 Section through a Dewar flask. The lines represent glass walls, and the region between the two glass walls is evacuated. The inside surface of the glass is silvered.

---

[11] Invented by James Dewar (1842–1923), Scottish physicist, who was the first to liquefy hydrogen.

11.8 Photon Statistics and the Planck Radiation Law

While it is possible to go further using only thermodynamic arguments (see Footnote 19 in this chapter), it is simpler to switch to the statistical mechanical approach at this point. We use the fact that photons are bosons[12] to derive $u_s(v, T)$.

We first show that the chemical potential ζ is zero for a photon gas (that is, electromagnetic radiation) in equilibrium. We cannot use the usual expression $\zeta = \left(\frac{\partial F}{\partial N}\right)_{V,T}$ for ζ, since one cannot increase N (that is, add photons to the system) at constant volume and at the same time keep the temperature constant; $\left(\frac{\partial F}{\partial N}\right)_{V,T}$ does not exist for a photon gas. Instead, we use the free enthalpy $G = F + pV$. The free energy $F = fV = -pV$ from Equation (11.3). Hence, $G = 0$, and it follows from $G = N\zeta$ (Equation 8.10) that $\zeta = 0$. Thus, photons in equilibrium have zero chemical potential, even though their free energy F is non-zero. This result can be obtained in a more intuitively appealing way by considering what happens if the volume of the cavity increases by dV, the temperature and hence the pressure remaining constant. This expansion increases the number of photons by an amount proportional to dV. For a change at constant pressure, G is a minimum in equilibrium, so that $dG = 0$. Hence, from Equation (8.9),

$$\zeta = \left(\frac{\partial G}{\partial N}\right)_{p,T} = 0.$$

The occupancy of a particular photon state of energy $E = hv$ is given by the Bose-Einstein distribution with $\zeta = 0$, which is, from Equation (9.10),

$$f(E, T) = \frac{1}{e^{\beta E} - 1}, \tag{11.9}$$

where $\beta = \frac{1}{k_B T}$. The density of states per unit volume with respect to photon frequency, $g(v)$, can be obtained from Equation (7.37). Since photons have two possible polarizations while phonons have three, we multiply by 2/3 to obtain:

$$g(v) = \frac{8\pi v^2}{c^3}. \tag{11.10}$$

The number of photons per unit volume with frequency between v and $v + dv$ is $g(v)f(E)\,dv$, and since each photon carries energy $E = hv$, the spectral energy density is:

$$\boxed{u_s(v, T) = hvg(v)f(E) = \frac{8\pi h}{c^3}\frac{v^3}{e^{\beta hv} - 1}.} \tag{11.11}$$

Equation (11.11) is Planck's radiation law,[13] and is represented by the continuous line in Figure 11.4. Also shown are the data obtained from a satellite measurement of the cosmic microwave background radiation mentioned in Section 6. The agreement between theory and experiment is perfect; the *only*

---

[12] One can obtain the spectral density function $u_s(v, T)$ by statistical mechanics without invoking the concept of photons, in a way exactly analogous to Debye's theory of the heat capacity of solids, discussed in Section 7.8. However, such a derivation is unsatisfactory, since it leads to an infinite zero point energy which must be subtracted out. While this problem is to a certain degree dealt with by quantum electrodynamics (see Footnote 8 of Chapter 7), we can avoid it altogether by treating the radiation as a boson gas.

[13] It is customary to express u_s as $u_s(\lambda, T)$; that is, as a function of wavelength λ rather than of frequency v: see Problem 11.7. Simple spectrometers usually measure the energy per unit wavelength interval and so give $u_s(\lambda, T)$ rather than $u_s(v, T)$. The more sophisticated spectrometer used in the satellite measurements shown in Figure 11.4 measures $u_s(v, T)$.

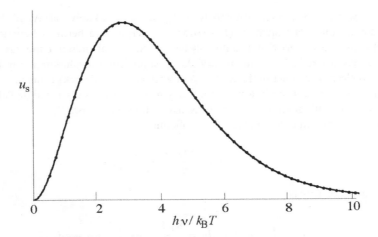

FIGURE 11.4 Line: Planck's radiation law (Equation 11.11). Points: observed energy flux of the cosmic microwave background radiation as a function of frequency in units of $k_B T/h$, with $T = 2.735$ K. (Adapted from Schwarzschild, B., *Physics. Today* **43**, 17–20, 1990.)

fitting parameter is the temperature, taken to be 2.735 K. Surprisingly, these are by far the most accurate spectral data ever obtained on black body radiation from any source.[14]

At low frequencies or high temperatures ($\beta h\nu \ll 1$), $e^{\beta h\nu} - 1 \approx \beta h\nu$, and Equation (11.11) becomes (see Equation (A.11))

$$u_s(\nu, T) = \frac{8\pi \nu^2}{c^3}\left(k_B T - \frac{1}{2}h\nu\right). \tag{11.12}$$

The temperature dependent term in Equation (11.12) is a purely classical result (note that it does not contain Planck's constant h) and can be obtained directly from equipartition (see Problem 11.4). It is known as the Rayleigh-Jeans radiation law[15] and is valid at low frequencies ($h\nu \ll k_B T$). It was assumed to hold (modified to take into account the lower dimensionality) in our calculation of electrical noise (Section 7.11). If it held at all frequencies, as required by classical theory, it would imply infinite $u(T)$, an obvious impossibility. This paradox of classical statistical mechanics is known as the ultraviolet catastrophe.

At high frequencies ($h\nu \gg k_B T$), the 1 in the denominator can be neglected and Equation (11.11) becomes,

$$u_s(\nu, T) \propto \nu^3 e^{-\beta h\nu}. \tag{11.13}$$

---

[14] Nevertheless, the measurements of Rubens and Kurlbaum in 1900, which stimulated Planck to develop his quantum theory of radiation, were remarkably accurate for the time. By fitting their data to Equation (11.11), Planck obtained values of h and k_B within 2.5% of the now accepted values.

[15] John William Strutt, Lord Rayleigh (1842–1919) and Sir James Jeans (1877–1946), English physicists. In 1900, a few months before Planck published Equation (10.9), Rayleigh published Equation (11.12) in the form $u_s \propto \nu^2 T$. Rayleigh recognized that equipartition could only apply at low frequencies and assumed an arbitrary cut off at high frequencies. Jeans' contribution was merely to correct a numerical error in Rayleigh's later (1905) derivation of the constant of proportionality (See Pais' article referred to in Footnote 17). Even as late as 1911, the ineffable Jeans was trying to save classical theory by an elaborate mechanical model, and earned a devastating response: "It is obvious that by giving suitable dimensions to the communicating tubes between his reservoirs and giving suitable values to the leaks, Jeans can account for any experimental results whatever. But this is not the role of physical theories..." Henri Poincaré in *La Théorie du rayonnement et les quanta* (Solvay Conference, Paris, 1911, eds. Langevin and de Broglie), quoted at greater length in *A Random Walk in Science*, eds. R. L. Weber and E. Mendoza (Inst. of Physics, 1973), p. 92.

Equation (11.13) was discovered experimentally by Wien[16] and is known as Wien's law. Planck first devised his radiation formula [Equation (11.11)] as an interpolation between Equations (11.12) and (11.13). His subsequent discovery that it could only be explained by abandoning the classical idea of continuous energy levels ushered in quantum theory, the most profound revolution in physics since Galileo overturned Aristotelean mechanics in the early Seventeenth Century. Planck's route to his discovery, and Einstein's recognition of its extraordinary significance five years later, is discussed by Pais.[17] A brief and illuminating, but historically slightly inaccurate, discussion is given by Born.[18]

If we substitute $x = \beta h\nu$ in Equation (11.11), we obtain:

$$u_s(x, T) = \frac{8\pi (k_B T)^3}{h^2 c^3} \frac{x^3}{e^x - 1}. \tag{11.14}$$

The integrated energy density is:

$$u(T) = \int_0^\infty u_s(\nu, T)\, d\nu = \frac{8\pi (k_B T)^4}{h^3 c^3} \int_0^\infty \frac{x^3}{e^x - 1}\, dx$$

$$= \frac{8\pi^5 k_B^4}{15 h^3 c^3} T^4. \tag{11.15}$$

(See Problem 7.7 for the evaluation of the integral).

Substituting Equation (11.15) in Equation (11.7), we obtain the radiative flux from a black body at temperature T:

$$J_u = \sigma_B T^4, \tag{11.16}$$

where $\sigma_B = \frac{2\pi^5 k_B^4}{15 h^3 c^2}$. Equation (11.16) is the Stefan-Boltzmann law [Equation (11.7)], with a value for σ_B in excellent agreement with experiment.

Differentiating Equation (11.13) with respect to ν shows that u_s is a maximum when $\nu = \nu_m \approx \frac{3k_B T}{h}$ (see Problem 11.7). Putting in numbers, we find that at 300 K, the peak emission occurs at a wavelength $\frac{c}{\nu_m} \sim 15\,\mu\mathrm{m}$, in the middle infrared. At 3000 K, the temperature of a typical incandescent lamp, the peak is at $1.5\,\mu\mathrm{m}$, in the near infrared, and the emission extends well into the visible. At the surface temperature of the sun ($\sim$ 6000 K), the peak is at 750 nm and about 40% of the radiation is in the visible. The proportionality of ν_m to T is a special case of Wien's displacement law, which states that:

$$u_s = \nu^3 \phi\left(\frac{\nu}{T}\right), \tag{11.17}$$

where ϕ is an unspecified function.[19]

Because of the variation of emissivity with wavelength, the radiation spectra of most real bodies deviate from the Planck curve. However, whatever the emissivity, Wien's displacement law ensures that the whole radiation curve always shifts to shorter wavelengths with increasing temperature. The "color temperature" of a hot object is defined as the temperature of a black body with the same variation of emission with

---

[16] Wilhelm Wien (1864–1928), German physicist, Nobel Prizewinner in 1911. Wien believed, incorrectly, that Equation (11.13) could be derived classically. Wien was a fine experimentalist who near the end of his career severely damaged his reputation for good judgment by proposing that Werner Heisenberg's doctoral thesis be rejected. Fortunately for the future of physics, Wien was overruled by his colleagues [D. C. Cassidy, *Uncertainty* (Freeman, 1992), p. 153].

[17] A. Pais, "Einstein and the quantum theory," *Revs. Mod. Phys.* **51**, 863 (1979).

[18] M. Born, *Natural Philosophy of Cause and Chance* (Oxford University Press, Oxford, UK, 1949), pp. 78–82.

[19] Wien's displacement law was originally derived by a purely classical argument: see Appendix 33 of Born. It can also be obtained from the Stefan-Boltzmann law by a dimensional argument if one assumes the relation $E = h\nu$ (see Problem 11.12).

wavelength (usually in the visible) as the object. It is an important parameter of a light source, particularly one used in color photography or television, where correct color balance is important.

Example 11.2: Spin Waves

Photons and phonons (see Footnote 3 of Chapter 7) are not the only non-conserved bosons. The lowest magnetic excitations of a ferromagnet (see Chapter 15) are *spin waves*, whose quanta are analogous to phonons, and are called *magnons*, in which the tilt of a spin relative to the average spin direction takes the place of atomic displacement. In zero external field, the dispersion relation for spin waves in an isotropic ferromagnet is $v \propto k^2$ for small k, where $\mathbf{k}$ is the wavevector. Find the temperature dependence of the mean number of magnons thermally excited at low temperature.

Since magnons, like photons, are not conserved, their chemical potential is zero. Because they have the same dispersion relation (apart from a multiplicative constant) as a non-relativistic particle, their density of states has the same dependence on frequency: $g(v) \propto v^{1/2}$. The number of magnons excited per unit volume is:

$$n(T) = \int g(v) f(v, T) \, dv -$$

where $f(v, T)$ is the Bose-Einstein distribution function with $E = hv$. As in the case of the Debye model of a solid, at low temperature the upper limit of the integral can be taken as infinity.

Hence, at low T,

$$n(T) \propto \int_0^\infty \frac{v^{1/2}}{e^{\beta h v} - 1} \, dv = \beta^{-3/2} \int_0^\infty \frac{x^{1/2}}{e^x - 1} \, dx$$

where $x = \beta h v$.

Hence, since the integral is just a number,

$$n(T) \propto T^{3/2}. \tag{11.18}$$

Each spin wave[20] reduces the magnetization by the same amount, and the predicted $T^{3/2}$ dependence of the decrease in magnetization as the temperature increases from zero is generally observed in ferromagnets (see Chapter 15).

11.9 The Greenhouse Effect

As mentioned before, the absorptivity of any substance varies with frequency in a way characteristic of that substance. The study of such variation is called *spectroscopy* and is an enormous field in itself. An environmentally important consequence of this variation is the *greenhouse effect*. A greenhouse is a shed made of glass so that the sun can shine in. Most of the sun's radiant energy is concentrated in the visible and near infrared (wavelengths less than about 3 μm), where glass is transparent, so the radiation enters the greenhouse and is absorbed by the plants and other objects within it. Their temperature, while raised above the ambient by the sun's heat, is still of course low relative to that of the sun, so that they emit radiation at much longer wavelengths, beyond 10 μm. Glass is opaque to these wavelengths (that is, it absorbs or reflects them), so that the radiation cannot escape. The interior of the greenhouse warms up until the heat gained by radiation from the sun is balanced by the losses by conduction.

---

[20] See L. R. Walker, "Spin waves and other magnetic modes," in *Magnetism*, Vol. 1, eds. G. T. Rado and H. Suhl (Academic Press, New York, 1963), Chapter 8.

In the earth's atmosphere, carbon dioxide (CO_2) is one of the "greenhouse" gases that play the role of the glass, being transparent at visible and near-infrared wavelengths, but strongly absorbing in the 10 to 15 μm range where the earth's thermal radiation is concentrated. Thus, an increase in the CO_2 content of the atmosphere raises the earth's average temperature (see Problem 11.1b). The actual amount of heating due to a given change in concentration of CO_2 is difficult to calculate, since other factors in the heat balance of the earth also change. In particular, the amount of solar radiation that reaches the earth's surface and is absorbed is strongly dependent on the average cloud cover, on the humidity, on the presence of pollutants in the atmosphere, and on the fraction of the earth's surface covered by snow or ice, all of which are affected by the CO_2 concentration.[21]

Another application of the variation of emissivity with wavelength, closely related to the greenhouse effect, is the *Welsbach mantle*, widely used in the days of gas lamps and still seen in kerosene-fueled hurricane lamps. The Welsbach mantle is made of a mixture of rare-earth oxides, which have strong absorption (and therefore emission) in the visible, but absorb only weakly in the infrared. When heated in a flame, the mantle cannot radiate until it is hot enough to emit in the visible, emitting a white light that seems cool because of the relative absence of infrared radiation.

11.10 Envoi

The solution of the "radiation problem" by Planck's introduction of the quantum of energy in 1900 is now seen as the beginning of the quantum revolution in physics.[22] However, the historical importance of Planck's radiation law should not obscure its relevance to current physics research and engineering practice. It is one of the few results of quantum theory that is relevant in the macroscopic domain where the atomic structure of matter can be ignored.

11.11 Problems

11.1 (a) If one neglects the small contribution of rock radioactivity and other terrestrial sources of heat, the earth's mean surface temperature in the steady state is determined by the balance between the radiation coming in from the sun and that reradiated out by the earth. Calculate this temperature, assuming that the sun and the earth are black bodies at all relevant wavelengths (in fact, this assumption is approximately correct for the sun, but not for the earth). The sun's average surface temperature is 5800 K, its radius 7×10^8 m, and the average earth-sun distance is 1.5×10^{11} m. Qualitatively, how would the result be affected by 10% cloud cover? Assume that the clouds are 100% reflective to the sun's radiation, but are black in the middle infrared and are close to the earth's surface temperature

     (b) By how much would the mean temperature of the earth be changed by the presence of enough CO_2 in the atmosphere to absorb 10% of the radiation emitted by the earth? Neglect all mechanisms of energy transfer between the earth and the atmosphere other than radiation, and assume that the atmosphere remains completely transparent to the sun's radiation

     (c) Now consider a solid planet with no atmosphere or oceans to transfer heat from one point on the surface to another. Find an expression for the noontime surface temperature as a function of latitude, at the equinox, in terms of the sun's surface temperature, and its angular diameter

---

[21] See, for example, R. A. Houghton and G. M. Woodwell, "Global climate change," *Sci. Am.* **260** #4, pp. 36–44 (April 1989); J. W. Firor, "Resource letter GW-1: Global warming," *Am. J. Phys.* **62** 490–495 (1994); R. S. Knox, "Physical aspects of the greenhouse effect and global warming," *Am. J. Phys.* **67** 1227 (1999).

[22] Planck's work was not seen as revolutionary at the time, and it was left to Einstein—who, unlike the conservative Planck was a true revolutionary—to take the quantum seriously with his introduction of the photon in 1905 [see Pais's article referred to in Footnote 17, and S. G. Brush, *Statistical Physics and the Atomic Theory of Matter, from Boyle and Newton to Landau and Onsager* (Princeton, University Press, Princeton, NJ 1983), Section 3.1.]

as seen from the planet. Assume that the only source of energy is the sun, that the surfaces of sun and planet are black at all wavelengths, and that the heat capacity and thermal conductivity of the ground are negligible

11.2 Radiation at temperature T_i fills a thermally insulated cavity of volume V. If the cavity expands isentropically to volume rV, how much work is done by the radiation, as a fraction of its initial energy? How much work would be done if the expansion were isothermal?

11.3 Extend the brightness theorem to the case of a surface whose normal is at an angle θ to the direction of viewing, and thus show that the radiant energy flux in this direction varies as $\cos \theta$

11.4 (a) Show that applying the equipartition theorem to the modes of cavity radiation leads to the Rayleigh-Jeans law for the spectral energy density:

$$u_s(\nu, T) = \frac{8\pi \nu^2}{c^3} k_B T \qquad (11.19)$$

(b) Explain why Equation (11.19) differs from the low frequency limit (Equation (11.12)) of Planck's radiation law

11.5 A *cryostat* (a device for holding experimental equipment at a low constant temperature) is cooled by a refrigerator with a capacity of 25 mW; that is, it can extract heat from the cryostat at this rate. The cryostat is insulated from its surroundings by a vacuum shield like the Dewar flask described in Example 11.2. The surface area of the shield is 100 cm^2, and the outer wall is held at 300 K

(a) What must the reflectivity of the walls in the relevant wavelength range be if the contents of the cryostat are to be held at a temperature of 10 K? Neglect all sources of heat input other than radiation

(b) Suppose that the silvering of the walls deteriorates and the reflectivity drops to 0.98. What temperature can be achieved now? Assume for simplicity that the refrigerator performance is independent of temperature (this is not in fact the case; see Chapter 3)

11.6 (a) A *heat shield* is a layer of material inserted between the inner and outer walls of an evacuated region such as that in the Dewar flask (see Example 11.2), but not touching them. Assuming that the only energy transfer to and from the heat shield is by radiation, and that the walls are plane and parallel, what temperature does the shield attain, and by what factor is the energy flux reduced from that in the absence of the shield? For simplicity, assume that the walls and shield are black

(b) Now suppose that there are N heat shields between the inner and outer walls. Show that the transfer is reduced by a factor $(N + 1)^{-1}$. This is the basis of *superinsulation*

11.7 (a) Show that at a given temperature T, the maximum of the Planck curve [Equation (11.11)] for $u_s(\nu)$ occurs when $\nu = \nu_m \approx \frac{3k_B T}{h}$. Estimate the percentage error in this expression for ν_m

(b) Show that the spectral density of black body radiation, expressed as a function of wavelength $\lambda = \frac{c}{\nu}$ rather than frequency (see Footnote 13), is:

$$u_s(\lambda, T) = 8\pi hc \frac{\lambda^{-5}}{e^{\beta hc/\lambda} - 1}. \qquad (11.20)$$

Show that Equation (11.20) has its maximum when $\lambda = \lambda_m \approx \frac{hc}{5k_B T}$

(c) Consider two spectrometers (a spectrometer is a device to measure the intensity of radiation as a function of wavelength). One spectrometer measures the radiation in the range λ to $\lambda + \Delta\lambda$, $\Delta\lambda$ being kept constant as λ varies, while the other measures the radiation in the range ν to $\nu + \Delta\nu$, $\Delta\nu$ being kept constant. Both are used to measure radiation which, like thermal radiation, is spread out over a range of wavelengths. Explain in physical terms why the outputs of the two spectrometers peak at different wavelengths

11.8   Show that in a ferromagnet at low temperatures, where spin waves are the only magnetic excitation (see Example 11.2) the magnetic heat capacity C_m (that is, the contribution of spin waves to the heat capacity) varies as $T^{3/2}$

11.9   (a) As any child with a magnifying glass knows, a much higher temperature can be reached with focused than with unfocused sunlight. Explain how this is consistent with the brightness theorem

       (b) Radiation from a filament at 3000 K is focused on to one side of a small thin disk suspended in a vacuum. The focusing lens is $f/2$ (this means that its diameter is one half its focal length), and the image of the filament completely covers the disk. The surroundings are at 300 K. What temperature does the disk reach? Assume that the imaging is perfect and lossless, and that the disk and filament are black at all wavelengths

       (c) Silicon is a semiconductor that is opaque in the visible but is almost transparent (at least in thin sections) to wavelengths longer than about 1.2 μm. If the disk in (b) were made of silicon, would you expect its temperature to be higher or lower than that calculated in (b)? Estimate the temperature T of the disk, assuming that it is a black body for wavelengths less than 1.2 μm and is perfectly transparent for longer wavelengths. Note that $\sim$ 50% of the radiation from the filament is below 1.2 μm, and that since the disk temperature $T \ll h\nu/k_B$ at 1.2 μm ($h\nu \approx 1$ eV), the rate at which energy is radiated by the disk varies very rapidly with temperature, so that a very rough estimate of the power yields quite an accurate value for T

11.10 (a) In Chapter 2, we obtained a relation between the heat capacities of an ideal gas at constant pressure and constant volume. Why would it make no sense to do the same thing for the photon gas?

       (b) Does a photon gas have an expansion coefficient α_b and a bulk modulus B_T? Explain.

11.11 It is customary in astronomy to estimate the surface temperature of a star by comparing the intensity of the star's image observed through different colored filters. Explain qualitatively why this works

       Astronomical magnitude is a negative logarithmic scale of luminosity, in which an increase of one magnitude represents a *decrease* in luminosity by a factor of 2.5 (approximately e). Thus, a star of apparent magnitude 5 is a factor $(2.5)^5 \approx 100$ more luminous (that is, it appears to be a factor $\sim$ 100 brighter) than a star of magnitude 10. If star A has an apparent magnitude equal to star B when viewed through a yellow filter (mean wavelength 550 nm) and has one higher apparent magnitude when viewed through a blue filter (mean wavelength 440 nm), what is the surface temperature of star A, if that of star B is known to be 6000 K? Assume that both stars are black bodies, that Wien's law applies in this spectral region, and that the *bandpass* of each filter (that is, the spread of wavelength transmitted) is the same, and is small enough that the spectral density can be taken to be that at the mean wavelength[23]

11.12 Obtain Wien's displacement law [Equation (11.17)] from the Stefan-Boltzmann law [Equation (11.6)] by a dimensional argument, assuming the relation $E = h\nu$

11.13 The maximum value of the curve in Problem 11.7 can also be determined graphically, either by hand or by a suitable routine in MatLab, IDL, Mathematica, etc. Show this

11.14 Why is there no sense in calculating the spectral energy density in Equation (11.11) using a 2D or 1D density of states? (While the answer to this question may seem obvious, nonetheless this is one of those times to point out that while something may be calculated mathematically, it is not necessarily physical.)

---

[23] Because stars are not black bodies, and for other reasons, the process of determining the surface temperature by this means is much more complex than is implied here. For an elementary account, see M. Zeilik and S. A. Gregory *Introductory Astronomy & Astrophysics* (Saunders, Philadelphia, PA, 1998), pp. 227–231.

12

The Perfect Bose Gas: Bose-Einstein Condensation

From a certain temperature on, the molecules 'condense' without attractive forces; that is, they accumulate at zero velocity. The theory is pretty, but is there some truth to it?

Albert Einstein, 1924

Superfluidity is a macroscopic single quantum state.

Fritz London

12.1 Gedanken

We pause to examine something odd that we may have, in our haste, passed over as we described the two types of particles, boson and fermions. Bosons correspond to the symmetric wave function under the exchange of two identical bosons. Let's repeat that with a slight modification: the multi-particle wave function is symmetric under the exchange of *any* two bosons. Imagine a picture that was identical under the exchange of any two people in the photo. Not just two, but any pairing of people in the picture. The implication is not just that there is a correlation between any two particles but there is a correlation over every pair of particles. If all of these identical bosons were then to all inhabit the same energy state, namely, the ground state, recalling that our designation of the ground state as the zero energy state is an arbitrary choice for our mathematical benefit and that ground states do have non-zero energy—do we expect such a dynamical state to look different from any other?

12.2 Conserved Bosons

We now turn to the other type of boson: a particle, such as an atom or molecule, with integer or zero spin, whose number is conserved. As in the case of the ideal gas, we integrate the distribution function over all the states to obtain a relation between particle density n and the chemical potential ζ. The density is, from Equation (9.4),

$$n = \int_0^\infty g(E)f(E)\, dE,$$

where, from Equation (9.10),

$$f(E) = \frac{1}{e^{\beta(E-\zeta)} - 1}.$$

Hence,

$$n = \int_0^\infty \frac{g(E)}{e^{\beta(E-\zeta)} - 1} \, dE, \tag{12.1}$$

or, in terms of the activity α,

$$n = \int_0^\infty \frac{g(E)}{\alpha^{-1} e^{\beta E} - 1} \, dE. \tag{12.2}$$

Note that the integral is finite only if $\alpha^{-1} \geq 1$, so that ζ must be negative or zero for a boson gas.

For three-dimensional non-relativistic particles with spin S, the density of states per unit volume is given by Equation (7.19):

$$g(E) = \frac{2S+1}{4\pi^2} \left(\frac{2m}{\hbar^2} \right)^{3/2} E^{1/2}.$$

Substituting $g(E)$ from Equation (7.19) in Equation (12.1), and putting $x = \beta E$, we obtain:

$$n = \frac{2S+1}{4\pi^2} \left(\frac{2mk_B T}{\hbar^2} \right)^{3/2} \int_0^\infty \frac{x^{1/2}}{e^{(x-\beta\zeta)} - 1} \, dx. \tag{12.3}$$

Equation (12.3) differs from the corresponding ideal gas result because -1 can no longer be neglected in the denominator, so that $e^{\beta\zeta}$ (that is, α) can no longer be taken outside the integral.

Since $\zeta \leq 0$, the maximum possible value of the integral in Equation (12.3) is obtained when $\zeta = 0$ and is (see Problem 12.1):

$$\int_0^\infty \frac{x^{1/2}}{e^x - 1} \, dx = 1.306\pi^{1/2}. \tag{12.4}$$

Thus, according to Equation (12.3), there is at any given temperature a maximum possible value of the density, called the *critical density* n_{cr}. Substitution of Equation (12.4) for the integral in Equation (12.3) gives, after a little algebra,

$$n_{cr} = 1.306\pi^{1/2} \frac{2S+1}{4\pi^2} \left(\frac{2mk_B T}{\hbar^2} \right)^{3/2} = 2.612 n_q, \tag{12.5}$$

where n_q is the quantum concentration, which varies as $T^{3/2}$ [Equation (7.31)].

12.3 Bose-Einstein Condensation

What happens if we go on adding particles until the actual density n exceeds n_{cr}, or, what comes to the same thing, reducing the temperature at a given density n until n_{cr} drops below n? The extra particles have to go somewhere, but there seems to be nowhere for them to go.

The reason why our calculation has no place for these extra particles is that we have used the continuum approximation, which is implicit in Equation (12.1), from which Equation (12.3) derives. Let us reexamine the derivation of Equation (12.1). For simplicity, we assume that the gas is in a cubic container of side L. Even though the volume L^3 of the container may be very large, it is still finite. At very low energies (that is, when $E \sim \frac{\hbar^2}{mL^2}$), we have to take into account the discrete nature of the quantum states. In particular, it is incorrect to say, as implied by Equation (7.19) for the density of states, that there are *no* states at

$E = 0$. There is in fact *one* state, the ground state, whose wave function (in a cubic container) is given by Equation (7.15) with $\kappa_x = \kappa_y = \kappa_z = \pi/L$. We define the energy of this state as our zero of energy.

Now let us examine the mean occupancy $f(0)$ of the ground state when ζ approaches zero from below. Since T is non-zero, we can take $|\zeta| \ll k_B T$, so that:

$$f(0) = \frac{1}{e^{-\beta\zeta} - 1} \approx \frac{1}{-\beta\zeta} = \frac{k_B T}{|\zeta|} \gg 1. \tag{12.6}$$

Thus as $|\zeta| \to 0, f(0) \to \infty$; this means that the ground state can accommodate *any number* of particles. In particular, it can take all those that are left over when $n > n_{cr}$. In general, once the condition $n > n_{cr}$ is satisfied, the occupancy of the ground state ceases to be a number of order 1. Instead it is given by $f(0) = (n - n_{cr})V$, where V is the volume of the system. This value of $f(0)$ we call N_c, the number of "condensed" particles. Note that N_c is of the same order as N, the number of particles in the system, and the occupancy of the ground state is macroscopic when $n > n_{cr}$.

What about the occupancy of low-lying excited states? If we substitute $f(0) = N_c$ in Equation (12.6), we obtain $\zeta \sim -\frac{k_B T}{N_c}$. Because N_c is a macroscopic number, $|\zeta|$ is very small indeed, much less than the separation E_1 between the ground state and the first excited state (see Problem 12.2). Let $f(E_1)$ be the occupancy of the first excited state. E_1 is also very small, and we are considering non-zero temperature, so that $\beta(E_1 - \zeta) \ll 1$. Hence, the mean occupancy of this state is:

$$f(E_1) = \frac{1}{e^{\beta(E_1 - \zeta)} - 1} \approx \frac{1}{\beta E_1 + N_c^{-1}} < \frac{k_B T}{E_1}, \tag{12.7}$$

which is the occupancy that one would obtain if one used the continuum approximation with $\zeta = 0$. Thus, the occupancy of the low lying excited levels remains microscopic and is small relative to that in the ground state: *all* the excess particles are in the ground state. This phenomenon is called *Bose-Einstein condensation*.

If a perfect boson gas of density n is cooled, at a certain temperature T_{BE} particles begin to condense into the ground state. T_{BE} is called the Bose-Einstein condensation temperature and is obtained by solving Equation (12.5) for T:

$$T_{BE} = 2\pi \frac{\hbar^2}{mk_B} \left(\frac{n}{2.612(2S + 1)} \right)^{2/3}. \tag{12.8}$$

Note that, apart from a numerical factor of order unity, T_{BE} is equal to the Fermi temperature T_F of a Fermi gas with density n, where $T_F \equiv E_F/k_B$ (see Section 13.2).

It is important to realize that although Bose-Einstein condensation is a phase transition having many similarities to familiar phase transitions like the condensation of a gas into a liquid (see Chapter 15), it is an entirely different phenomenon, not only in its cause, but in its consequences.[1] One essential difference is that in the gas to liquid transition, which is due to interparticle attraction, the liquid phase occupies a different region of space from the gas (consider, for example, the droplets of condensed water in fog or steam). Each particle in the Bose-Einstein condensate, on the other hand, has a wave function that fills the entire volume available and cannot be localized in any part of the container. The condensed bosons

---

[1] The existence of a phase transition without any overt interaction between the particles was regarded as "quite mysterious" by Einstein, until Paul Dirac ["On the theory of quantum mechanics," *Proc. Roy. Soc.* **112**, 661–677 (1926)] pointed out that quantum mechanics requires that the many-particle wave function of identical bosons be symmetric under particle exchange (see Footnote 4 of Chapter 10). This requirement produces the correlations between particles which are responsible for the condensation.

differ from the uncondensed ones in having essentially zero momentum (that is, their wave vector **k** is as small as the size of the container permits); the phase separation occurs in k-space (see Chapter 7), not in real space.

Bose-Einstein condensation was predicted theoretically by Einstein in 1924, but it has only recently been unambiguously observed experimentally in a perfect gas. The reason is that for ordinary gases at the lowest temperatures accessible by normal cryogenic techniques (~ 1 K), the critical density given by Equation (12.5) is so high that intermolecular interactions are extremely important, and the gas cannot be treated as perfect. For example, helium is the most promising candidate since it is light, monatomic, spinless, and the interaction between He atoms is exceptionally weak. If we take n to have its ideal gas value at atmospheric pressure we find $T_{BE} \approx 3$ K. However, interatomic interactions cause helium at this density to condense to a liquid at 4.2 K, so it obviously cannot be treated as a perfect gas. Nevertheless, liquid helium does undergo a transition at 2.18 K (called the λ-*point* because of the shape of the heat capacity curve at that point) to a new phase called He II. Elegant experiments by Petr Kapitza,[2] J. F. Allen,[3] and others have demonstrated that He II behaves as if it consists of two interpenetrating components, one of which, the "normal fluid," has normal viscosity and entropy, while the other, the "superfluid," has no viscosity, and when it flows, it carries no entropy.[4] The experimental fact that the transition to the superfluid state occurs only in the isotope He^4, which is a boson, and not in He^3, which is a fermion,[5] strongly suggests that the transition is related in some way to Bose-Einstein condensation, even though it occurs in a liquid, not a perfect gas. It is natural to identify the superfluid with the Bose-Einstein condensate. Being in its ground state, the condensate has no entropy. The zero viscosity comes from the fact that the atoms in the condensed state cannot be displaced from it by a small perturbation, just as an electron in the ground state of an atom cannot be so displaced.

Example 12.1: The Fountain Effect in Superfluid Helium

Two vessels, A and B, containing liquid helium at slightly different temperatures below the λ-point (the temperature at which the transition to superfluidity takes place) are connected by a narrow tube called a *superleak*. A superleak is so small that viscosity prevents any significant flow of normal fluid through it; the superfluid, on the other hand, having no viscosity, can flow freely. Liquid helium is essentially incompressible, and we can neglect any small changes in density.

1. Show that, if the normal fluid cannot flow, equilibrium is established (that is, superfluid flow ceases) when the chemical potentials in the two vessels are equal, even if the temperatures are different[6]

---

[2] Petr Leonidovich Kapitza (1894–1984), Russian physicist. After the 1917 revolution, he worked with Ernest Rutherford in Cambridge, where he established the Royal Society Mond Laboratory for low temperature physics. In 1934, on a visit to the USSR, he was detained on Stalin's orders, and was made head of the newly established Institute for Physical Problems in Moscow. It was here that he discovered the superfluidity of liquid helium, for which he won the Nobel Prize in 1978. See *Kapitza in Cambridge and Moscow: Life and Letters of a Russian Physicist*, eds. J. W. Boag, P. R. Rubinin, and D. Shoenberg (Elsevier, Amsterdam, the Netherlands, 1990).

[3] John Frank Allen (1908–2001), Canadian physicist, who succeeded Kapitza as director of the Mond Laboratory.

[4] The remarkable properties of superfluid helium are described in P. V. E. McClintock, D. J. Meredith, and J. K. Wigmore, *Matter at Low Temperatures* (John Wiley & Sons, New York, 1984). For a thorough discussion of the observable effects of Bose-Einstein condensation, and of how far one can understand the properties of superfluid helium in terms of it, see Goodstein pp. 127–139.

[5] He^3 does in fact undergo a transition to a very complex superfluid state at a much lower temperature. Robert Richardson, David Lee, and Douglas Osheroff of Cornell University won the 1996 Nobel Prize for Physics for this discovery. The transition is analogous to the superconducting transition of an electron gas in which two fermions pair up to form a boson.

[6] Note that $\zeta \neq 0$, so that superfluid helium differs from a true Bose-Einstein condensate; it is in equilibrium with low density helium vapor (which approximates to an ideal gas, so that its chemical potential is negative) and must have the same chemical potential. The chemical potential in the liquid is negative because of the attractive interaction between the atoms.

2. Hence, show that a pressure difference builds up proportional to the temperature difference, and find the constant of proportionality

Suppose that the temperature in vessel A is higher than in B by dT.

In an isolated system, U is minimized:

$$dU = T \, dS - p \, dV + \zeta \, dN = 0.$$

Suppose that a small amount dN of superfluid transfers from A to B. Then $dN_A = -dN_B = dN$. Flow of the superfluid involves no transport of entropy and the process is reversible, so that $dS = 0$; furthermore, $dV = 0$ if the density of the fluid in vessels A and B is the same.

Hence, $dU = (\zeta_A - \zeta_B) \, dN = 0$ so that $\zeta_A = \zeta_B$.

This is true even though the temperatures are different (it is not true if the normal fluid can flow, since then the flow involves an irreversible process and $dS \neq 0$)

3. A difference in temperature produces a difference in ζ. This must be compensated by a difference in pressure in order to keep ζ the same in the two vessels.

$$d\zeta = \left(\frac{\partial \zeta}{\partial T}\right)_p dT + \left(\frac{\partial \zeta}{\partial p}\right)_T dp = 0.$$

There are Maxwell relations:[7]

$$\left(\frac{\partial \zeta}{\partial p}\right)_{T,N} = \left(\frac{\partial V}{\partial N}\right)_{T,p} = \frac{V}{N}$$

and:

$$\left(\frac{\partial \zeta}{\partial T}\right)_{p,N} = -\left(\frac{\partial S}{\partial N}\right)_{T,p} = -\frac{S}{N},$$

where we have used the fact that V, S, and N are all extensive quantities and are proportional to each other at constant temperature and pressure. Hence,

$$\frac{dp}{dT} = \frac{S}{V},$$

and the pressure must rise in vessel A (the hotter one) to keep ζ constant. The only way this can happen is for superfluid to flow from A to B, raising the level and providing hydrostatic pressure. The effect can be quite dramatic (see Problem 12.3 and the cover picture of the book by McClintock *et al.* referred to in Footnote 4).

12.4 Experimental Observation of Bose-Einstein Condensation in Perfect Gases

There are two routes to obtaining Bose-Einstein condensation in a perfect gas; that is, one whose density is sufficiently low that interparticle interactions are insignificant. One is to reduce the mass of the particle, thus reducing n_{cr} to a low value at a reasonable temperature. The only elementary particle with a small enough mass is the electron. This is, of course, a fermion, but in solids it is possible to create a composite boson called the *exciton*, which consists of an electron bound to "hole" (a "missing" electron and so also a fermion: see Problem 9.1, and Chapter 14). In semiconductors, excitons have a mass comparable to

---

[7] These can be obtained from $dG = -S \, dT + V \, dp + \zeta \, dN$ by equating cross derivatives (see Appendix E).

that of the electron. Such excitons are very large particles on an atomic scale and interact even at quite low densities. However, in a high magnetic field at low temperature, the excitons are in a triplet state, in which the electron and hole spins are parallel. Excitons in the triplet state tend to repel each other, so that they keep apart and the interactions are minimized. Another problem with excitons is that they are usually short-lived and are created by light so that it is difficult to ensure that thermal equilibrium has been attained. However, these problems have been overcome and Bose-Einstein condensation of very long-lived triplet excitons in the semiconductor cuprous oxide has been observed.[8]

The other route is to lower the temperature to the point where the critical density n_{cr} is so low that the particles are too far apart to interact with each other. Recently, techniques have been developed for bringing neutral atoms to temperatures below 10^{-8} K by a combination of laser and evaporative cooling.[9] It has recently been shown that a very low density gas ($n \sim 10^{18}$ m^{-3}, less than 10^{-7} of the density of an ideal gas at room temperature and pressure) of rubidium, or of certain other neutral atoms, undergoes a transition to a new condensed phase, with the characteristics of a Bose-Einstein condensate, at a temperature ~ 20 nK, the critical temperature predicted by Equation (12.8).[10]

12.5 Envoi

Einstein's prediction that a gas of conserved bosons could condense without any overt interaction between the molecules, coupled with Dirac's explanation in terms of quantum-mechanically induced correlations (see Footnote 1) was the first application of the new quantum mechanics outside atomic physics. Unlike most predictions of quantum theory, Bose-Einstein condensation had to wait nearly sixty years before the development of new techniques for achieving extremely low temperatures made possible unambiguous proof of its existence. The remarkable properties of the Bose-Einstein condensate are now the subject of active research.

12.6 Problems

12.1 Prove Equation (12.4) using $\int_0^\infty x^{1/2} e^{-x}\, dx = \Gamma\left(\frac{3}{2}\right)$ and $\sum_1^\infty n^{-3/2} = \zeta\left(\frac{3}{2}\right)$, where $\Gamma(u)$ is the gamma function and $\zeta(u)$ is the Riemann zeta function. Both are tabulated in Abramowitz and Stegun

12.2 N atoms of a perfect Bose gas are contained in a cubic box at a temperature well below the Bose-Einstein condensation temperature T_{BE}. Show that if $N \gg 1$, $|\zeta| \ll E_1$, where E_1 is the separation between the ground and first excited states

12.3 The normal fluid of helium consists of phonon-like excitations of the superfluid and its thermal properties can be approximated by a Debye model with kmolar heat capacity $C_V \approx 80$ T^3 J K^{-1} kmole^{-1} (see Problem 7.12). The particle density of liquid helium is 2×10^{28} m^{-3} and the mass of the helium atom is $4M_H$

Estimate the pressure difference Δp in the fountain effect for a temperature difference $\Delta T = 0.2$ K if $\langle T \rangle = 2$ K. If the vessels are open to the atmosphere, to roughly what height will the helium level in the hotter vessel rise?

---

[8] J. L. Lin and J. P. Wolfe, "Bose-Einstein condensation of paraexcitons in stressed Cu$_2$O," *Phys. Rev. Lett.* **71**, 1222–1225 (1993); D. W. Snoke, "Bose-Einstein condensation of excitons," *Comments on Condensed Matter Physics* **17**, 217–238, 325–347 (1995). See also *Bose-Einstein Condensation*, eds. A. Griffin, D. W. Snoke, and S. Stringari (Cambridge University Press, New York, 1995).

[9] Steven Chu, Claude Cohen-Tannoudji, and William Phillips won the 1997 Nobel Prize for developing these techniques.

[10] M. H. Anderson, J. R. Ensher, M. R. Matthews, C. E. Wieman, and E. A. Cornell, "Observation of Bose-Einstein condensation in a dilute atomic vapor," *Science* **269**, 198–201 (1995); C. E. Wieman, "Bose-Einstein condensation in an ultra-cold gas," *Am. J. Phys.* **64**, 847–855 (1996). Rubidium has an odd number of electrons and its nucleus is a fermion, so that the neutral atom is a composite boson. This is true of any alkali metal and also of hydrogen. Carl Wieman, Eric Cornell, and Wolfgang Ketterle shared the 2001 Nobel Prize for this achievement.

12.4 (a) Show that a two-dimensional perfect Bose gas in a box (that is, the gas is confined to a certain area of a plane by a potential which is uniform within that area and infinite outside it) does not undergo Bose-Einstein condensation

(b) Now suppose that the Bose gas is still confined to a plane (so that it is two-dimensional), but within the plane it is confined by a potential of the form $\phi(r) = ar^2$, where r is measured from some point in the plane. Show that Bose-Einstein condensation is now possible and find the critical density

12.5 Find the fractional mean square fluctuation in the particle density $\frac{\langle \Delta n^2 \rangle}{\langle n \rangle^2}$ of a given volume V of a perfect degenerate boson gas (that is, one in which $|\zeta| \ll k_B T$), and show that the fluctuation diverges when the condensation density is approached

12.6 Strictly speaking for statistical mechanics to adhere to the axioms from which we started, the number of particles, N, must be much larger than 1 and the number of available states, M, must be much larger than N. However, this is a constraint that we can, and have, pushed quite a bit. Consider a simple systems of only two levels, duplicated by the presence of N particles. Given the non-interacting nature of the bosons, we could consider this synonymous to N independent systems of two levels

(a) Write the partition function for a single particle in the two-level system

(b) Now consider the N = 2 case, writing out an exact expression for Z

(c) Compare this to situation where we simply raise the answer in (a) to the power of 2, considering the non-interacting bosons to simply be two independent versions of the answer found in part (a)

(d) Consider the observation in (c) as N grows large and comment on the constraint for M and N

13

The Perfect Fermi Gas

We shall now work out...the equation of state of the (perfect) gas on the assumption that the solution with antisymmetrical eigenfunctions is the correct one.

Paul Dirac, 1926

Ignorance is never better than knowledge.

Enrico Fermi

13.1 Gedanken

In the last chapter we considered the implication of all of the identical particles in a system being symmetric under the exchange of any pair of the particles. What made that thought experiment interesting was that all of the particles could simultaneously occupy a single quantum state in the system. Here, as we consider the particles that are antisymmetric, we are constrained against considering the single state situation by the Pauli Exclusion Principle. The occupation by any fermion is either 0 or 1, meaning conversely to the boson wavefunction that can overlap in a single quantum state, we're forced for fermions to consider a minimization wherein they must occupy N_A states per mole of the system. No matter how the system is excited (or not) stressed (or not) heated (or not), we're always going to have to consider N_A states to describe the entire system, even at $T = 0$.

We now turn to the perfect gas of fermions, which obeys the Pauli principle and whose distribution function is given by Equation (9.3):

$$f(E, T) = \frac{1}{e^{\beta(E-\zeta)} + 1},$$

where the chemical potential ζ is a function of the particle density. We are interested in the *degenerate* gas; that is, one whose density is so high that $n > n_q$, so that 1 cannot be neglected in the denominator of Equation (9.3). We have seen in Chapter 9 that for such a gas ζ is positive, and for most of this chapter we confine our attention to the limit in which ζ is close to its $T = 0$ value, the Fermi energy E_F.

The most important degenerate Fermi gas is the electron gas in metals and in white dwarf stars. Another case is the neutron star, whose density is so high that the neutron gas is degenerate (see Problem 13.8e). Here, we concentrate on the electron gas.

13.2 Electrons in Metals

We start from the Drude[1] model of a metal. This model, which preceded quantum theory, accounted for the high electrical and thermal conductivity of a metal by assuming that when the constituent atoms are brought together, their outer electrons break away and can move freely through the solid. Thus, a metal contains a large number of free electrons, of order one or more per atom. They are prevented from escaping from the metal by the net Coulomb attraction of the positive ions (charged atoms) that they leave behind; the energy required for an electron to escape is typically a few eV and is called the *work function*. The model assumes that these electrons form a perfect Fermi gas confined within impenetrable walls.

We are immediately faced with a problem: What right have we to treat this dense gas as perfect? There are two objections: Coulomb interaction between the electrons is extremely strong, and in a solid, the electrons move in the strong electric fields of the positive ions. The full answer to the first objection lies in Landau's[2] Fermi liquid theory, an advanced topic in the many-body theory of solids, which is beyond the scope of this book.[3] Briefly, the high conductivity of the electron gas screens the Coulomb interaction so that it has a very short range, making the electron-electron interaction effectively weak. Ziman[4] has an apposite quote from *Alice through the Looking Glass* at the head of his chapter on this subject:

> Oh I was thinking of a plan
> To dye my whiskers green,
> But always use so large a fan
> That they would not be seen.

The answer to the second objection is that while the field of the positive ions in a simple metal alters the density of states and the effective mass (a concept which will be explained in Chapter 14) of the electrons, it does not otherwise affect the validity of the perfect gas approximation in a metal.[5] We shall find that we can understand a great deal about electrons in metals without going into these deep problems of solid state physics, and that we can use the perfect gas model to calculate with reasonable accuracy the contribution of the electrons to the heat capacity and to the magnetic susceptibility of the metal, as well as other properties. In Chapter 14, we shall see that the perfect Fermi gas provides a starting point for understanding semiconductors.

The density of states per unit volume for a three-dimensional free electron gas is, from Equation (7.20),

$$g(E) = \frac{1}{2\pi^2}\left(\frac{2m}{\hbar^2}\right)^{3/2} E^{1/2},$$

where m is the mass of the electron and E its energy.

The number of electrons per unit volume (that is, the particle density), is, from Equation (9.5),

$$n = \int_0^\infty g(E)f(E)\,dE.$$

---

[1] Paul Drude (1863–1906), German physicist. His model, developed immediately after the discovery of the electron in 1897, is the ancestor of the modern theory of metals.

[2] Lev Davidovich Landau (1908–1968), Russian theoretical physicist, winner of the 1962 Nobel Prize.

[3] See, for example, N. W. Ashcroft and N. D. Mermin, *Solid State Physics* (Saunders, College, Philadelphia, PA, 1976), Chapter 17.

[4] J. Ziman, *Electrons and Phonons* (Oxford University Press, Oxford, UK, 1960).

[5] This is shown in any textbook of solid state physics.

If $T = 0$, $f(E) = 0$ for $E > \zeta$, and $f(E) = 1$ for $E < \zeta$ (see Figure 9.1), so that, as in Equation (9.7),

$$n = \int_0^{E_F} g(E)\, dE,$$

where we write E_F for the chemical potential of the Fermi gas at absolute zero. E_F is called the Fermi energy and is a function only of the mass m and the particle density n. We now use Equation (9.7) to find E_F from n. Substituting Equation (7.20) in Equation (9.7), we obtain:

$$n = \frac{1}{2\pi^2} \left(\frac{2m}{\hbar^2} \right)^{3/2} \int_0^{E_F} E^{1/2}\, dE \tag{13.1}$$

$$= \frac{1}{3\pi^2} \left(\frac{2mE_F}{\hbar^2} \right)^{3/2}, \tag{13.2}$$

or:

$$\boxed{E_F = \frac{\hbar^2}{2m} \left(3\pi^2 n \right)^{2/3}.} \tag{13.3}$$

Putting in numbers for a typical metal such as copper, which has one free electron per atom, $n \sim 8 \times 10^{28}$ m^{-3}, $m \sim 10^{-30}$ kg, we find $E_F \sim 10^{-18}$ J ~ 6 eV. It is sometimes convenient to express E_F in terms of the Fermi temperature $T_F \equiv E_F/k_B$. A Fermi gas is degenerate when $T \ll T_F$. In copper, $T_F \approx 6 \times 10^4$ K, a typical number for most metals and well above the boiling point, so that the electrons in a solid or liquid metal are always degenerate, in the sense that $\zeta \approx E_F$ at all accessible temperatures.

Most of the electrons in a metal are moving at a very high velocity, of the order of the Fermi velocity v_F, where $v_F = (2E_F/m)^{1/2}$. The Fermi velocity is typically $\sim 10^6$ m/s, which is of the same order as the orbital velocities of the outer electrons in an atom. This is ~ 10 times the mean thermal velocity that an ideal (that is, non-degenerate) electron gas would have at room temperature and ~ 300 times the sound velocity in a typical metal.

What happens as we raise T from absolute zero, but keep $k_B T \ll E_F$ so that $\zeta \approx E_F$? We see from Equation (9.3) and Figure 9.2 that if $E - \zeta$ is greater than a few $k_B T$, $f(E) = 0$, while if $\zeta - E$ is greater than a few $k_B T$, $f(E) = 1$. Hence, all the variation in $f(E)$ occurs within a few $k_B T$ of E_F, and over this range of energy, the density of states $g(E)$ in most metals does not vary appreciably from $g(E_F)$, its value at the Fermi energy. The density of states at the Fermi energy is one of the most important properties of a metal, since it determines the number of states available to the only electrons which can change their state without the provision of a large amount of energy. Electrons with energies well below E_F cannot change their state since all states close in energy are already occupied. The effect of raising the temperature is to transfer a small number of electrons from just below E_F to just above it, as illustrated in Figure 13.1. The empty states below E_F are called holes.

So long as $k_B T \ll E_F$, the distribution function is symmetric about E_F, in the sense that:

$$1 - f(-\epsilon) = f(\epsilon), \tag{13.4}$$

where $\epsilon \equiv E - E_F$ (see Figure 13.2 and Problem 9.1). It follows that if the density of states is constant over the range of energy where $f(E, T)$ differs significantly from $f(E, 0)$, the shaded area above E_F in Figure 13.1 is equal to the unshaded area below it. The integral in Equation (13.1) is thus unchanged, so that ζ retains its $T = 0$ value, which is E_F. This cannot hold at all temperatures, since we know that if $k_B T \gg E_F$, the gas is ideal and $\zeta < 0$. The transition from the degenerate to the ideal regime occurs when $k_B T \sim E_F$ and is discussed at the end of this chapter. Here, we consider only the degenerate case ($k_B T \ll E_F$), and take $\zeta = E_F$.

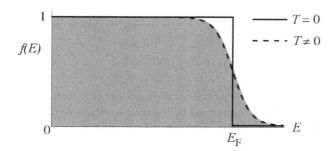

FIGURE 13.1   Fermi-Dirac distribution at $T = 0$ and at $0 < T \ll T_F$. Filled states at $T \neq 0$ are shaded.

Example 13.1: Fermi Energy of an Extreme Relativistic Fermi Gas

So far we have assumed that the fermions are non-relativistic, so that their dispersion relation is $E = \frac{\hbar^2}{2m}k^2$. If the density is high enough, some of the fermions become relativistic, and this relation ceases to hold. We now consider the opposite extreme: a Fermi gas so dense that most of the fermions are moving at a velocity close to c, the velocity of light. Find the Fermi energy of an electron gas with the extreme relativistic dispersion relation, $E = \hbar c k$. Note that the rest mass of the particle does not enter this relation.

From Equation (7.17), the number of electron states per unit volume of real space within the sphere in k-space of radius k is $G(k) = \frac{k^3}{3\pi^2}$, where the factor of 2 for spin degeneracy has been included. At 0 K, the states are all occupied up to the Fermi wave vector k_F, so that $n = G(k_F)$ and the Fermi energy is:

$$E_F = \hbar c k_F = \hbar c (3\pi^2 n)^{1/3}. \tag{13.5}$$

13.3 Heat Capacity of a Degenerate Electron Gas

We now calculate the energy per unit volume required to change the $T = 0$ distribution, shown by the full line in Figure 13.1, to the distribution for $0 < T \ll T_F$, shown by the shading. This is the energy required to raise electrons from the unshaded region just below the Fermi energy to the shaded region just above it. This energy makes a temperature dependent contribution to the internal energy U of the electron gas, and hence contributes to the heat capacity of the metal. As in Equation (13.4), we measure energy from E_F, writing $\epsilon \equiv E - E_F$. We assume that since $k_B T \ll E_F$, the density of states $g(E)$ can be taken as equal to $g(E_F)$ over the relevant range of ϵ. Consider the group of electrons between $-\epsilon$ and $-(\epsilon + d\epsilon)$, which are raised to an energy between ϵ and $(\epsilon + d\epsilon)$ when the temperature is raised from zero to T (see Figure 13.2). There are $g(E_F)f(\epsilon)\,d\epsilon$ electrons in this group, and each electron has increased its energy by 2ϵ. Hence, the internal energy per unit volume is increased by:

$$U - U_0 = 2g(E_F) \int \epsilon f(\epsilon)\, d\epsilon, \tag{13.6}$$

where U_0 is the energy at 0 K, which will be calculated in Section 13.4.

The lower limit of the integral over ϵ is zero; the upper limit can be taken as infinity without error, since $f(\epsilon)$ drops rapidly to zero for $\epsilon > k_B T$. Hence,

$$U - U_0 = 2g(E_F) \int_0^\infty \frac{\epsilon}{e^{\beta \epsilon} + 1}\, d\epsilon.$$

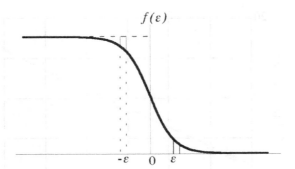

FIGURE 13.2 Fermi-Dirac distribution for $0 < T \ll T_F$, as a function of $E - E_F$. As T is raised from zero, electrons at $E = E_F - \epsilon$ are raised to $E = E_F + \epsilon$.

Substituting $x = \beta\epsilon$ and remembering that $\beta \equiv \frac{1}{k_B T}$, we find:

$$U - U_0 = 2\left(k_B T\right)^2 g\left(E_F\right) \int_0^\infty \frac{x}{e^x + 1}\, dx.$$

The integral is $\pi^2/12$ (see Problem 13.3c), so that the electronic contribution to the heat capacity *per unit volume* is:

$$C_e = \frac{dU}{dT} = \frac{\pi^2}{3} k_B^2 g(E_F) T. \tag{13.7}$$

Equation (13.7) shows that the electronic heat capacity varies linearly with temperature and is proportional to the density of states at the Fermi energy. The coefficient of T is often denoted γ (not to be confused with the ratio of the heat capacities of an ideal gas). Note that no assumption about $g(E)$ has been made here, other than that it does not vary appreciably over the range $\pm k_B T$, so that Equation (13.7) is valid for any degenerate electron gas, not necessarily free.

The heat capacity of the degenerate Fermi gas is much less than that of an ideal gas, as can be seen qualitatively as follows. From Equation (10.16), the heat capacity per unit volume of an ideal gas of density n is $\sim nk_B$. In unit volume of the metal there are n electrons, whose energies are spread out over a range $\sim E_F$. Each electron occupies one state, so that the density of states per unit volume at the Fermi energy $g(E_F) \sim n/E_F$. Hence, Equation (13.7) gives $C_e \sim nk_B^2 T/E_F$, which is less than that of the ideal gas by a factor of order $k_B T/E_F$. The small heat capacity is a direct consequence of the Pauli principle, which ensures that most of the electrons cannot change their energy when the temperature changes, since the states immediately above and below them are filled. Only those electrons within a few $k_B T$ of the Fermi energy have an empty state nearby into which they can move, and the fraction of the electrons in this energy range is of order $k_B T/E_F$. The restriction to electrons near E_F applies to all the properties of the metal; for example, the paramagnetic susceptibility (see Example 13.2) and the electrical and thermal conductivity (see Chapter 16). The derivation of Equation (13.7) by Sommerfeld[6] in 1928 resolved a serious problem with the classical Drude model of a metal, which assumed an ideal gas of electrons and was unable to account for their small contribution to the heat capacity.

The electronic heat capacity given by Equation (13.7) is generally much less than the lattice heat capacity due to the vibrations of the atoms, which is $3nk_B$ for most solids at ordinary temperatures, where n is now the number of atoms per unit volume [compare Equation (6.17)]. Only at very low temperatures, where the lattice heat capacity varies as T^3 (see Equation 7.44), can the electronic contribution be detected against the lattice background. Since the two contributions to the total heat capacity $C(T)$ vary differently

---

[6] Arnold Sommerfeld (1868–1951), German theoretical physicist.

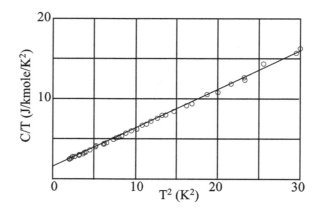

FIGURE 13.3 Heat capacity of sodium at low temperature, plotted as C/T against T^2 in order to separate the lattice and electronic contributions (Adapted from Lois, M. R., *Proc. Phys. Soc.*, **70B**, 744, 1957.)

with T, they can be separated out. A simple way to do this is to plot $C(T)/T$ against T^2; the slope gives the coefficient of T^3 in the lattice contribution, while the intercept is $\gamma = \frac{\pi^2}{3} k_B^2 g(E_F)$, the coefficient of T in C_e [see Equation (13.7)]. Such a plot is shown in Figure 13.3. In the early days of solid state physics (up to around 1954), $g(E_F)$ determined in this way was virtually the only experimental information available concerning the fundamental properties of electrons in metals.[7]

Example 13.2: Pauli Paramagnetism

An important application of the theory developed in this chapter is to the magnetic suscepti-bility χ due to the spins in a degenerate Fermi gas. We shall see that so long as $k_B T \ll E_F$, χ is independent of temperature and is much smaller than the classical value predicted by Curie's law [Equation (6.39)]. The calculation is an illuminating application of the result obtained in Chapter 8, that the chemical potentials of two systems in diffusive equilibrium are equal.

The effect of a magnetic field B on an electron gas is to displace all spin-up ↑ states downwards in energy by μB and all spin-down ↓ states upwards by μB, where μ is the spin magnetic moment (see Appendix B). For any reasonable field, $\mu B \ll E_F$. In equilibrium, the spin-up and spin-down populations must have the same chemical potential and to achieve this some spins must flip; that is, they must reverse their direction relative to the field. We use this fact to calculate the net excess of spin-up over spin-down electrons in a metal when a field B is applied.

Let $E_{F\uparrow}$ and $E_{F\downarrow}$ be the Fermi energies that up and down spins would have in the presence of the magnetic field if no spins flipped (see Figure 13.4). Since all ↑ spin states are shifted down in energy by μB and all ↓ spin states shifted up by μB,

$$E_{F\downarrow} - E_{F\uparrow} = 2\mu B. \tag{13.8}$$

When the up and down spins are in equilibrium, ζ must be the same for both. To achieve this, some spins must flip, as indicated by the shaded regions in Figure 13.4. The number of electrons with a given spin and with energies between E and $E + \mu B$ is $\frac{1}{2}g(E)\mu B$. The factor $\frac{1}{2}$ is neces-sary because $g(E)$ includes both spin states. Since $\mu B \ll E_F$, $g(E)$ is almost constant over these

---

[7] The paramagnetic (Pauli) susceptibility gives the same information in principle (see Example 13.2), without requiring low temperature measurements, but in practice its accuracy is vitiated by the difficulty of correcting for the Landau dia-magnetism of the electron gas [see R. E. Peierls, *Quantum Theory of Solids* (Oxford University Press, London, 1955), p. 144].

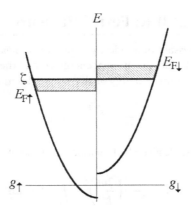

FIGURE 13.4   Density of states $g_\uparrow(E)$ and $g_\downarrow(E)$ for spin-up and spin-down electrons, respectively, plotted in opposite horizontal directions. $E_{F\uparrow}$ and $E_{F\downarrow}$ are the Fermi energies that would obtain if the spins did not flip, and ζ is the actual chemical potential. The $\downarrow$ spins in the shaded region on the right flip to fill the shaded $\uparrow$ spin region on the left.

narrow regions. The number of spins that flip is $\frac{1}{2}g(E_F)\mu B$, and the net number of up spins is $N_\uparrow - N_\downarrow = g(E_F)\mu B$.

The susceptibility per unit volume is:

$$\chi \equiv \frac{M}{B} = \mu \frac{N_\uparrow - N_\downarrow}{B} = \mu^2 g(E_F). \tag{13.9}$$

This is called the Pauli[8] paramagnetic susceptibility, and it does not depend on T so long as our assumption that $\zeta \approx E_F$ in zero B is valid; that is, so long as $k_B T \ll E_F$. Like Equations (13.7), Equation (13.13) is valid for any degenerate electron gas, not necessarily free.

For free non-relativistic electrons, $g(E) \propto E^{1/2}$, so that:

$$n = \int_0^{E_F} g(E)\,dE = \frac{2}{3}E_F g(E_F). \tag{13.10}$$

Hence, for a degenerate free electron gas,

$$\chi = \frac{3}{2}\frac{\mu^2 n}{E_F}. \tag{13.11}$$

Thus, the Pauli susceptibility of a free electron gas is less than the classical Curie susceptibility per unit volume, which is $\frac{\mu^2 n}{k_B T}$ [Equation (6.39)], by a factor $\frac{3k_B T}{2E_F} \ll 1$. This factor is roughly the same as the reduction factor in the heat capacity and has the same cause: only electrons within $\sim k_B T$ of the Fermi energy can respond to the field by flipping spin. For all other electrons, the spin-flipped state is already occupied.

The derivation of Equation (13.9) by Pauli in 1927 was one of the earliest applications of Fermi-Dirac statistics. It resolved a serious problem which arose in the Drude model of a metal when the electron spin was discovered by Goudsmit and Uhlenbeck, since the model assumed a classical electron gas whose spin magnetic moment should produce a large and strongly temperature dependent Curie susceptibility. Experimentally, the susceptibility is found to be small and independent of temperature, in agreement with Equation (13.9), but quantitative comparison requires that one take the diamagnetism of the electron gas into account. Landau found that for free electrons, this is $-1/3$ of the Pauli paramagnetism (see the reference in Footnote 7).

---

[8] Wolfgang Pauli (1900–1958), Austrian-Swiss theoretical physicist, who won the Nobel Prize in 1945 for his discovery of the exclusion principle.

13.4 Internal Energy at 0 K: Fermi Pressure

While the internal energy and pressure of an ideal gas goes to zero as $T \to 0$, this is not the case for a degenerate Fermi gas. We now calculate the internal energy U_0 and the pressure at 0 K.

We obtain $U_0(V)$ by summing over the energy of all the particles in volume V:

$$U_0 = V \int_0^{E_F} E g(E) \, dE. \tag{13.12}$$

We assume that the density of states is given by Equation (7.20), so that:

$$U_0 = \frac{V}{2\pi^2} \left(\frac{2m}{\hbar^2} \right)^{3/2} \int_0^{E_F} E^{3/2} \, dE$$

$$= \frac{V}{5\pi^2} \left(\frac{2m}{\hbar^2} \right)^{3/2} E_F^{5/2}. \tag{13.13}$$

Substituting for E_F from Equation (13.3) and putting the number of particles $N = nV$, we obtain:

$$U_0 = \frac{3}{5} N E_F, \tag{13.14a}$$

which can also be written:

$$U_0 = \frac{1}{5\pi^2} \frac{\hbar^2}{2m} (3\pi^2 N)^{5/3} V^{-2/3}. \tag{13.14b}$$

At $T = 0$ there is no distinction between the free energy F and the internal energy U_0, so that we can write $p = -\left(\frac{\partial U_0}{\partial V} \right)_N$. Differentiation of Equation (13.14b) with respect to V gives, after some algebra, the *Fermi pressure*:

$$p = \frac{2}{5} n E_F. \tag{13.15}$$

Note that even though the internal energy density and the pressure differ greatly from their ideal gas values, their ratio $\frac{pV}{U_0} = \frac{2}{3}$, just as in an ideal gas. This ratio is a direct consequence of the quadratic relation between energy and momentum [Equation (B.1)] and of the assumption that interparticle interaction can be neglected. In Appendix H, the ratio is derived by kinetic theory, making no assumption about the distribution function [see Equation (H.7)].

In a degenerate Fermi gas $E_F \gg k_B T$, so that the Fermi pressure given by Equation (13.15) is much greater than that of an ideal gas of the same density, which is $n k_B T$ (Equation (10.11)). Furthermore, in a metal, the electron density is much larger than in a typical ideal gas, and putting in numbers for (say) copper we find that $p \sim 3 \times 10^{10}$ Pa ≈ 300 kbar, compared with atmospheric pressure (about 1 bar). In a metal, this enormous pressure is counteracted by the Coulomb attraction of the electrons to the positive ions. In a white dwarf or neutron star, the Fermi pressure is even larger, and as long as the star is not too massive, this pressure prevents it from collapsing under gravity and becoming a black hole.

Example 13.3: White Dwarfs and Neutron Stars: The Chandrasekhar Limit

When a star has used up most of the hydrogen that fuels its thermonuclear reactions, it collapses under its own gravitation. As the density increases, the electrons become degenerate, and their Fermi pressure prevents further collapse if the star is not too massive. The resulting star is called a *white dwarf.*

The star can be modeled as a spherical body of radius R, mass M, and electron density n. In a real star, the density increases towards the center, but here, for simplicity, we assume that n is uniform. Most stars are also rotating, but we ignore this complication. Since all the hydrogen has been used up, the elements that predominate in a white dwarf have nuclear masses of approximately $2ZM_H$, where Z is the atomic number of the element and M_H is the mass of the proton. Since the matter in the star is completely ionized, Z is the number of electrons in the gas per nucleus. Hence, $M \approx 2NM_H$, where $N = \frac{4\pi}{3}R^3 n$ is the total number of electrons.

1. Find the gravitational potential energy U_g of a non-rotating spherical star by a dimensional argument.

Newton's equation for the gravitational potential ϕ of two objects of mass m_1 and m_2 separated by a distance R is $\phi(R) = -\frac{Gm_1 m_2}{R}$, where G is the gravitational constant. Like U_g, ϕ is an energy. U_g can depend only on M, G, R, and a dimensionless function of the mass distribution in the star and is negative because the force is attractive. Hence, the only dimensionally consistent equation for U_g is:

$$U_g = -C_g \frac{GM^2}{R},$$

where C_g is a positive numerical constant of order 1, which depends on the radial dependence of the density. It is shown in many elementary mechanics texts that $C_g = 3/5$ if the density is uniform

2. In the degenerate limit the electron gas, if non-relativistic, has internal energy U_0 given by Equation (13.14). Find an expression for the equilibrium radius R_m as a function of M by minimizing the total energy $U \approx U_g + U_0$ with respect to R, and show that the equilibrium is stable against small changes in R.

The number of electrons N is proportional to the number of atoms and hence to the mass M, while from Equation (13.3), the Fermi energy of a non-relativistic electron gas is proportional to $n^{2/3}$, where $n = \frac{N}{V} \propto \frac{M}{V}$. If the density n is uniform, Equation (13.14a) gives $U_0 \propto Nn^{2/3} \propto M^{5/3}V^{-2/3}$. Hence, since $V \propto R^3$:

$$U_0 = C_1 \frac{M^{5/3}}{R^2},$$

where C_1 is a constant.

The total energy $U = U_g + U_0$, so that:

$$\frac{dU}{dR} = C_g \frac{GM^2}{R^2} - 2C_1 \frac{M^{5/3}}{R^3}.$$

Putting $\frac{dU}{dR} = 0$ and solving for R gives the equilibrium radius:

$$R_m = \frac{2C_1}{C_g G} M^{-1/3}.$$

When $R = R_m$, $\frac{d^2 U}{dR^2} = \frac{C_g GM^2}{R_m^3} > 0$, so that the equilibrium is stable.

Thus, as the mass increases, the star gets smaller, but if our assumptions are correct, a star of any given mass can reach equilibrium, where the Fermi pressure balances the gravitational attraction. For a star with the same mass as the Sun, R_m is roughly equal to the radius of the Earth (see Problem 13.7a)

3. What happens if the electrons are relativistic?

If the density is high enough, E_F exceeds the rest energy of the electron and the electrons reach relativistic velocities (see Example 13.1 and Problem 13.7c). In the relativistic limit, the

Fermi energy is given by Equation (13.5) so that $E_F \propto n^{1/3}$. U_0 is still proportional to NE_F (see Problem 13.8), and $N \propto M$, so that we have $U_0 \propto Nn^{1/3} \propto M^{4/3}V^{-1/3}$; that is:

$$U_0 = C_2 \frac{M^{4/3}}{R},$$

where C_2 is a constant. Hence,

$$\frac{dU}{dR} = \frac{M^{4/3}}{R^2}\left(C_g M^{2/3}G - C_2\right).$$

There is now no stable minimum in U. If $M < M_{Ch}$, where $M_{Ch} \equiv \left(\frac{C_2}{C_g G}\right)^{3/2}$, $\frac{dU}{dR} < 0$ and the radius will grow, reducing the density (and thus the Fermi velocity) until the deviation from the relativistic limit for U_0 is sufficient to introduce a stabilizing term proportional to R^{-2}, as in the non-relativistic case. On the other hand, if $M > M_{Ch}$, $\frac{dU}{dR} > 0$ and the star is unstable against collapse, since the Fermi pressure can no longer overcome gravity. As the radius gets smaller, the electrons become ever more relativistic, $\frac{dU}{dR}$ increases, and there is no stable radius. A white dwarf with $M > M_{Ch}$ collapses until the density approaches the density of the nucleus; the electrons then combine with the nuclear protons by inverse β-decay to form a neutron star. M_{Ch} is called the Chandrasekhar[9] mass, and is about 1.4 solar masses (see Problem 13.9).

Since neutrons are also fermions, neutron stars can suffer the same collapse, though the problem is more complicated and the numbers are slightly different (see Problem 13.10). The limiting mass is about three solar masses, and a star more massive than that collapses to a black hole.

13.5 Fermi Gases at Non-zero Temperature: The Changeover to the Ideal Gas

Up till now, we have assumed that $k_B T \ll E_F$, so that the chemical potential $\zeta \approx E_F$, which is positive. If we use Equations (13.3) and (9.14) to relate the density n to E_F and $k_B T$ to the quantum concentration n_q, we find that, for $S = \frac{1}{2}$,

$$\frac{n}{n_q} = \frac{4}{3\pi^{1/2}}\left(\frac{E_F}{k_B T}\right)^{3/2}. \tag{13.16}$$

Thus, as pointed out in Chapter 9, the condition $k_B T \ll E_F$ is equivalent to $n \gg n_q$. On the other hand, we know that if $n \ll n_q$, the gas is ideal and its chemical potential is given by $\zeta = -k_B T \ln\left(\frac{n_q}{n}\right)$ (Equation 9.15), so that ζ is then negative. It follows that, for a given density, as T increases ζ must change sign when $n_q \sim n$.

---

[9] Subrahmanyan Chandrasekhar (1910–1995), Indian/American theoretical astrophysicist, who won the 1983 Nobel Prize for this work. When Chandrasekhar first predicted this collapse in 1935, the eminent British astronomer Sir Arthur Eddington (1882–1944) refused to believe it, declaring that "there ought to be a law to forbid such ridiculous behavior" and describing the theory as an "unholy alliance (between) relativistic classical mechanics and non-relativistic quantum mechanics." As a result, in spite of the privately expressed support of physicists such as Dirac, Pauli, and R. H. Fowler, it was many years before Chandrasekhar's theory was generally accepted by astronomers. The existence of neutron stars (pulsars) and black holes is now well established (see any modern textbook of astronomy). The exact mass-radius relation for a white dwarf, calculated allowing for the fact that the density is not uniform and that not all the electrons are relativistic, is shown in S. Chandrasekhar, *Eddington, the Most Distinguished Astrophysicist of his Time* (Cambridge, 1983), p. 51. Chandrasekhar's biographer, K. C. Wali, remarks that the title may be a conscious echo of Anthony's "Brutus is an honorable man" in *Julius Caesar*.

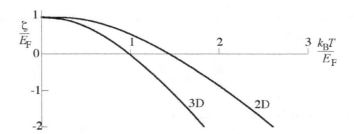

FIGURE 13.5 Temperature dependence of the chemical potential (Fermi level) of a perfect Fermi gas in two and three dimensions, showing the changeover from the degenerate gas ($\zeta \approx E_F$) at low T to the ideal gas $\zeta < 0$ at high T. The 3D curve is obtained from Equation (13.20) for $k_B T < 0.2E_F$ and from the Joyce-Dixon approximation (see Footnote 12) for $k_B T > 0.2E_F$. On the scale shown this curve is indistinguishable from the exact numerical result. The 2D curve is the analytic function obtained in Problem 13.2.

The chemical potential of a Fermi gas at any temperature can be obtained from the particle density n, by solving Equation (9.6):

$$n = \int_0^\infty \frac{g(E)}{e^{\beta(E-\zeta)} + 1} \, dE.$$

From Equation (13.3), n can be expressed in terms of E_F. For a given form of $g(E)$, it can easily be shown dimensionally that $\frac{\zeta}{E_F}$ is a universal function of $\frac{k_B T}{E_F}$, valid for any perfect Fermi gas, regardless of the particle's mass or spin (see Problem 13.1);[10] all these parameters are subsumed in the single parameter E_F. However, the form of the function depends on the dimensionality through $g(E)$. This function is shown for a free particle in two and three dimensions in Figure 13.5.

In three dimensions, $\zeta(T)$ can only be found numerically, but analytic expressions exist in the limits $k_B T \ll E_F$ and $k_B T \gg E_F$. The initial temperature dependence of ζ when $k_B T \ll E_F$ is most easily derived from the heat capacity[11] with the aid of the Maxwell relation[Equation (E.9)]

$$\left(\frac{\partial \zeta}{\partial T}\right)_N = -\left(\frac{\partial S}{\partial N}\right)_T,$$

where S is the entropy.

When $k_B T \ll E_F$, the electronic heat capacity per unit volume C_e is given by Equation (13.7), so that:

$$S = \int_0^T \frac{C_e}{T} \, dT = \frac{\pi^2}{3} k_B^2 g(E_F) T, \tag{13.17}$$

and:

$$\left(\frac{\partial S}{\partial N}\right)_T = \frac{\pi^2}{3} k_B^2 T \frac{dg(E_F)}{dn}. \tag{13.18}$$

The dimensionality enters through the factor $\frac{dg}{dn}$. In two dimensions, $g(E)$ is a constant, so that $\frac{dg(E_F)}{dn} = 0$ so long as Equation (13.7) holds; that is, so long as the integral in Equation (13.6) can be taken to infinity

---

[10] That the function is independent of spin can be seen as follows. Spin only enters through the factor $(2S+1)$ in the density of states [see Equations (7.9), (7.17), and (7.20)]. From Equation (9.5), spin only appears in the expression for ζ or E_F in the form $\frac{n}{2S+1}$. Since n does not appear explicitly in the function relating ζ/E_F to $k_B T/E_F$, it follows that S cannot either. It is, of course, included implicitly through E_F.

[11] A logically more direct, but mathematically more complex, derivation is described in Problem 13.3.

without error. In three dimensions, on the other hand, algebraic manipulation of Equations (7.20) and (13.3) gives $\frac{dg(E_F)}{dn} = \frac{1}{2E_F}$, so that:

$$\frac{d\zeta}{dT} = -\frac{\pi^2}{6} \frac{k_B^2 T}{E_F}. \tag{13.19}$$

Integrating Equation (13.19) and using $\zeta(0) = E_F$, we obtain:

$$\frac{\zeta}{E_F} = 1 - \frac{\pi^2}{12} \left(\frac{k_B T}{E_F}\right)^2, \tag{13.20}$$

so that initially ζ falls quadratically as T is raised from zero. Since Equation (13.7) for C_e was obtained on the assumption that $\zeta = E_F$, Equation (13.20) is accurate only to second order in $\frac{k_B T}{E_F}$.

When $k_B T \gg E_F$, ζ is given by the ideal gas relation (Equation 9.15). Substituting $\frac{n}{n_q}$ from Equation (13.16) gives:

$$\frac{\zeta}{E_F} = -\frac{k_B T}{E_F} \left[\ln\left(\frac{3\sqrt{\pi}}{4}\right) + \frac{3}{2}\ln\left(\frac{k_B T}{E_F}\right)\right]. \tag{13.21}$$

A series expansion called the Joyce-Dixon approximation[12] extends Equation (13.21) to intermediate temperatures. It is very accurate for $\frac{k_B T}{E_F} > 0.4$ and provides a good interpolation between Equations (13.20) and (13.21). This was used to obtain the 3D curve in Figure 13.5.

13.6 Envoi

The discovery by Sommerfeld and Pauli in the late 1920s that the new quantum theory could account for the observed low heat capacity and magnetic susceptibility of metals was the first time that quantum theory had explained data from a field outside atomic physics and chemistry. It established the theory of solids as a distinct branch of physics, which was, however, at first considered "somewhat less respectable than other branches of theoretical physics.[13]" This suspicion was directed primarily at the drastic approximations that were initially needed for progress to be made. In particular the free electron model, with its neglect of the strongly varying potential acting on electrons and of the Coulomb interactions between them, seemed unconscionable. The latter approximation was justified by Landau in 1957 (see Footnotes 2 and 3). The former was made plausible by the development in the 1950s of pseudopotential theory,[14] which showed that canceling terms in the self-consistent potential made its effect on observable properties much less than one might expect. However, accurate so-called "first principles" calculations (that is, calculations which do not require empirical data as input) of the energy-momentum relation of an electron in a real solid (its *band structure*) had to await the advent of powerful computers. Nevertheless, solid state physics (now called "condensed matter physics," since it has subsumed the theory of liquids) has turned out to be one of the most successful branches of physics, both in theoretical understanding that has been achieved and in the practical applications that it has generated.

---

[12] W. B. Joyce and R. W. Dixon, "Analytic approximations for the Fermi energy of an ideal Fermi gas," *Applied Physics Letters* **31**, 354 (1977). In this title, "Fermi energy" means "chemical potential," a confusing usage common in the semiconductor literature, and "ideal" means "perfect" in our sense.

[13] R. E. Peierls, *Quantum Theory of Solids* (Clarendon Press, Oxford, UK, 1955), preface.

[14] See W. A. Harrison, *Pseudopotentials in the Theory of Metals* (W.A. Benjamin, New York, 1966).

13.7 Problems

13.1 Use dimensional arguments to prove the following for a non-relativistic d-dimensional Fermi gas
(a) $E_F \propto \hbar^2 m^{-1} n^{2/d}$, where n is the number of particles per unit length, area or volume for $d = 1$, 2, and 3, respectively, and the constant of proportionality is dimensionless
(b) $\frac{\zeta}{E_F} = f\left(\frac{k_B T}{E_F}\right)$ where $f(x)$ is a dimensionless function which cannot be determined by dimensional arguments alone

13.2 (The two-dimensional Fermi gas; an exactly soluble model.)
(a) Find an analytic expression for the chemical potential of a two-dimensional Fermi gas as a function of temperature and of Fermi energy E_F (which is a function of density only). Assume that the areal density is independent of T. Hint: Use the fact that in two dimensions $g(E)$ is independent of E. The integral is most easily evaluated by making the substitution $x = e^{\beta(\zeta - E)}$. Confirm that your expression for ζ goes to the degenerate limit $\zeta = E_F$ as $T \to 0$ (see Problem 9.2) and to the ideal gas limit (Equation 9.15), modified for two dimensions, as $T \to \infty$
(b) What is the fractional deviation of ζ from E_F when $k_B T = 0.5 E_F$?
(c) Find an expression for the value of $\frac{k_B T}{E_F}$ at which ζ changes sign.

13.3 (a) Use the fact that $f(\epsilon) - 1/2$ is an odd function of ϵ (see Problem 9.1a) to show that Equation (9.4) can be put in the form:

$$n = \int_0^\infty [g(\epsilon) - g(-\epsilon)] f(\epsilon)\, d\epsilon + \int_0^\zeta g(E)\, dE,$$

where $\epsilon \equiv E - \zeta$
(b) Show that if $k_B T \ll E_F$,

$$\int_0^\zeta g(E)\, dE \approx n + (\zeta - E_F) g(E_F)$$

(c) Hence show that, to lowest order in $\frac{k_B T}{E_F}$, the chemical potential of a three-dimensional Fermi gas is given by Equation (13.20)
Hint: The first integral in (a) can be evaluated by approximating $g(\epsilon)$ as a linear function of ϵ for small ϵ, expanding the denominator as a series in $e^{-\beta\epsilon}$, and using the fact that $\eta(2) = \frac{\pi^2}{12}$, where $\eta(x) \equiv \sum_{n=1}^\infty (-1)^{(n+1)} n^{-x}$ [Abramowitz and Stegun, p. 808]
(d) Explain in physical terms why the range of temperature over which $\zeta \approx E_F$ is narrower in the three-dimensional than in the two-dimensional case (Problem 13.2b)

13.4 Show that the chemical potential of a three-dimensional Fermi gas is zero when $k_B T = 0.99 E_F$. You will need the result $\eta\left(\frac{3}{2}\right) = 0.765$ [$\eta(x)$ is defined in Problem 13.3c]

13.5 Calculate the contribution of the Fermi pressure to the bulk modulus $B = n\frac{dp}{dn}$ at 0 K. Evaluate B for Li ($E_F = 4.7$ eV) and compare your result with the observed value (1.2×10^{10} N m^{-2}). What does this comparison tell you about the contribution of the electrons, regarded as a perfect gas (that is, without Coulomb interactions), to the stability of the solid?

13.6 Find the enthalpy H of an electron gas at low, but non-zero temperature ($T \ll T_F$). Hence, find the free enthalpy G and show that the enthalpy term in G exceeds the entropy term. Compare your result with Equation (13.20)

13.7 (a) For a white dwarf with mass equal to the solar mass (2×10^{30} kg), estimate the equilibrium radius R_m, the corresponding density n_m, and E_F in the non-relativistic approximation. For simplicity, assume that the density is uniform, and that the atoms in the star are all fully ionized and have mass $2 Z M_H$
(b) The surface temperature of a white dwarf is in the region 10^4 to 5×10^4 K, compared to 5800 K for the Sun. The interior temperature is higher, but never exceeds 10^8 K. Is the zero

temperature approximation valid for the electron gas? Explain. Show that if the star remains gaseous,[15] the gas of (mostly) helium nuclei is non-degenerate and that the contribution of the kinetic energy of the nuclei to U is small relative to U_0

(c) Using the non-relativistic approximation, estimate $\frac{v_F}{c}$, the ratio of the Fermi velocity of the electrons in the white dwarf to the velocity of light, as a function of its mass (in units of the solar mass). Is the non-relativistic approximation valid for a white dwarf with the solar mass?

13.8   Find the internal energy U_0 and the Fermi pressure of the extreme relativistic electron gas described in Example 13.1. Express your result in the same form as Equations (13.14) and (13.15). What is the ratio $\frac{pV}{U_0}$? Why does it differ from the non-relativistic case?

13.9   Estimate the Chandrasekhar mass M_{Ch}, taking $C_g \approx 1$. Express your result in terms of the mass of the proton M_H and the Planck mass $M_P \equiv \left(\frac{\hbar c}{G}\right)^{1/2} \approx 1.3 \times 10^{19} M_H \approx 20\ \mu g$. The result should be quite close to Chandrasekhar's value given in the text, which was obtained from a more realistic model in which the density is not uniform

13.10 In a neutron star, collapse has proceeded further, the electrons combining with the protons in the helium nuclei (inverse beta decay) to form a degenerate Fermi gas of neutrons

(a) Making the same assumptions as Example 13.3b, estimate the equilibrium radius and mean density of a neutron star with the mass of the sun (note that in this case, almost all the mass is in the form of neutrons). Compare the density with that of the sun and with that of nuclear matter (the effective radius of a spherical nucleus with mass number A is about $1.2 \times 10^{-15} A^{1/3}\ m$, where A is the mass of the nucleus in atomic mass units)[16]

(b) Making the same approximations as Problem 13.9, estimate the limiting mass (analogous to the Chandrasekhar mass of a white dwarf) of a neutron star. Because our model neglects the nuclear interactions and general relativistic effects, this is not such an accurate estimate as in the case of the white dwarf.

13.11 Suppose that you want to cool a metal, starting from a temperature T_1 which is so low that the heat capacity of the metal is purely electronic and is proportional to T [Equation (13.7)]. You have an ideal reversible refrigerator that discharges heat to a heat bath at T_1. Calculate the work needed to cool an unit volume of the metal from T_1 to T_2, where $T_2 \ll T_1$, giving your result in terms of the density of states at the Fermi energy $g(E_F)$

13.12 Similar to Problem 12.6, let's examine the rigor of demanding that the number of particles in a system, N, and the number states in the system, M, be larger than 1 and M much larger than N. Consider a system comprised of four levels with energies E_1, $2E_1$, $3E_1$, and $4E_1$ that are occupied by 3 fermions

(a) Write the Helmholtz Free Energy for this system

(b) Since we are allowed to set the zero of energy once in a problem, demonstrate that by a convenient and sly choice of where the zero energy is set in this system that the answer in (a) is completely and physically equivalent to the Free Energy if one thinks of the state of the system as being due to the unoccupied level (hole)

---

[15] The center of a white dwarf is now thought to be a solid well below its Debye temperature. This does not alter the result that the nuclear contribution to U is small relative to U_0.

[16] See *American Institute of Physics Handbook*, 3rd ed., ed. D. E. Gray (McGraw-Hill, New York 1972) pp. 2–91, 8–96.

IV

Modern Thermostatistical Applications

14

Electrons and Holes in Semiconductors

The technology which has transformed practical existence is largely an application of what was discovered by... allegedly irresponsible (natural) philosophers.

Sir Cyril Hinshelwood

It has today occurred to me that an amplifier using semiconductors rather than vacuum is in principle possible.

Entry in William Shockley's laboratory notebook, December 29, 1939.

14.1 Gedanken

If we think of our picture of phonons in a solid, and for simplicity consider the contradictory picture of a 1-D solid – a chain of springs connecting N_A masses. With the exception of the masses at the two ends of the chain, which would represent the surface of the solid, our concept of a real solid is that every atom (mass) in an homogeneous solid would experience the very same forces from each of its immediate nearest neighbors (on the left and on the right, so to speak) in 1-D. Let's take advantage of that symmetry to imaging the entire potential that each mass sits in, extending out essentially forever to the left and to the right. That potential would be the same for each atom to the immediate left and right, so we'd be replicating the potential over and over and over again offset by one lattice spacing for each atom. Now imagine looking on a graph used to represent the location of one of the atoms relative to the infinitely wide potential for each of the atoms (see Figure 14.1). The information for each adjacent atom, identical to the one you are looking at, is graphed further and further above the one atom relative to how further and further away each neighbor is, coming in symmetrically from the left and from the right. This means that all of the information for all N_A identical atoms is present on the graph representing each atom. So let us reduce our work by $1/N_A$ and consider only the information in one identical box and its atom due to all the other atoms, which is just the potential for that atom of interest folded back within the unit cell of the atom using periodic boundary conditions.

14.2 Metals, Insulators, and Semiconductors

In Chapter 13, we treated the electrons in a metal as if they were free, except for the confining potential which keeps them within the metal. This is a drastic assumption, which is valid as a rough approximation in most metals, but is not valid, even qualitatively, for an insulator or semiconductor. In a real solid,

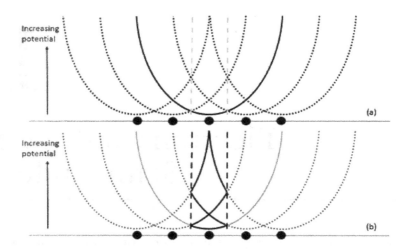

FIGURE 14.1 Shifted potentials for atoms in a 1-D solid. In (a) the middle atom's potential is represented by the dark solid line, adjacent atoms' potentials are dotted lines. The potential for the middle atom due to each adjacent atom can be visualized as the information contained between the two vertical dashed lines in light gray. By symmetry, this is also the potential for every adjacent atom due to the middle atom. In (b) this information is redrawn as the solid lines between two vertical dashed lines in black. Electron-electron repulsion in the crossing points of the solid black lines will lead to gaps in the band structure (see Figure 14.4).

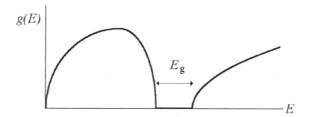

FIGURE 14.2 Schematic density of states $g(E)$ in a solid. E_g is the width of the bandgap.

the electrons move in the rapidly varying potential of the atoms, and the density of states $g(E)$ is more complicated than the free electron value given by Equation (7.20). In general, there are regions of energy, called *bandgaps* or *energy gaps* (or *gaps*), where there are no states.[1] The regions of energy where $g(E) \neq 0$ are called *energy bands*, which are separated by gaps where $g(E) = 0$. A density of states with such a gap is illustrated schematically in Figure 14.2. The width of the gap is denoted E_g.

In a metal at $T = 0$, the states are filled up to the Fermi energy E_F, and there are empty states immediately above it, as illustrated in Figure 14.3. An electron at the Fermi energy can be raised to a slightly higher energy by, for example, an electric field. For this reason, metals conduct electricity regardless of the exact form of $g(E)$, and, as seen in Chapter 13, most of the observable properties of a metal depend only on $g(E_F)$.

In an insulator, on the other hand, the Fermi energy lies in a bandgap, where $g(E) = 0$, so that all states up to the lower edge of the bandgap are filled, and those above the upper edge are empty, as in Figure 14.4. These edges are labeled by their energies E_v and E_c, respectively. The range of energy in

---

[1] See, for example, C. Kittel, *Introduction to Solid State Physics*, 7th edition (John Wiley & Sons, New York, 1996), p. 176.

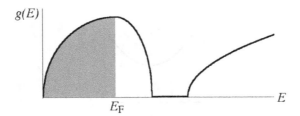

FIGURE 14.3 Schematic density of states in a metal, showing the states filled up to the Fermi energy E_F.

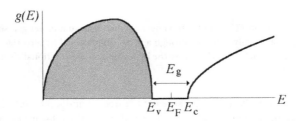

FIGURE 14.4 Schematic density of states in an insulator or semiconductor, showing the states filled up to the top of the valence band at E_v, and empty states in the conduction band whose bottom is at E_c. $E_g \equiv E_c - E_v$.

which the states are filled is called the *valence band*, and the range with the empty states, the *conduction band*.[2] No conduction is possible at 0 K in equilibrium since there are no states to which an electron in the valence band can be excited by a small electric field. A semiconductor is an insulator in which the bandgap is relatively small (e.g., in silicon it is 1.1 eV, compared with $\sim$ 10 eV in a "good" insulator such as silicon dioxide). As is explained later in this chapter, a semiconductor, which is an insulator when pure, can be made to conduct at room temperature by the addition of rather small amounts of impurity.

In a pure insulator or semiconductor, electrons can be excited from the valence band to the conduction band in two ways: by thermal excitation or by the absorption of a photon with energy greater than E_g. Such excitation leaves a *hole* (that is, an unfilled electron state) in the valence band. A hole behaves as a particle whose only qualitative difference from an electron is its positive charge; just as a bubble in beer *rises* under the influence of gravity, so in an electric field a hole moves in the opposite direction to an electron. When an electron meets a hole, they mutually annihilate or recombine, releasing an energy of about E_g. This process may be radiative, in which case a photon with $h\nu \approx E_g$ is emitted, or non-radiative, the energy released being dissipated as heat.

14.3 Density of States Near a Band Edge

Just as in metals, we are primarily interested in states near the Fermi energy, so in semiconductors we are primarily interested in states close to the bandgap. We confine our attention to the simple case where the band edges—that is, the maximum in the valence band (VBM) and the minimum in the conduction

---

[2] The reason for these names is that the valence band contains the electrons that hold the material together; these are the valence electrons in the chemical sense. A photon with energy above E_g can excite an electron from the valence band to a higher band where it is free to move through the crystal and conduct electricity; hence the name conduction band.

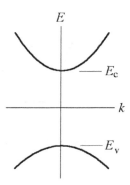

FIGURE 14.5 Dispersion relation $E(k)$ for electrons near the bandgap in a direct gap semiconductor. Since **k** is a vector, the lines in this figure are really sections through a three-dimensional hypersurface. Most semiconductors of technical interest have cubic symmetry, in which case the hypersurfaces are rotationally symmetric about the E axis if the bands are non-degenerate.

band (CBM)—are both at $k = 0$; this is the so-called "direct" bandgap case,[3] illustrated in Figure 14.5. Unlike the case of metals, the density of states varies rapidly over the relevant range of energy, and we can no longer treat it as even approximately constant.

At a band edge, the function $E(k)$ has a turning point; a minimum at E_c and a maximum at E_v, as in Figure 14.5. Hence, the Taylor expansion of $E(k)$ about $k = 0$ has no linear term, and for small k the energy is a quadratic function of k. In the conduction band, the electron energy measured upwards from E_c is:

$$\epsilon_e(k) \equiv E(k) - E_c = b_c k^2 + \text{higher powers of } k, \qquad (14.1)$$

where b_c is a positive constant.

Since a hole is the *absence* of an electron in the valence band, the further it is below the valence band edge, the higher its energy (again, consider the analogy with a bubble in beer: the further it is below the surface, the higher its potential energy). Hence, the hole energy is measured *downwards* from E_v and is:

$$\epsilon_h(k) \equiv E_v - E(k) = b_v k^2 + \text{higher powers of } k, \qquad (14.2)$$

where b_v is another positive constant. For sufficiently small k, the higher terms are negligible. To bring out the analogy between Equations (14.1) and (14.2) and the free particle $E(k)$ relation (Equation B.1), it is customary to write these two equations in the form:

$$\boxed{\epsilon_{e,h} = \frac{\hbar^2 k^2}{2m^*_{e,h}},} \qquad (14.3)$$

where m^*_e is the *effective mass* of the electron and m^*_h that of the hole, while the zero of ϵ is the appropriate band edge (E_c for electrons and E_v for holes). The effective mass for electrons is $m^*_e \equiv \frac{\hbar^2}{2b_c}$ and for holes is $m^*_h \equiv \frac{\hbar^2}{2b_v}$. The effective masses are usually less than the free electron mass m_0, often substantially less in the case of the electron; for example, in gallium arsenide, the material on which most semiconductor

---

[3] The bandgap in semiconductors used for lasers, such as gallium arsenide, is always direct. Silicon and germanium on the other hand, have indirect bandgaps, with VBM and CBM at different values of k. The discussion in the text applies, with some minor modifications, to the indirect case too (see any textbook on semiconductors). We also ignore the effect of band degeneracy, which complicates the valence band of most semiconductors considerably; see, for example, G. Bastard, *Wave Mechanics Applied to Semiconductor Heterostructures* (Editions de Physique, Paris, France, 1988), Chapter 2.

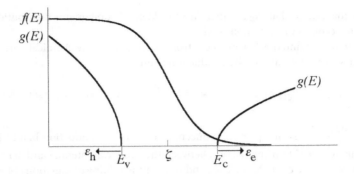

FIGURE 14.6 Density of states $g(E)$ and distribution function $f(E)$ for $T \neq 0$, in the vicinity of the band gap. In a pure semiconductor, the chemical potential ζ is within a few $k_B T$ of the middle of the gap (see text). The electron and hole energies ϵ_c and ϵ_v are measured from the respective band edges E_c and E_v.

lasers are based, $m_e^* = 0.067 m_0$. Since electrons and holes are electrically charged and can carry current, they are collectively referred to as *carriers*.

The density of states near the band edge as a function of ϵ is identical in form to that for a free electron obtained in Chapter 7, the only difference being the replacement of the mass m in Equation (7.20) by the effective mass m^*. Thus we have, from Equation (7.20),

$$g(E) = \frac{1}{2\pi^2}\left(\frac{2m^*}{\hbar^2}\right)^{3/2}\epsilon^{1/2}, \tag{14.4}$$

where m^* is the effective mass (m_e^* or m_h^*) for the appropriate carrier. The density of states at given ϵ varies as $m^{*3/2}$ and goes to zero at the band edge. When the effective mass is small, the density of states near a band edge can be orders of magnitude smaller than that of a free electron of the same energy.

The density of states in the vicinity of the band gap is shown schematically in Figure 14.6, together with the Fermi-Dirac distribution function $f(E)$ at a temperature such that $k_B T \sim 0.2 E_g$. It may be asked what meaning can be attached to the distribution function $f(E)$ in the gap, where there are no states to be occupied. That $f(E)$ is still meaningful will become clear later in the chapter, when we consider a semiconductor containing a small number of impurities, which produce states in the gap. The occupancy of such states is then determined by $f(E)$. We shall see later that the chemical potential ζ (called the *Fermi level* in the semiconductor literature, a usage that we will follow[4]) is near the middle of the band gap in a pure material, so that only the tails of the Fermi distribution reach into the energy region where the density of states is non-zero. We shall find that ζ usually varies much more rapidly with temperature than it does in a metal. It is easily shown (see Problem 9.1b) that if ζ is near the middle of the bandgap, and if $k_B T \ll E_g$, the distribution functions for the electrons and holes, f_e and f_h, take the Maxwell-Boltzmann form:

$$f_e(\epsilon_e) = e^{\beta(\zeta - E_c - \epsilon_e)}, \qquad f_h(\epsilon_h) = e^{\beta(E_v - \zeta - \epsilon_h)}. \tag{14.5}$$

Here we write $f_h(\epsilon_h)$ for $1 - f(E)$ in the valence band, since we are interested in number of holes in the valence band, rather than the number of electrons. Equation (14.5) shows that $\zeta - E_c$ and $E_v - \zeta$, both of which are negative, play the role that ζ plays in the ideal gas [compare Equation (9.12)]. They can be regarded as "effective" chemical potentials for electrons and holes respectively, but we will not use this term since it can cause confusion: in equilibrium there can be only one true chemical potential, the Fermi level ζ. The condition for the validity of Equation (14.5) is that both $\zeta - E_c$ and $E_v - \zeta$ be negative and

---

[4] It is important to distinguish between the Fermi *level* $\zeta(T)$ and the Fermi *energy* E_F; they are only equal at $T = 0$. This distinction is not always observed in the literature (see, for example, Footnote 12 of Chapter 13).

much greater in magnitude than $k_B T$; otherwise the Maxwell-Boltzmann distribution function is not applicable and life becomes more complicated.

As in Chapter 9, we can find the density of electrons n_e in terms of the chemical potential by integrating over the occupancy of all the states in the conduction band:

$$n_e = \int_0^\infty g(\epsilon_e) f(\epsilon_e) \, d\epsilon_e = e^{\beta(\zeta - E_c)} \int_0^\infty g(\epsilon_e) e^{-\beta \epsilon_e} \, d\epsilon_e = n_c e^{\beta(\zeta - E_c)}, \qquad (14.6)$$

where $n_c = 2\left(\frac{m_e^* k_B T}{2\pi\hbar^2}\right)^{3/2}$ is the quantum concentration for the conduction band.[5] Equation (14.6) is equivalent to Equation (9.15) for the relation between the chemical potential and density of an ideal gas, if we substitute n_c for n_q and $\zeta - E_c$ for ζ. The condition for the validity of Equation (14.6) is $E_c - \zeta \gg k_B T$; that is, $n_e \ll n_c$. Similarly, the density of holes is

$$n_h = \int_0^\infty g(\epsilon_h) f(\epsilon_h) \, d\epsilon_h = e^{\beta(E_v - \zeta)} \int_0^\infty g(\epsilon_h) e^{-\beta \epsilon_h} \, d\epsilon_h = n_v e^{\beta(E_v - \zeta)}, \qquad (14.7)$$

where $n_v = 2\left(\frac{m_h^* k_B T}{2\pi\hbar^2}\right)^{3/2}$ is the quantum concentration for the valence band. Equation (14.7) only holds if $\zeta - E_v \gg k_B T$; that is, $n_h \ll n_v$.

The product of the electron and hole densities is independent of ζ, but not of temperature:

$$n_e n_h = n_c e^{\beta(\zeta - E_c)} n_v e^{\beta(E_v - \zeta)} = n_c n_v e^{-\beta E_g}, \qquad (14.8)$$

since $E_g = E_c - E_v$.

Equations (14.6) through (14.8) hold only if ζ is within the band gap, and not so close to a band edge that Maxwell-Boltzmann statistics cease to apply. This is true for a pure semiconductor so long as $k_B T \ll E_g$, as is usually the case, but we shall find in Section 14.5 that ζ depends critically on the chemical purity of the semiconductor, and that Maxwell-Boltzmann statistics only apply if the impurity concentration is much less than n_c or n_v.

14.4 Intrinsic Conduction in a Pure Semiconductor

In a pure semiconductor or insulator, every electron promoted to the conduction band leaves a hole in the valence band, so that in this case, the electron and hole densities are equal: $n_e = n_h$. Hence, in thermal equilibrium,

$$n_c e^{\beta(\zeta - E_c)} = n_v e^{\beta(E_v - \zeta)},$$

or:

$$k_B T \ln\left(\frac{n_v}{n_c}\right) = 2\zeta - (E_c + E_v). \qquad (14.9)$$

When $T = 0$, the left-hand side of Equation (14.9) is zero, so that $E_F \equiv \zeta(0) = \frac{1}{2}(E_c + E_v)$; the Fermi energy lies halfway between the band edges. If $n_c \neq n_v$ (as is usually the case), ζ varies with T, and is given by:

$$\zeta(T) = E_F + \frac{1}{2} k_B T \ln\left(\frac{n_v}{n_c}\right). \qquad (14.10)$$

---

[5] The quantum concentration for the conduction or valence bands is often called the "effective density of states" in the semiconductor literature; however, this terminology is very misleading (concentration and density of states have different dimensions), and we do not use it.

As T increases, ζ moves towards the band with the smaller quantum concentration, in order to keep the electron and hole densities equal. However, the second term in Equation (14.10) is of order $k_B T$, and so long as $k_B T \ll E_g$, ζ remains near the center of the band gap, so that the condition for the Maxwell-Boltzmann approximation (Equation (14.5)) to $f(\epsilon)$ continues to be satisfied. Note the difference from a metal, where $\zeta(T) \approx E_F$ and does not change appreciably with temperature so long as $k_B T \ll E_F$. The reason for the difference is that a metal has a non-zero (usually large) density of states at E_F, while in a semiconductor $g(E_F) = 0$.

We can use Equation (14.8) to find the temperature dependence of the intrinsic electron (or hole) density n_i:

$$n_i = (n_v n_c)^{1/2} e^{-\beta E_g / 2}. \tag{14.11}$$

The factor $1/2$ in the exponent comes from the fact that *two* particles, an electron and a hole, are created in the excitation process, rather than one (compare Chapter 8).

The density of electrons and holes n_i given by Equation (14.11), being a property of the pure material, is called the *intrinsic density*. Since the bandgap E_g is usually large relative to $k_B T$, the intrinsic density is very small at ordinary temperatures. e.g. in silicon at room temperature, $n_i \sim 10^{16}$ m^{-3}, which is a factor $\sim 10^{12}$ smaller than the electron density in a typical metal. However, we shall see in the next section that the addition of a small number of impurities can drastically alter the electron and hole densities.

14.5 Impurities in Semiconductors: Extrinsic Conduction

Technologically, the most useful property of a semiconductor is the fact that its conductivity is strongly affected by the presence of minute quantities of impurity. The reason for this can be seen by considering a simple example.

Suppose that we substitute a phosphorus atom for a silicon atom in an otherwise pure silicon crystal (the process of deliberately adding impurities is called *doping*). Phosphorus has five valence electrons, while silicon has four. Four electrons are needed to satisfy the bonds to neighboring silicon atoms and thus fill the valence band, so that for each phosphorous atom, there is an electron left over. This electron can only go into the conduction band, leaving a positively charged phosphorus atom behind. An impurity that gives up an electron in this way is called a *donor*. Except at very low temperatures, the large dielectric constant of the semiconductor makes the Coulomb interaction energy between the electron and the positively charged atom small relative to $k_B T$, and the electron moves freely through the material. Such a donor-doped semiconductor is referred to as *n-type* since the current carriers are electrons which are negatively charged.[6] Since the number of electrons is equal to the number of impurity atoms, and the intrinsic concentration of electrons is usually extremely small, it only takes a minute concentration of impurity to dominate the conductivity. For example, the presence of one phosphorus atom for every 10^{10} silicon atoms is sufficient to increase the room temperature conductivity of otherwise pure silicon by more than two orders of magnitude.

Now suppose that we dope with gallium instead of phosphorus. Gallium has three valence electrons, one short of the number needed to satisfy the silicon bonds. Thus, there is an electron missing in the valence band; that is, there is a hole. Such an impurity, which creates holes, is called an *acceptor*. Like the electron, the hole can move freely through the material at normal temperatures, leaving the gallium atom with a negative charge. An acceptor-doped semiconductor is referred to as *p-type* since the current carriers are holes, which are positively charged.

---

[6] The names *n*-type and *p*-type preceded the understanding of semiconductors in terms of the sign of the carriers, and were based on the observed sign of the thermoelectric power relative to a metal. This is negative for electrons and positive for holes (see Example 14.1).

We now calculate the chemical potential in an n-type semiconductor. Suppose that there are n_D donor atoms per unit volume, each of which gives up one electron to the conduction band. For simplicity, we assume that $n_i \ll n_D \ll n_c$. The density of electrons is now $n_e = n_D$. Substituting in Equation (14.6) gives:

$$\zeta - E_c = k_B T \ln \left(\frac{n_D}{n_c} \right). \tag{14.12}$$

Since $n_D \ll n_c$, the logarithm is negative, so that $\zeta < E_c$; that is, the Fermi level is in the band gap. As n_D gets larger, ζ approaches the conduction band edge. In a typical moderately doped n-type semiconductor at room temperature, $n_D \sim n_c$, so that $\zeta \approx E_c$. The exact relation between ζ and n_D is now more complicated, since the Maxwell-Boltzmann distribution no longer applies and Equation (14.12) is not strictly accurate.[7] However, it remains true that the Fermi level is close to the conduction band edge in a moderately doped n-type semiconductor. We saw in Equation (14.8) that in equilibrium, the product of the electron and hole densities is independent of ζ; combining Equations (14.8) and (14.11) gives:

$$n_e n_h = n_i^2. \tag{14.13}$$

Since the intrinsic carrier density n_i is small at normal temperatures, a large electron density implies that, in equilibrium, the hole density is very small indeed, and vice versa.

Similarly, if the semiconductor is p-type with n_A acceptors per unit volume, making the same approximations we find:

$$E_v - \zeta = k_B T \ln \left(\frac{n_A}{n_v} \right). \tag{14.14}$$

A heavily doped p-type semiconductor has $n_A \sim n_v$, so that $\zeta \approx E_v$; that is, the Fermi level is close to the valence band edge.

In real semiconductor devices, both types of impurity are usually present, and in equilibrium, the number of carriers is $|n_D - n_A|$, since if both electrons and holes are present, they recombine (that is, they mutually annihilate) until only the majority species remains. The remaining carriers are electrons if $n_D - n_A$ is positive, holes if it is negative. If $|n_D - n_A| \ll n_D, n_A$, the semiconductor is said to be "compensated."

As remarked above, in real semiconductor devices, the assumption that $|n_D - n_A| \ll n_c, n_v$ is not usually valid, and the electron or hole gas is not ideal. In the opposite limit, $n_D - n_A \gg n_c$, we have a degenerate Fermi gas of electrons and ζ is in the conduction band. Similarly, if $n_A - n_D \gg n_v$, we have a degenerate Fermi gas of holes and ζ is in the valence band. In practice, $n_A \gg n_v$ is more difficult to achieve at room temperature than $n_D \gg n_c$, since in most semiconductors holes are much heavier than electrons, so that $n_v \gg n_c$. The degenerate case is treated in Problem 14.5.

Example 14.1: Thermoelectricity

The fact that, in lightly doped semiconductors, the chemical potential depends strongly on temperature, manifests itself in the thermoelectric effect, which is in general much more pronounced in semiconductors than in metals. Thermoelectricity is most easily understood in terms of a temperature-dependent work function ϕ_w, which is defined as the difference between the chemical potential of an electron outside and inside the solid (see Example 8.1). If two conductors with work functions ϕ_{w1} and ϕ_{w2} are brought into contact, a contact voltage V_c develops between them. It was shown in Chapter 8 that $qV_c = \phi_{w2} - \phi_{w1}$, where $q = 1$ if ϕ_{w1} and ϕ_{w2} are measured in eV.

---

[7] The Joyce-Dixon approximation, which gives a more accurate value of ζ (see Figure 13.5 and Footnote 12 of Chapter 13), was derived to deal with this case.

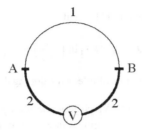

FIGURE 14.7 Schematic of a circuit with junctions A and B between different materials (labeled 1 and 2). The two junctions are at different temperatures. The thermoelectric voltage is measured by the meter V.

In Figure 14.7, two wires of different composition, 1 and 2, are joined at two points, A and B, to make a loop. If A and B are at the same temperature, the contact voltages at A and B cancel and no voltage is detected by the voltmeter V. On the other hand, if the temperature of A exceeds that of B by ΔT, the contact voltages, which depend on the chemical potentials and hence on temperature, are in general no longer equal. Since the cancellation is no longer perfect, a thermoelectric voltage V_t develops across the voltmeter.

1. Assuming that ΔT is small, express the thermoelectric power (thermopower for short) $Q \equiv \frac{V_t}{\Delta T}$ in terms of the charge on the electron $-q$ ($q > 0$) and the temperature derivatives of the chemical potentials, ζ_1 and ζ_2, of the two materials.

From Example 8.1, the contact potential at each junction is V_c, where $qV_c = \phi_{w2} - \phi_{w1} = \zeta_1 - \zeta_2$. Since the electron is negatively charged, V_c is positive in conductor 1 if $\zeta_1 > \zeta_2$, since electrons then transfer from conductor 1 to conductor 2 in order to equalize the electrochemical potential.

If there is a small temperature difference ΔT between A and B, the net voltage is, to first order in ΔT,

$$V_t = V_c(A) - V_c(B) \approx q^{-1}\left(\frac{\partial \zeta_1}{\partial T} - \frac{\partial \zeta_2}{\partial T}\right)\Delta T,$$

so that the thermopower is:

$$Q = q^{-1}\left(\frac{\partial \zeta_1}{\partial T} - \frac{\partial \zeta_2}{\partial T}\right).$$

2. Now suppose that conductor 1 in Figure 14.7 is a lightly doped n-type semiconductor with an electron density n_D such that $n_i \ll n_D \ll n_c$. Conductor 2 is a metal, for which $\frac{\partial \zeta}{\partial T}$ is less than that in the semiconductor by a factor of order $\frac{k_B T}{E_F}$ (see Problem 14.1), so that its contribution to the thermopower can be neglected. Find the magnitude and sign of Q. How will the result be different if the semiconductor is lightly p-type?

In a lightly doped n-type semiconductor, the chemical potential is, from Equation (14.12),

$$\zeta = E_c - k_B T \ln\left(\frac{n_c}{n_D}\right),$$

where $n_c \gg n_D$. The temperature dependence of E_c is usually relatively small, so that

$$\frac{\partial \zeta}{\partial T} \approx -k_B \ln\left(\frac{n_c}{n_D}\right) - k_B T \frac{\partial \ln(n_c)}{\partial T}$$

$$= -k_B\left[\ln\left(\frac{n_c}{n_D}\right) + \frac{3}{2}\right],$$

since $n_c \propto T^{3/2}$.

Since the contribution of the metal to Q is small,

$$|Q| = q^{-1}k_B\left[\ln\left(\frac{n_c}{n_D}\right) + \frac{3}{2}\right],$$

so that the thermopower increases in magnitude as the doping decreases, as long as the conduction remains extrinsic.

The sign of Q needs some thought. Which junction goes positive, A or B? Since $n_c > n_D$, $\frac{\partial \zeta}{\partial T} < 0$ in the semiconductor; that is, ζ moves down (away from the conduction band) as T increases. Thus, the thermopower is negative: the hotter junction becomes negative from the point of view of the voltmeter. This is the origin of the term n-type. A more physical way of seeing this is to suppose that the voltmeter has infinite impedance so that charge can move only in the semiconductor. Since ζ decreases in the semiconductor at the hot junction, electrons will flow from the cold junction until an electrostatic potential V_t, negative at the hot junction, has built up to stop them. Thus, the hot junction goes negative. If the voltmeter has finite impedance, electrons flow through it counterclockwise; that is, positive current flows clockwise.

Hence, the final result for an n-type semiconductor is:

$$Q = -q^{-1}k_B\left[\ln\left(\frac{n_c}{n_D}\right) + \frac{3}{2}\right]. \tag{14.15}$$

We can proceed in the same way for a lightly doped p-type semiconductor; that is, one in which $n_i \ll n_A \ll n_v$. Since $\frac{\partial \zeta}{\partial T} > 0$, the hot junction is now attractive to holes. The sign of Q is now positive; hence the name p-type. Semiconductors were classified as n- or p-type long before the mechanism was understood, and checking the sign of the thermopower with a hot soldering iron is still often the quickest and most convenient way of determining the type of an unknown sample.

The picture of thermoelectricity given here is oversimplified, but it gives the correct sign and approximate magnitude. The first term inside the bracket in Equation (14.15) is correct, but the second is only qualitatively accurate. To be quantitative, one has to take into account the different mobilities (see Chapter 16) of electrons with different energies. There is also a third term due to the interaction of electrons with the phonons, which move under the temperature gradient and tend to drag the electrons with them.[8]

14.6 The $p-n$ Junction

The key advance that made possible the development of semiconductor devices such as the transistor and microprocessor was the achievement of precise control of impurity content. For example, one can dope one region of a semiconductor n-type and another p-type, with an abrupt transition between the two regions (see Figure 16.4 for a common method of achieving this by solid state diffusion). The transition is called a $p-n$ junction. Let us consider what happens when an n-type and a p-type semiconductor are brought into contact. For simplicity, we assume that the two materials differ only in doping.

When the two materials are not in contact, the energy (relative to the energy of a free electron at rest in the vacuum) of any particular state (e.g., the conduction band energy E_c) is the same in each material; this energy is determined only by the properties of the semiconductor itself, and E_c and E_v are not affected by the presence of small amounts of impurity. However, as shown in the previous section, the chemical

---

[8] See, for example, K. Seeger, *Semiconductor Physics: An Introduction*, 3rd ed. (Springer-Verlag, Berlin, Germany, 1985), pp. 79–85; R. W. Ure, *Thermoelectric Effects in III-V Compounds in Semiconductors & Semimetals*, eds. R. K. Williamson and A. C. Beer (Academic Press, New York, 1972) **8**, 67–102. For a different approach to thermoelectricity, more appropriate to metals, see R. G. Chambers, "Thermoelectric effects and contact potentials," *Physics Education* **12**, 374–380 (1977). The thermoelectric effect described here (properly called the Seebeck Effect) is closely related to the Peltier Effect, which is widely used in thermoelectric coolers, and the Thomson Effect; see M. W. Zemansky and R. H. Dittman, *Heat and Thermodynamics*, 6th ed. (McGraw-Hill, New York, 1981), pp. 431–442.

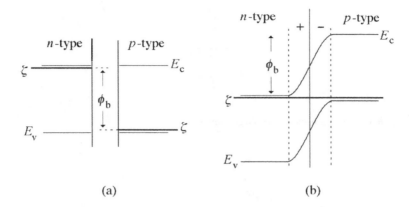

FIGURE 14.8 Schematic energy diagrams for an *n*- and a *p*-type semiconductor. The ordinate is electron energy. The abscissa represents position, the junction between the two materials being at the center of the diagram. ζ is the electrochemical potential. (a) No contact between the *n*-type and *p*-type material; the chemical potentials differ by ϕ_b. (b) The two materials are brought into contact, forming a *p–n* junction. The region between the two dotted lines is depleted of carriers, leaving a space charge which is positive in the *n*-type region and negative in the *p*-type. In diffusive equilibrium, there is an electric field in the depletion layer, which makes the electrochemical potential ζ the same everywhere.

potential in the *n*-type material is close to the conduction band edge, while in the *p*-type material, it is close to the valence band edge. Hence, ζ is higher in the *n*-type material than in the *p*-type by the *barrier height* ϕ_b, which is close to E_g.

Figure 14.8a shows the band edge energies, relative to the energy of an electron at rest outside the semiconductor, as a function of position for an electron in the *n*-type and *p*-type materials when they are not in contact. Note that electron energy is measured upwards, and that the charge on the electron is $-q$. When the two materials are brought into contact, as in Figure 14.8b, the difference of chemical potential makes electrons flow from the *n*-type to the *p*-type material. In doing so, they leave behind positively charged donor atoms which are no longer neutralized by the electrons. In the *p*-type material, the electrons flowing in combine with the holes already present, leaving acceptor atoms that are negatively charged since there are no longer holes to neutralize them. The resulting double layer of charge produces an electrostatic potential difference between the *n*- and *p*-type regions. An electrostatic potential V_b builds up, and equilibrium is reached when $-qV_b$ exactly cancels the chemical potential difference; that is, when $V_b = -\phi_b/q$ (where $q = 1$ if ϕ_b is measured in electron volts). The electrochemical potential (that is, the sum of the chemical potential and the electrostatic potential energy) is now the same everywhere, as shown in Figure 14.8b. The region where there is an electric field contains no electrons or holes and is called the *depletion layer*. The width of the depletion layer depends on the doping and can be found by solving Poisson's equation for the electrostatic potential. For our present purposes, all that matters is that in equilibrium, the electrochemical potential is the same on both sides of the junction, so that there is an electrostatic potential difference ϕ_b across the junction. This potential difference is called the *barrier height* or *built-in voltage*. Typical depletion layer widths are between 0.1 and 1 μm, so that if $\phi_b \sim E_g \sim 1$ eV, the built-in electric field in the junction is in the range 10^6 to 10^7 V/m.

So far we have assumed that the two pieces of semiconductor are not connected to an external circuit. Now suppose that they are connected to a voltage source such as a battery, as in Figure 14.9. The junction is no longer in equilibrium, since there is now an external source of energy, and the chemical potential need not be the same everywhere. Equilibrium between electrons and holes requires recombination, which is relatively slow. On the other hand, the strong Coulomb interaction between the carriers ensures that thermal equilibrium is rapidly established among the electrons and holes separately. Consequently, each type of carrier can be described by its own chemical potential, commonly called the *quasi-Fermi level*,

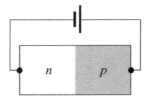

FIGURE 14.9 A *p–n* junction connected to a voltage source.

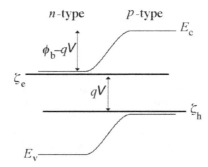

FIGURE 14.10 Effect of a forward bias V on the energy diagram of a *p–n* junction. ζ_e and ζ_h are the quasi-Fermi levels for electrons and holes, respectively.

ζ_e for electrons and ζ_h for holes. If the electrical resistance of the contacts is small, the carriers on each side of the junction are in local equilibrium with the respective battery terminal. Hence, the difference between the quasi-Fermi levels in a *p–n* junction is qV, as indicated in Figure 14.10.

Suppose that V is positive on the *p*-type side of the junction. For reasons that will become clear, this is called *forward bias*. Since a positive voltage is attractive to electrons, its effect is to partially cancel ϕ_b, lowering the energy barrier to electrons from ϕ_b to $\phi_b - qV$, so that they diffuse into the *p*-type material, producing a *diffusion current* proportional to the probability of overcoming a barrier of this height, which is given by the Boltzmann factor.[9] Hence, the diffusion current is:

$$I = I_0 e^{-\beta(\phi_b - qV)}, \tag{14.16}$$

where I_0 is independent of V and varies only slowly with T.

In most semiconductors $k_B T \ll \phi_b$, so that $\beta \phi_b$ is a large number, and the current only reaches an appreciable value when the forward bias reaches a certain voltage, called the *turn-on voltage* V_{to}, of roughly $\frac{\phi_b - 10 k_B T}{q}$, which is ~ 0.6 V for silicon. When the electrons have diffused across the depletion layer and reach the *p*-type region, they recombine with holes. Similarly, holes diffuse into the *n*-type region and recombine with electrons. Electrons in *p*-type material and holes in *n*-type are called *minority carriers*. In silicon, this recombination process is slow, so that the minority carriers diffuse an appreciable distance before recombining. In a direct band gap material such as gallium arsenide, the electron-hole

---

[9] There is also a small current $I_0 e^{-\beta \phi_b}$ in the reverse direction, called the *generation-recombination* current, which is due to thermal excitation of electron-hole pairs in the depletion region. The large electric field in this region sweeps out these thermally generated carriers as soon as they are created, producing a current in the opposite direction to the diffusion current given by Equation (14.16). This current ensures that when $V = 0$, the net current is zero, as it must be. In a small diode, the generation-recombination current is typically less than 10^{-9} A in the dark at room temperature, and can be neglected for most purposes. Photons with $h\nu > E_g$ can greatly enhance the generation-recombination current, so that it is no longer cancelled out by the diffusion current (see Example 14.2).

recombination is more rapid and is radiative; that is, a photon of energy $h\upsilon \approx E_g$ is emitted, rather than the energy of recombination being dissipated as heat. This radiative recombination is the basis of the light-emitting diode (LED) and of the semiconductor laser.

If the voltage is applied in the reverse direction (*reverse bias*), the situation is quite different, since the barrier height now increases with voltage and diffusion is suppressed. Even though the chemical potentials in the n- and p-type regions differ, only the tiny generation-recombination current (see Footnote 9) can flow, because the conducting regions are separated by the insulating depletion layer. Thus, the p–n junction acts as a rectifier. As the reverse voltage increases, the electric field in the depletion layer rises until it is so large that quantum mechanical tunneling[10] of electrons through the depletion region occurs, a phenomenon called Zener breakdown.

Example 14.2: The Photovoltaic Effect

When light is absorbed in the depletion region of a p–n junction, electrons and holes are created which are immediately swept out of the region by the built-in field. If there is no external electrical connection to the p- and n-type layers, these carriers will remain in the respective layers and neutralize the space charge. Complete neutralization would produce the *flat band* condition where the conduction and valence bands are at the same energy on both sides of the junction, and the output voltage would be ϕ_b/q. However, for the reason given in (b) below, this condition cannot be achieved in practice. The process of generating a voltage with light is called the *photovoltaic effect* and is the basis of the semiconductor photodetector and of the solar cell. If the terminals of the device are connected to an external circuit, a current called the *photocurrent* will flow.

1. If the device is connected to a low impedance circuit, so that no voltage is generated across its terminals, what photocurrent will be generated if r photons per second are absorbed in the depletion layer? Neglect any recombination that might occur before the carriers are swept out of the depletion layer

If charge is not to build up, current must flow. If r electron-hole pairs are generated per second, a current rq flows. Note that it is *not* $2rq$: To see this, suppose that the photogenerated holes stay in the p-layer, while the electrons traverse the entire circuit, neutralizing the extra holes when they reach the p-layer. Only r charges pass through the circuit in unit time. Now imagine neutralization occurring anywhere in the circuit; only r charges (which may be electrons going one way or holes going the other) pass any point in unit time

2. If a circuit has non-zero impedance, a voltage V develops across the junction. If r electron-hole pairs are generated per second and the current drawn from the device is $I < rq$, what will be the sign and magnitude of the voltage? What is the maximum power that can be drawn from the device?

The current generated by the light is in the opposite sense to the diffusion current given by Equation (14.16). Thus, the net photocurrent is:

$$I = rq - I_0 e^{-\beta(\phi_b - qV)},$$

whence:

$$V = q^{-1}\left[\phi_b - k_B T \ln\left(\frac{I_0}{rq - I}\right)\right].$$

This voltage is positive on the p-type side of the junction.

In practice $I_0 \gg rq$, and V is insensitive to the net current drawn, remaining close to the turn-on voltage of the junction, V_{to}, so long as $I < rq$. Hence, the maximum value of IV is a little less than rqV_{to}

---

[10] See any quantum mechanics textbook.

3. A solar cell consists of a flat p–n junction situated as close as practicable to the surface of the cell. In a silicon solar cell illuminated with solar radiation, what is the maximum possible efficiency, defined as $\frac{\text{electric power out}}{\text{radiative power in}}$? Assume that the sun is a black body at temperature 5800 K, and that all radiation with $h\nu > E_g$ is absorbed in the p–n junction. Neglect atmospheric absorption and reflection of light at the surface of the cell (which can be minimized by appropriate coating). For silicon, $V_{to} \approx 0.6$ V.

If every photon with $h\nu > E_g$ produces an electron-hole pair, and the cell has unit area, $r = J_n$, where J_n is the flux of such photons. To find r for a given incident radiative power, we rearrange Equation (11.11) and integrate over the range of photons that can be absorbed, using Equation (H.4) to obtain:

$$r = \frac{c}{4} \int_{E_g/h}^{\infty} g(\nu) f(E)\, d\nu$$

$$= \frac{2\pi}{c^2} \int_{E_g/h}^{\infty} \frac{\nu^2}{e^{\beta h\nu} - 1}\, d\nu.$$

Substituting $x = \beta h\nu$ gives:

$$r = \frac{2\pi}{c^2} \left(\frac{k_B T}{h} \right)^3 \int_{\beta E_g}^{\infty} \frac{x^2}{e^x - 1}\, dx,$$

where the integral can be evaluated numerically. If $T = 5800$ K and $E_g = 1$ eV, $\beta E_g = 2$. The integral is approximately $0.6 \int_0^{\infty} \frac{x^2}{e^x - 1}\, dx = 1.2\zeta(3) = 1.44$, where $\zeta(n)$ is the Riemann zeta function (see Problem 7.7). The radiative flux is $\sigma_B T^4$, where σ_B is Stefan's constant (see Chapter 11). Putting in numbers, we find that the maximum possible (ideal) efficiency is 26%.

Actual solar cells have efficiencies < 20%. The reduction from the ideal value has several causes.[11] The output voltage is always less than V_{to}. Not every photon with $h\nu > E_g$ which falls on the cell is absorbed in the depletion layer of the p–n junction; in silicon, which has an indirect gap (see Footnote 3), the absorption near the band gap is weak and most photons with $h\nu$ close to E_g pass right through the junction, while at large $h\nu$, where the absorption is strong, a substantial number are reflected at the surface or are absorbed before reaching the junction. There are also various internal losses: in particular, recombination at defects in the junction, and internal resistance.

14.7 Envoi

In the early days of research on semiconductors, their sensitivity to impurity content was responsible for their most puzzling and frustrating feature: that the conductivity could vary by many orders of magnitude from one nominally identical sample to another, and that even the type (p or n) seemed to vary almost at random.[12] However, once the problem of accurately controlling the impurity content had been solved, the fact that the conductivity can be varied at will over an enormous range, and can be made p-type or n-type, became the basis of the vast semiconductor device industry.

---

[11] See S. J. Fonash, *Solar Cell Device Physics* (Academic, Press, Burlington, MA, 1981), pp. 127–130.

[12] Partly because of this irreproducibility, research on semiconductors had a bad reputation amongst physicists. When I was a physics graduate student, in the early days of the transistor, metals were considered (at least in Cambridge) to be the only respectable subject of inquiry in solid state physics. If you studied ionic crystals, you were a chemist; if semiconductors, you were an engineer. See also the section on the semiconductor laser in M. D. Sturge, "Impossibility in technology: When the expert says no," in *No Way: On the Nature of the Impossible*, eds. D. Park and P. J. Davis (W. H. Freeman, New York, 1986), p. 111.

This chapter treats the statistical mechanics of semiconductors at the minimum level needed to understand many of their remarkable properties, but it only skims the surface. For a much more thorough treatment, see J. S. Blakemore, *Semiconductor Statistics* (Dover, New York, 1987).

14.8 Problems

14.1 Prove the statement in the text (Example 14.1) that the contribution of the metal to the thermopower Q is smaller than that of a lightly doped semiconductor by a factor of order $\frac{k_B T}{E_F}$ (see Section 13.5)

14.2 (The heterojunction.) Two semiconductors with different bandgaps E_{g1} and E_{g2} (in eV) are brought into contact with each other to form a *heterojunction*. Take $E_{g1} > E_{g2}$. Both are moderately n-type, so that $n_e \approx n_c$. Draw an energy-position diagram (analogous to the diagram of the p–n junction in Figure 14.8), assuming that the pure semiconductors have the same work function. Show the electrochemical potential ζ and the conduction and valence band edges on either side of the junction. Indicate the depleted region. What is the built-in voltage difference barrier height between the valence band edges?

14.3 An n-type semiconductor, with electron density n_e and effective mass m_e^*, contains an impurity which can bind an electron with energy E_0. This means that an electron in the bound state has an energy $E_c - E_0$, where E_c is the energy of the conduction band edge. Because of the electron spin, the state is doubly degenerate. If the bound state is 50% occupied at temperature T, what is n_e?

14.4 (Semi-insulators.) No material can be made absolutely pure. Suppose that a semiconductor unavoidably contains n_D shallow donors and n_A shallow acceptors, where "shallow" means that the energy level is within $k_B T$ of the respective band edge. If $n_D < n_A$, we can deliberately dope with n_m "deep" donors, whose energy level is near the center of the bandgap, so that it takes an energy of roughly $E_g/2$ to remove the electron from the deep donor and put it in the conduction band. Show that if $n_m > n_A - n_D$, charge neutrality requires that the deep donor level be partially occupied. Hence, show that the doping lowers the carrier density from $|n_A - n_D|$ to a value $\sim n_i$. A semiconductor doped in this way (or, if $n_D > n_A$, doped with deep acceptors[13]) is called *semi-insulating*. Assume that $k_B T \ll E_g$

14.5 (Degenerate semiconductor.) In the text we assumed that $n_D \ll n_c$ and $n_A \ll n_v$, so that the electrons and holes could be treated as ideal gases. As $T \to 0$, n_c and $n_v \to 0$, so that if there is any doping at all, this criterion cannot be satisfied

   (a) On a figure like Figure 14.6, indicate where the chemical potential (Fermi energy) is at 0 K in a semiconductor which is (i) strongly n-type and (ii) strongly p-type. Show that if $n_e \gg n_c$, the electrons form a degenerate Fermi gas and that the chemical potential is given by:

$$\zeta = E_c + \frac{\hbar^2}{2m_e^*}(3\pi^2 n_e)^{2/3}. \tag{14.17}$$

   (b) Write down the chemical potential in the p-type material in which $n_h \gg n_v$

   (c) In practice, doping levels are often used that are greater than the quantum concentrations even at room temperature. Estimate the minimum value of n_e required for Equation (14.17) to be a reasonable approximation in GaAs at 300 K, given that $m_e^* = 0.067 m_0$.

---

[13] In practice, since the unwanted impurities may cause the material to be either n- or p-type, it is customary to dope with an amphoteric impurity, which can act as a donor or as an acceptor according to the position of the Fermi level. For a thorough discussion of the phenomenon of semi-insulation, see D. C. Look, "The electrical and photoelectronic properties of semi-insulating GaAs," in *Semiconductors & Semimetals*, eds. R. K. Williamson and A. C. Beer (Academic Press, New York, 1983), **19**, 76–170.

14.6 (Optical absorption in a doped semiconductor; the Burstein shift.) When light is absorbed in a pure semiconductor, an electron makes a transition from an initial state at energy E_1 in the valence band to a final state at energy $E_2 = E_1 + h\nu$ in the conduction band, where $h\nu$ is the photon energy. The photon cannot be absorbed if $h\nu < E_g$, since if there is a state at E_1 there cannot be one at E_2. If $h\nu > E_g$, the photon can be absorbed (subject to optical selection rules) if there is a filled initial state at E_1 and an empty final state at E_2. This is the case for all states if ζ is within the gap. Thus if we increase $h\nu$ from less than E_g, a pure semiconductor will have an absorption edge at $h\nu = E_g$, where the absorption coefficient increases abruptly from zero to a high value.

Show that in a heavily doped n-type semiconductor at 0 K, the absorption edge is shifted to a higher energy by $\frac{\hbar^2}{2m_e^*}(3\pi^2 n_e)^{2/3}$. This is called the Burstein shift. What is the Burstein shift in a heavily doped p-type semiconductor? What will happen to the shape of the absorption edge and its position relative to the bandgap as the temperature is raised from 0 K?

14.7 (Optical gain in a semiconductor: quasi-Fermi levels and the lasing condition). In a non-equilibrium situation, Equation (14.17) gives the quasi-Fermi level ζ_e for electrons if $n_e \gg n_c$. Similarly, the answer to Problem 14.5b gives the quasi-Fermi level ζ_h for holes if $n_h \gg n_v$.

Consider optical transitions of an electron between the valence and conduction bands in a forward biased p–n junction in which the quasi-Fermi levels are ζ_e and ζ_h. The occupancy of the valence band is determined by ζ_h, and that of the conduction band by ζ_e. These occupancies are given by the Fermi-Dirac distributions $f(E - \zeta_h)$ and $f(E - \zeta_e)$ respectively, where $f(x) = \frac{1}{e^{\beta x}+1}$.

(a) Show that the probability of an upwards transition from a valence band state at E_1 to a conduction band state at E_2 is proportional to $f(E_1 - \zeta_h)[1 - f(E_2 - \zeta_e)]$.

In addition, there are stimulated transitions[14] downwards from E_2 to E_1, whose probability is proportional to $f(E_2 - \zeta_e)[1 - f(E_1 - \zeta_h)]$. Einstein's theory of stimulated emission shows that the coefficients of proportionality for these two processes are equal. The absorption coefficient for light with photon energy $h\nu = E_1 - E_2$ is proportional to the difference between these two probabilities; that is, to the net upwards transition rate.

(b) Show that the absorption coefficient is proportional to $f(E_1 - \zeta_h) - f(E_2 - \zeta_e)$. Hence, show that the condition for the absorption coefficient to be negative, so that light is amplified rather than attenuated as it passes through the semiconductor, is:

$$h\nu < \Delta\zeta,$$

where $\Delta\zeta \equiv \zeta_e - \zeta_h$.

Since transitions are only possible if $h\nu > E_g$, this condition is equivalent to:

$$\Delta\zeta > E_g. \tag{14.18}$$

Equation (14.18) is called the transparency condition and lasing can only occur if it is satisfied. In practice, the difference of quasi-Fermi levels $\Delta\zeta$ must be somewhat larger than this in order for the gain to be sufficient to overcome losses.[15] The chemical potential difference, $\Delta\zeta$, is often rather confusingly called the chemical potential in the semiconductor literature.

---

[14] See any introductory atomic physics text.

[15] See O. Svelto and D. C. Hanna, *Principles of Lasers*, 4th ed. (Plenum Press, New York, 1998). Equation (13.18) was first published by the Russian physicists N. G. Basov, O. N. Krokhin, and Y. M. Popov, "Production of negative-temperature states in p-n junctions of degenerate semiconductors," *J. Exper. Theor. Physics* **40**, 1879–1880 (1961), [Engl. transl. in *Soviet Physics JETP* **13**, 1320–1321]. An equivalent result was derived 8 years earlier by the Hungarian-American mathematician John von Neumann (1903-1957), but was not published in his lifetime [see "Notes on the photon-disequilibrium amplification scheme," *in Collected Papers of J. von Neumann*, ed. A. H. Taub, **5**, 420 (Pergamon, New York, 1963) and C. H. Townes, *How the Laser Happened* (Oxford University Press, Oxford, UK, 1999), p. 71].

(c) In GaAs $m_e^* = 0.067m_0$, and $m_h^* \gg m_e^*$. Consider a forward biased *p-n* junction, in which current flowing through the junction produces an electron density n_e in the *p*-type layer. Use Figure 12.5 to find the minimum value of n_e for lasing to be possible (i) at 77 K, (ii) at 300 K. Neglect losses, and assume that ζ_h coincides with the valence band edge. Assume that in the *p*-type layer, $n_e \ll n_h \sim n_v$, and neglect the shift of ζ_h from its equilibrium value when current flows. Justify this neglect.

14.8 The Fermi energy is a convenient shorthand for characteristic quantities in a system. Assuming the non-relativistic Fermi energy is known for a material, show how other characteristic quantities can immediately be determined such as the Fermi temperature, Fermi momentum, and the Fermi velocity.

15

Phase Transitions

You can swim (uncomfortably) in water at a temperature slightly above freezing; a tiny drop in temperature—or a miracle—allows you to walk on water.

Bohren and Albrecht

Modern physics often advances only by sacrificing some of our traditional philosophical convictions.

Fritz London and Edmond Bauer, 1939

15.1 Gedanken

There's a certain irony about two of our most common examples we use in thermodynamics. We began this book motivating temperature by the expansion or contraction of a column of mercury whether the temperature is raise or lowered, respectively. For much of our beginning discussion we also use the ubiquitous example of water to appeal to the vast amount of personal experiences students have with that substance thermodynamically or otherwise. However, there's something extremely quirky about the world around us. When water freezes, it doesn't contract, as most substances do, but it expands. Perhaps no other single detail in the universe is ultimately so necessary for this existence we call life: if water behaved as most other substances do, we would not have ice floats on top of great oceans at the pole(s), but rather our existence, if there was any, would be confined to the edges of great ice caps that sank to the ocean floor until all the water on the planet froze. What a radically different world that would be to imagine.

15.2 First Order Transitions

With a few exceptions, the systems that we have considered so far are perfect gases; that is, systems in which the interactions between particles can be neglected. While nitrogen and oxygen at room temperature are good examples of such gases, they constitute only a tiny fraction of the substances which we come across in nature; furthermore, the intermolecular interaction makes all gases deviate greatly from ideal behavior at high pressure or low temperature. The most dramatic effect of this interaction is condensation of the gas to a liquid or solid. This occurs when the intermolecular attraction is strong enough, and the temperature low enough, that the energy gained by bringing the molecules into close proximity overcomes the reduction in entropy due to condensation.

We will make that general statement more precise later, but first we need to review the significance of the free enthalpy $G \equiv U + pV - TS$. As pointed out in Chapter 6, at a fixed temperature and volume, equilibrium is attained when the free energy $F \equiv U - TS$ is minimized. However, most processes are carried out at constant pressure rather than at constant volume; this is particularly true of phase transitions, many of which are accompanied by large changes in volume, and would produce dangerously high pressures if they were to occur at constant volume. Hence, the free enthalpy G is usually a more useful quantity than F, although the simple relation [Equation (6.29)] between F and the many-particle partition function makes F a more natural quantity to calculate in statistical mechanics. From Equation (8.9), the chemical potential of the i'th species is related to G by:

$$\zeta_i = \left(\frac{\partial G}{\partial N_i}\right)_{T,p},$$

and from Equation (8.12),

$$G = \sum_i N_i \zeta_i. \tag{15.1}$$

We also saw that at given temperature and pressure, equilibrium is attained when the free enthalpy G is minimized, so that in equilibrium:

$$dG = -S\,dT + V\,dp + \sum_i \zeta_i\,dN_i = 0. \tag{15.2}$$

In this chapter, we assume that there is only one chemical substance present, which can exist in two different phases. It is convenient to treat the phases as different species subject, of course, to conservation of the total number of particles. We label the two phases 1 and 2, where phase 1 is stable at high temperature and phase 2 at low. For definiteness, we take phase 1 to be a gas and phase 2 to be a condensed (that is, liquid or solid) phase, although the treatment applies equally well to other transitions; for instance, solid-liquid.

If dN particles condense at constant temperature and pressure, so that the change in the number N_1 of particles in the gas phase is $-dN$ and the change in the number N_2 in the condensed phase is $+dN$, the change in G is:

$$dG = \zeta_1\,dN_1 + \zeta_2\,dN_2 = (\zeta_2 - \zeta_1)\,dN \tag{15.3}$$

$$= 0 \quad \text{in equilibrium.}$$

Hence, if the two phases are to coexist, $\zeta_1 = \zeta_2$. If $\zeta_1 > \zeta_2$, G is decreased by condensation and the condensed phase is the stable one, while if $\zeta_1 < \zeta_2$, the gaseous phase is stable.

Since $F \equiv U - TS$, we can write:

$$\zeta_i = \left(\frac{\partial F}{\partial N_i}\right)_{T,V} = \left(\frac{\partial U}{\partial N_i}\right)_{T,V} - T\left(\frac{\partial S}{\partial N_i}\right)_{T,V}, \tag{15.4}$$

where $i = 1$ for the gas phase, and $i = 2$ for the condensed phase. As we saw in Chapter 10, in an ideal gas $\frac{\partial U}{\partial N_1}$ is small and positive, coming only from the additional kinetic energy carried by the particle, while $T\frac{\partial S}{\partial N_1}$ is large and also positive. Both terms increase with increasing temperature. Hence, ζ_1 is negative, and it becomes less negative as the temperature decreases [see Equation (9.15)]. Small deviations from ideality do not change this, and ζ_1 approaches zero from below as the gas cools from high temperature. In a liquid or solid, on the other hand, the interparticle potential, which is attractive and therefore negative, dominates U, so that $\frac{\partial U}{\partial N_2}$ is negative and is rather insensitive to temperature. In a liquid, $\frac{\partial S}{\partial N_2}$ is small relative to the gas, and it is smaller still in a solid, since there is little choice as to where an additional molecule can go. Hence ζ_2, while still negative, varies more slowly with temperature than ζ_1. If we cool from a high

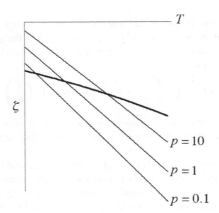

FIGURE 15.1 Schematic $\zeta(T)$ relations for a representative gas-liquid system (note that ζ is always negative in such a system, and that the origin is false for both ζ and T). The thin lines are for the gaseous phase (ζ_1) at three different pressures (in an arbitrary unit), while the thick line is for the liquid (ζ_2). The pressure dependence of ζ_2 is too small to see on this scale. For any given pressure p, ζ is the same for both phases at the point where the light and heavy lines cross, so that these crossing points define the coexistence line $T_t(p)$.

temperature, where $\zeta_1 < \zeta_2$ so that the gas is the stable phase, sooner or later we reach a temperature T_t, the transition temperature, where the $\zeta_1(T)$ curve crosses the $\zeta_2(T)$ curve; that is, $\zeta_1(T_t) = \zeta_2(T_t)$. At a given pressure, the two phases can coexist in equilibrium only at this temperature. If the temperature is reduced further, the condensed phase is the stable one. For reasons that will become clear later, such a transition is called a *first order phase transition*.

In general, ζ depends on pressure p as well as temperature, and the functions $\zeta_1(p, T)$ and $\zeta_2(p, T)$ can be represented by surfaces in three dimensional (p, T, ζ) space. Figure 15.1 shows sections through these surfaces at three different pressures for a representative gas-liquid system. Since the liquid is almost incompressible relative to the gas, the pressure dependence of ζ_2 is small and is omitted for clarity. At the intersection of the ζ_1 (gas) and ζ_2 (liquid) lines for a given pressure [that is, where $\zeta_1(p, T) = \zeta_2(p, T)$], the two phases can coexist in equilibrium. The points of intersection thus define the coexistence line $T_t(p)$.

Example 15.1: Gas-Solid Phase Transition

We can illustrate the above discussion more quantitatively by calculating the coexistence line $p(T_t)$ for a monatomic gas of spinless atoms in equilibrium with the Einstein solid described in Section 6.5 (solids are easier to model than liquids). We assume that the gas is ideal, that the volume of the solid is negligible compared to that of the gas, and that, relative to an atom at rest in the gas phase, an atom at rest in the solid phase has an energy $-E_0$, independent of temperature. To simplify the mathematics, we also assume that the transition temperature T_t is above the solid's Einstein temperature θ_e, which is defined in Example 6.3.

We first calculate the chemical potentials, ζ_1 and ζ_2, of the gas and of the solid as functions of pressure and temperature. The calculation for the gas is easy, since it is ideal. Combining Equation (9.15) with the equation of state $p = nk_BT$ gives

$$\zeta_1 = -k_BT \ln\left(\frac{n_q}{n}\right) = -k_BT \ln\left(\frac{k_BTn_q}{p}\right),$$

in which $n \ll n_q$ since the gas is assumed to be ideal.

The free energy of the solid has two terms: the potential energy $-NE_0$ and the vibrational free energy of $3N$ oscillators, which is $-3Nk_BT\ln Z$, where Z is the partition function of a single oscillator. From Equation (6.14), $Z \approx T/\theta_e$ when $T > \theta_e$, so that the chemical potential of the solid is

$$\zeta_2 = \left(\frac{\partial F}{\partial N_2}\right)_{T,V} \approx -E_0 - 3k_BT\ln\left(\frac{T}{\theta_e}\right).$$

The second term, which becomes increasingly negative as T increases because of the entropy's contribution to the free energy, is smaller than the corresponding term in ζ_1, since n_q/n is very large while T/θ_e is not.

At low temperature, $\zeta_2 \to -E_0$, while $\zeta_1 \to 0$, so that $\zeta_2 < \zeta_1$ and the solid is the stable phase. However, while both ζ_2 and ζ_1 decrease with increasing temperature, ζ_1 decreases more rapidly than ζ_2 because the entropy of the gas is much greater than that of the solid, so that ultimately the two curves cross. At the transition temperature T_t, $\zeta_2 = \zeta_1$ and the solid and gas coexist.

Equating ζ_1 and ζ_2 using Equation (7.31) to obtain n_q in terms of T_t, and solving for p, we find for the gas-solid coexistence curve:

$$p(T_t) = \left(\frac{mk_B^2\theta_e^2}{2\pi\hbar^2}\right)^{3/2} (k_BT_t)^{-1/2}e^{-E_0/k_BT_t}. \tag{15.5}$$

Equation (15.5) is usually called the *vapor pressure curve* (a "vapor" being a condensible gas), and gives the pressure of the gas in equilibrium with the solid at temperature T_t. While the details change from substance to substance, virtually all vapor pressure curves approximate to this exponential form. We shall see that this is a very general consequence of the requirement that the chemical potentials remain the same along the coexistence curve.

15.3 Superheating and Supercooling

At any temperature and pressure, the lower value of ζ at the corresponding point in Figure 15.1 gives the chemical potential of the stable phase. However, physical meaning can still be attached to the higher ζ (that is, ζ_1 when $T < T_t$, ζ_2 when $T > T_t$), even though this corresponds to the phase that is not stable at that temperature. Because there is usually an energy barrier that prevents a transition occurring from the higher ζ to the lower ζ phase, the higher ζ phase may be metastable (see Figure 6.5). For example, if one takes a gas at a given pressure and cools it below T_t, it does not immediately condense, since surface energy makes the formation of very small drops of liquid energetically unfavorable (this is made more quantitative in Example 15.2). Under such conditions, the gas phase is metastable and may survive for a long period if not disturbed; a phenomenon known as *supercooling*. Similarly, if a liquid is heated above T_t it does not immediately vaporize, and the metastable liquid is said to be *superheated*. Once bubbles of gas form in a superheated liquid, they may expand explosively, as every cook knows.

A familiar example of supercooling is the sodium acetate hand warmer. When a hot concentrated solution of sodium acetate in water is cooled to room temperature, the solution is metastable and the sodium acetate "wants" to crystallize out of solution. However, because the solid-liquid interface has a surface energy, there is an energy barrier to crystallization and the solution supercools. If undisturbed, the solution remains in this supercooled condition indefinitely, but if it is subjected to sudden shear (for instance, by squeezing the package containing it), the barrier is lowered and crystallization occurs, releasing the latent heat of solution. The sodium acetate can be redissolved by heating the package.

Supercooling of a gas below its condensation point was used by C. T. R. Wilson[1] to display the tracks of energetic charged particles in the *cloud chamber*. Air saturated with water vapor (that is, the partial

---

[1] Charles T. R. Wilson (1869–1959), Scottish meteorologist and physicist, won the 1927 Nobel Prize for his invention of the cloud chamber.

FIGURE 15.2 A cloud chamber photograph showing the ionization tracks of α-particles emitted by a polonium source. (From Rasetti, F., *Elements of Nuclear Physics*, Prentice Hall, New York, 1936.)

pressure of the vapor is equal to its condensation pressure at ambient temperature) is cooled by isentropic expansion, so that its temperature falls below T_t for water. Even though the bulk liquid now has a lower free enthalpy G than the gas and is therefore the stable phase, in clean air a small water droplet has a higher G than the same mass of gas because of its surface energy. Hence, a droplet cannot form and the vapor becomes supercooled. When a charged particle passes through the chamber, it ionizes some of the air molecules along its path; that is, it knocks electrons off the atoms, leaving behind a string of positively charged ions. These ions are attractive to liquid water, reducing G sufficiently to overcome the surface tension and permitting droplets to form (see Problem 15.12). Thus, a track of water droplets marks the passage of the particle. A photograph is taken before more water vapor has had time to condense, which enlarges the droplets and blurs the track. A typical cloud chamber photograph of particle tracks is shown in Figure 15.2. The modern version of the cloud chamber is the bubble chamber,[2] which is based on the same principle: but uses superheated liquid hydrogen rather than supercooled water vapor.

Example 15.2: Nucleation of a Liquid Drop in a Supercooled Vapor.

A free surface of a liquid has an energy σ per unit area (see Problem 2.5). Suppose that a monatomic gas at a given pressure is cooled to a temperature $T_t - \Delta T$, where T_t is the condensation temperature at that pressure, and that $\Delta T \ll T_t$.

1. Find the difference ΔG between the free enthalpy of N atoms in the gas phase and that of the same number in the form of a spherical liquid drop of radius R. The latent heat is L and the kmolar volume in the liquid state is $V_{m\ell}$, which is considerably less than the kmolar volume in the gaseous state.

---

[2] Donald A. Glaser (1926–2013), American biophysicist, won the 1960 Nobel Prize for this invention.

It follows from $S = -\left(\frac{\partial G}{\partial T}\right)_p$ [Equation (5.27)], $G = N\zeta$ in the bulk [Equation (8.10)], and $L = T(S_g - S_\ell)$, where the subscripts g and ℓ refer to the gaseous and bulk liquid phase respectively, that:

$$\left(\frac{\partial(\zeta_g - \zeta_\ell)}{\partial T}\right)_p = -\frac{L}{N_A T}.$$

At $T = T_t$, $\zeta_g = \zeta_\ell$, so that at $T = T_t - \Delta T$, where $\Delta T \ll T$,

$$\zeta_g - \zeta_\ell \approx -\left(\frac{\partial(\zeta_g - \zeta_\ell)}{\partial T}\right)_p \Delta T = \frac{L \, \Delta T}{N_A T}.$$

A drop of radius R has a surface contribution to the free enthalpy given by:

$$G_s = 4\pi R^2 \sigma = \frac{3V\sigma}{R},$$

where V is the volume of the drop. For N atoms $V = \frac{NV_{m\ell}}{N_A}$, so that the difference in free enthalpy is:

$$\Delta G = N(\zeta_g - \zeta_\ell) - G_s = \frac{N}{N_A}\left(\frac{L \, \Delta T}{T} - \frac{3V_{m\ell}\sigma}{R}\right)$$

2. Show that the drop evaporates unless R exceeds a critical value R_c, and find R_c in terms of L, σ, $V_{m\ell}$, and ΔT. Find R_c for water, for which $L = 45$ MJ/kmole, $\sigma = 0.075$ J m^{-2}, $V_{m\ell} = 0.018$ m^3, $T = 300$ K, $\Delta T = 30$ K (a typical value for a Wilson cloud chamber). The system "wants" to go to the lowest G, so that if $\Delta G < 0$ the drop will evaporate. This will happen if

$$R < R_c = \frac{3V_{m\ell}\sigma T}{L \, \Delta T}. \tag{15.6}$$

For water under the specified conditions, $R_c \approx 10$ nm. Note that while this sized drop is submicroscopic, it still contains $\sim 10^5$ molecules, far too many to accumulate spontaneously, so that pure water vapor will not spontaneously condense even when supercooled by more than 30 K.

15.4 Properties of the Coexistence Curve: The Clausius-Clapeyron Equation

On the coexistence curve between phases 1 and 2, the chemical potentials of the two phases are equal, so that both potentials must change by the same amount as we move along the curve. For an infinitesimal move along the coexistence curve, $d\zeta_1 = d\zeta_2$. To simplify the notation (in this section only), we drop the subscript "t" and write the coexistence curve as $p(T)$. If the pressure change is dp and the corresponding change in the transition temperature is dT, it follows that:

$$\left(\frac{\partial\zeta_2}{\partial p}\right)_T dp + \left(\frac{\partial\zeta_2}{\partial T}\right)_p dT = \left(\frac{\partial\zeta_1}{\partial p}\right)_T dp + \left(\frac{\partial\zeta_1}{\partial T}\right)_p dT, \tag{15.7}$$

where it must be remembered that dp and dT are not independent because we are following the coexistence curve. Rearranging Equation (15.7) we obtain:

$$\frac{dp}{dT} = \frac{\left(\frac{\partial \zeta_1}{\partial T}\right)_p - \left(\frac{\partial \zeta_2}{\partial T}\right)_p}{\left(\frac{\partial \zeta_1}{\partial p}\right)_T - \left(\frac{\partial \zeta_2}{\partial p}\right)_T}. \tag{15.8}$$

From the fact that dG in Equation (15.2) is a perfect differential, we can obtain two Maxwell relations [Equations (E.9) and (10.39)]:

$$\left(\frac{\partial \zeta}{\partial T}\right)_{p,N} = -\left(\frac{\partial S}{\partial N}\right)_{p,T}$$

and:

$$\left(\frac{\partial \zeta}{\partial p}\right)_{T,N} = \left(\frac{\partial V}{\partial N}\right)_{p,T}.$$

Substituting these into Equation (15.8) gives:

$$\frac{dp}{dT} = \frac{\dfrac{\partial S}{\partial N_1} - \dfrac{\partial S}{\partial N_2}}{\dfrac{\partial V}{\partial N_1} - \dfrac{\partial V}{\partial N_2}}, \tag{15.9}$$

where all the derivatives are at constant p and T. For given p and T, S and V are linear functions of the N_i, so that $\frac{\partial S}{\partial N_i}$ and $\frac{\partial S}{\partial N_i}$ are independent of N_i. Hence, the numerator in Equation (15.9) is the change in S when one particle is transferred from phase 2 to phase 1, and the denominator is the corresponding change in volume. Thus, Equation (15.9) can be written:

$$\frac{dp}{dT} = \frac{\Delta S}{\Delta V}, \tag{15.10}$$

where ΔS and ΔV are, respectively, the change in entropy and volume for a given number of particles (it is customary to choose this number to be Avogadro's number or N_A). Integration of Equation (15.10) gives the coexistence curve; that is, the relation between p and T which must hold if the two phases are in equilibrium.

The energy which must be supplied to change one kmole (N_A particles) from phase 2 to phase 1 at constant pressure and temperature (that is, the enthalpy change ΔH) is the kmolar latent heat L. At constant pressure, $dH = T\,dS$, so that $L = T\,\Delta S$ and Equation (15.10) can be written:

$$\boxed{\frac{dp}{dT} = \frac{L}{T\,\Delta V_m},} \tag{15.11}$$

where ΔV_m is the change in volume per kmole. This relation, which gives the slope of the coexistence curve, is called the Clausius-Clapeyron equation. Since L is positive by definition, the sign of ΔV_m determines the sign of the slope. For a liquid-gas transition, ΔV_m, and hence $\frac{dp}{dT}$, is positive, but for other transitions, this may not be the case. For example, the density of liquid water is greater than that of ice, so for the ice-water transition ΔV_m and $\frac{dp}{dT}$ are negative. Raising the pressure *lowers* the melting point of ice.

In a transition from a condensed phase to a gas, the volume of the condensed phase is often negligible, in which case ΔV_m is simply the kmolar volume of the gas V_m. If the gas can be approximated as ideal,[3] $V_m = \frac{R_0 T}{p}$. Substituting in Equation (15.11) gives:

$$\frac{dp}{dT} = \frac{Lp}{R_0 T^2}.$$

This can be written:

$$\frac{d(\ln p)}{d(T^{-1})} = -\frac{L}{R_0}; \tag{15.12}$$

that is, the slope of an Arrhenius plot of $\ln(p)$ against T^{-1} is proportional to L. This can be regarded as an example of Arrhenius' law (Equation 6.44), with L/N_A representing the activation energy for escape of a molecule from the liquid.

If the temperature dependence of L is ignored, Equation (15.12) can be integrated to give:

$$p = p_0 e^{-L/R_0 T} \tag{15.13}$$

where p_0 is a constant. In general, L is a function of temperature, but Equation (15.13) is usually quite a good approximation over a wide range of temperature, although it must be used with caution as a means of determining L (see Problem 6.8). A typical vapor pressure curve is shown in Figure 15.3, where $\log_{10} p$ for water vapor in equilibrium with liquid water or ice is plotted against $10^3/T$ from 213 K to 647 K, which is the critical point where the liquid-gas distinction vanishes (see later in this section).

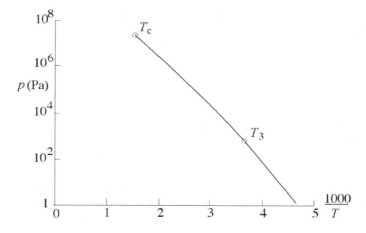

FIGURE 15.3 Arrhenius log-linear plot of vapor pressure against $10^3/T$ for H_2O. The curve stops at the critical point T_c (647 K), above which the distinction between gas and liquid disappears. Note the change in slope at the triple point T_3 (273 K) where ice, liquid water, and water vapor coexist (see Problem 15.13). At temperatures above T_3, the condensed phase is water, below T_3 it is ice.

---

[3] While this might seem to be a poor approximation, since the gas is close to turning into a liquid, it is usually quite accurate at pressures and temperatures which are not too close to the critical point. For example, at atmospheric pressure (1 bar), the density of steam at its condensation temperature (373 K) is only 1.5% greater than the ideal gas value, while C_p exceeds the ideal gas value of $4R_0$ (see Chapter 10) by only 10%, and varies little with temperature. On the other hand, at 80 bar (about 1/3 of the critical pressure), the density at the condensation temperature (568 K) exceeds the ideal gas value by 40% and C_p is about $10R_0$, varying strongly with T. These data are taken from R. W. Haywood, *Thermodynamic Tables in SI Units*, 2nd ed. (Cambridge University Press, Cambridge, UK, 1972).

While the Clausius-Clapeyron relation [Equation (15.11)] is a rigorous consequence of thermodynamics, Equation (15.13) depends not only on the assumption of ideal gas behavior, but also on the assumption that L is independent of temperature, which is certainly not the case. Furthermore, the constant p_0 is undetermined. One can do better by using a particular model for the condensed phase, such as the Einstein model that led to Equation (15.5).

Example 15.3: The Pressure Cooker

The boiling point of water at atmospheric pressure (1 bar) is 373 K, and the latent heat of vaporization at this temperature is 2.25 MJ/kg. Assuming the latent heat to be constant over the relevant temperature range, that water vapor is an ideal gas, and that the volume of the liquid is negligible relative to that of the vapor, find the temperature inside a pressure cooker operating at an absolute pressure of 2 bar.

Given these assumptions, Equation (15.13) applies, so that:

$$\Delta\left(\frac{1}{T}\right) = -\frac{R_0}{L}\,\Delta(\ln p).$$

where L is the kmolar latent heat and T is the absolute temperature.

The pressure is increased by a factor of 2, so that $\Delta(\ln p) = \ln 2 \approx 0.7$. $L = 18 \times 2.25$ MJ/kmole. Hence,

$$T = \left(\frac{1}{373} - 0.7\,\frac{8.3 \times 10^3}{18 \times 2.25 \times 10^6}\right)^{-1} = 394 \text{ K},$$

so that the temperature in the cooker is 394 K or 121 °C

The coexistence curve in the (T, p) plane is very useful, but it obscures one vital piece of information: the change in density at the transition. To reveal this, it is necessary to plot isotherms (lines of constant temperature) on a pressure-density (or volume) diagram. Two isotherms passing through a liquid-gas transition are shown schematically in Figure 15.4. Since we are varying p while keeping T constant, it is convenient to write p_t for the pressure at which the transition occurs when the temperature is T, and to write the coexistence curve as $p_t(T)$.

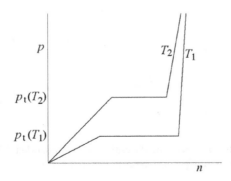

FIGURE 15.4 Pressure-density diagram for a condensible gas showing two typical isotherms. For simplicity we assume that the gas phase is ideal, so that $p \propto n$ until the gas begins to condense. The isotherms are flat in the coexistence region, within which n represents the average particle density N/V of the liquid-gas mixture.

We start with the gas at a low pressure and density, and compress at constant temperature T_1. When the transition pressure $p_t(T_1)$ is reached, the gas begins to turn to liquid (assuming that equilibrium is always maintained, so that there is no supercooling). As we continue to compress, liquid and gas coexist, so that the pressure remains constant at $p_t(T_1)$ until all the gas has condensed. In this coexistence region, the two phases are spatially separated; either a cloud of droplets forms, or else gravity causes the liquid to collect at the bottom of the vessel. Once the gas has condensed, one can reduce the volume further only by compressing the liquid, so that the pressure rises rapidly.

Now let's look at the curve for a higher temperature T_2. Since $\Delta V > 0$ for the liquid to gas transition, $\frac{dp_t}{dT} > 0$ from Equation (15.11). Usually $L \gg R_0 T$, so that raising the pressure increases the density of the gas on the coexistence curve much more rapidly than raising the temperature reduces it (see Problem 15.2). It follows that as one compresses the gas isothermally, one reaches the coexistence line at a *higher* density at a higher temperature. The molar volume V_m of the gas on the coexistence line is correspondingly smaller at the higher temperature. On the other hand, the liquid is not very compressible, and expands as the temperature is raised, so that its density at the transition is *less* at T_2 than at T_1, and its molar volume is greater at the higher temperature. Hence, the change of density Δn as one goes from gas to liquid, and the corresponding change in volume $|\Delta V|$, decreases as the temperature rises, as indicated in Figure 15.4.

The extra degrees of freedom (in the sense used in the equipartition theorem; see Appendix G) associated with intermolecular interactions make the heat capacity larger in the condensed than in the gaseous phase, so that as the temperature rises, the enthalpy of the liquid increases faster than that of the gas.[4] Hence for a gas-liquid transition the latent heat $L(= \Delta H)$ also decreases with temperature. This will be made more quantitative in Section 15.6, where we consider the van der Waals model of a condensible gas.

The locus of the points where the isotherm has an abrupt change in slope is shown by the dotted line in Figure 15.5. The gas and liquid coexist in the *coexistence region* enclosed by this line. As the temperature is raised, Δn decreases, so that the width of the coexistence region (that is, the length of the flat portion of the isotherm in Figures 15.4 and 15.5) decreases, and ultimately disappears at the *critical point* (T_c, p_c). At this point, the distinction between liquid and gas disappears and the latent heat goes to zero. The *critical isotherm* [that is, the $p(n)$ curve at the critical temperature T_c], has zero gradient and a point of inflection at the critical point. At higher temperatures still, the isotherms have no flat region, indicating that there is no region where two phases coexist. In fact, the phase transition has disappeared; as the pressure increases,

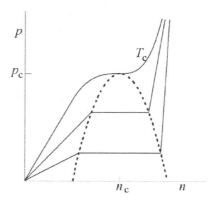

FIGURE 15.5 Pressure-density diagram including the critical isotherm (labeled T_c). The coexistence region is the region delineated by the dotted line. It disappears at the critical point (n_c, p_c).

---

[4] For example, the kmolar heat capacity at constant pressure of liquid water at room temperature is 75 kJ/kmole $\approx 9R_0$, while that of the vapor, treated as an ideal gas with three rotational degrees of freedom, is $4R_0$.

the gas steadily becomes more dense and more and more resembles a liquid. If the pressure is held at the critical value p_c and the temperature is varied, there is a transition at T_c, called a *second order phase transition*. A second order phase transition is a transition in which there is no latent heat, but there is a discontinuity in the heat capacity. Second order phase transitions are most easily introduced in the context of magnetism, which we discuss in the next section, returning to the liquid-gas case later.

15.5 Second Order Transitions: The Ferromagnet: Order-Disorder Transitions

A *ferromagnet* is a material that has a spontaneous magnetic moment even in zero field. We shall see that it is the simplest system exhibiting a second order phase transition. Our model of the ferromagnet is based on that of Weiss.[5] We first consider a collection of independent spin 1/2 elementary magnets (spins), each with magnetic moment μ. We discussed this system in Chapters 4 and 6, and found it to be a paramagnet. In a magnetic field B at temperature T, the magnetization (magnetic moment per unit volume), which is the vector sum of all the spins, is $m = n\langle\mu_z\rangle$ where n is the number of spins per unit volume and $\langle\mu_z\rangle$ is the average moment per atom, projected on the direction of the applied magnetic field B. This average is, from Equation (6.46),

$$\langle\mu_z\rangle = \mu\tanh(\beta\mu B)$$

(see Example 6.5 and Problem 6.3). Here, as always, $\beta = \frac{1}{k_B T}$.

Our model of a ferromagnet starts with the same collection of spins, but assumes that there is an interaction between them, so that the energy of a spin depends on its direction relative to the directions of neighboring spins.[6] In the Weiss model this interaction is represented by a molecular (or internal) magnetic field, which is assumed to be proportional to the magnetization m, where $m = n\langle\mu_z\rangle$. Thus, each elementary magnet is assumed to behave as if it were subject to a magnetic field $B_{int} = \lambda_w m$, where λ_w is a constant, called the *molecular field parameter*, which represents this hypothetical interaction and whose magnitude is chosen to fit experiment. The assumption that the molecular field is proportional to the magnetization, which is an average property of the system, is called the *mean field approximation*.

If no external field is applied, the average moment is given by Equation (6.46), with B_{int} substituted for B. Since $B_{int} = \lambda_w m$, m is given by the solution of the transcendental equation:

$$m = n\langle\mu_z\rangle = n\mu\tanh(\beta\mu B_{int}) = m_0\tanh(\beta\mu\lambda_w m), \tag{15.14}$$

where $m_0 \equiv n\mu$ is the *saturation magnetization*; that is, the magnetization that would be obtained if all the spins were lined up in the same direction. We can simplify Equation (14.13) by making the substitutions $t \equiv \frac{T}{T_c} - 1$, where $T_c \equiv \frac{\mu\lambda_w m_0}{k_B}$, and $x \equiv \frac{m}{m_0(1+t)}$. The significance of T_c and the reason for this choice of variables will become clear later. With these substitutions, Equation (15.14) becomes:

$$(1 + t)x = \tanh(x). \tag{15.15}$$

Equation (15.15) can be solved graphically, as shown in Figure 15.6, in which we plot the functions $y = \tanh(x)$ and $y = (1 + t)x$ on the same graph. The intersection of the two curves gives the solution to Equation (15.15), the value of y at that point being m/m_0. The function $(1 + t)x$ is shown for two cases: (a) $t > 0$ and (b) $t < 0$. Since the initial slope of $\tanh(x)$ is 1, the only solution for $t > 0$ is $x = 0$,

---

[5] Pierre Weiss (1864–1940), French physicist, introduced his model in "The molecular field hypothesis and ferromagnetism," *Journal de Physique*, ser.4, **6**, 661 (1907).

[6] This interaction is now known to arise from the exchange energy, which depends on the relative orientation of the spins (see any quantum mechanics text).

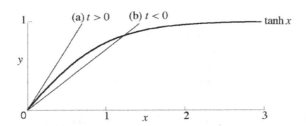

FIGURE 15.6 Graphical solution of Equation (15.15). The heavy line is the right-hand side of Equation (15.15), while the light lines are the left-hand side for (a) positive t and (b) negative t (corresponding to T greater than and less than T_c).

$y = 0$; that is, $m = 0$. This means that if $T > T_c$, the material has no magnetization in zero field and is a paramagnet. However, if $T < T_c$, so that $t < 0$, there is an additional solution with $x \neq 0$. It is easily seen that in this case the $x = 0$ solution is not stable: If m has small fluctuation from zero, B_{int} increases so that m increases further. The other solution, which is stable, gives non-zero y, so that there is a spontaneous magnetization $m = y m_0$. As $T \to 0$ (that is, $t \to -1$) the line $y = (1 + t)x$ becomes horizontal, so that $x \to \infty$ and $y \to 1$ at the point of intersection. The temperature T_c, where $t = 0$, above which the spontaneous magnetization vanishes, is called the Curie temperature.

Note that when $t = 0$, the two curves have the same slope at the origin; they touch, but do not cross. It follows that as $t \to 0$ from below, the point of intersection goes to the origin, so that the magnetization goes to zero and has no discontinuity at T_c. Unlike the case of a first order transition, nothing changes abruptly at a second order transition; it only begins to change. We shall see later that the same is true of the critical point of a condensible gas, mentioned in the previous section, and that in fact such critical phenomena are observed in a wide variety of systems and have many features in common.

When $T > T_c$, one must apply an external field B to obtain non-zero magnetization. Then Equation (15.14) becomes:

$$m = m_0 \tanh[\beta \mu (\lambda_w m + B)]. \tag{15.16}$$

If we assume that B and m are small, the argument of the hyperbolic tangent is small and, since $\tanh(x) \approx x$ for small x, we can write:

$$m \approx m_0 \beta \mu (\lambda_w m + B). \tag{15.17}$$

Solving for m, we obtain for the susceptibility:

$$\chi \equiv \frac{m}{B} = \frac{m_0 \beta \mu}{1 - m_0 \beta \mu \lambda_w} = \frac{n \mu^2}{k_B (T - \theta_w)}, \tag{15.18}$$

where $\theta_w \equiv \frac{\mu \lambda_w m_0}{k_B} = T_c$. Equation (15.18) is known as the Curie-Weiss law. The susceptibility in this model has the same form as the Curie susceptibility [Equation (6.39)] for non-interacting spins, but with $T - T_c$ substituting for T, so that χ diverges at T_c. This divergence is observed experimentally, and Equation (15.18) usually holds for T, not too close to T_c, but the fitted value of the Curie-Weiss temperature θ_w differs somewhat from T_c. Also, many materials have $\theta_w < 0$ and do not show spontaneous magnetization; these are known as *antiferromagnets* and can be qualitatively described in terms of the Weiss model

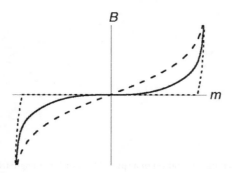

FIGURE 15.7 $B(m)$ isotherms in the Weiss model. Full line: $T = T_c$ (the critical isotherm); dashed line: $T > T_c$; and dotted line: $T < T_c$.

with $\lambda_w < 0$. In an antiferromagnet, neighboring spins are aligned antiparallel, rather than parallel as in a ferromagnet. This alignment can be detected by neutron diffraction.[7]

Equation (15.18) diverges when $T = \theta_w = T_c$, and the dependence of m on B ceases to be linear even at small B. It can be shown (see Problem 15.3) that the model predicts that at $T = T_c$, $m \propto B^{1/3}$ for small B. The critical isotherm $B(m)$, ($T = T_c$), is thus a cubic with a point of inflection and zero slope at $B = 0$. Isotherms of $B(m)$ for T less than, equal to, and greater than T_c are shown schematically in Figure 15.7. They are plotted this way to bring out the resemblance to the isotherms of a gas-liquid system in the vicinity of the critical point (compare Figure 15.5). We shall see in the next section that the van der Waals model of a condensible gas has a close analogy to the Weiss model of a ferromagnet.

We now take a closer look at the spontaneous magnetization in the region just below T_c. We expect m and therefore x to be small. However, the approximation $\tanh(x) \approx x$ in Equation (15.15) gives no solution, so that we have to extend the Taylor series for $\tanh(x)$ to the next non-zero term, which is $-x^3/3$. Substituting for $\tanh(x)$ in Equation (15.15), we obtain

$$x(1 + t) \approx x \left(1 - \frac{x^2}{3} \right),$$

with the only non-trivial solution, existing only if $t < 0$,

$$\frac{m}{m_0} \approx x = (3|t|)^{1/2}. \tag{15.19}$$

Thus, near T_c we expect $m \propto |t|^{1/2} \propto |T_c - T|^{1/2}$. Exactly at T_c there is no spontaneous magnetization, but as soon as T drops below T_c the magnetization increases rapidly.

The complete magnetization curve $m(T)$, obtained by solving Equation (15.18) numerically, is shown by the full line in Figure 15.8. Typical experimental data are shown by the open circles. They agree qualitatively with the theoretical curve, but deviate from it in two ways. One difference is that as T increases from zero, the observed m drops off more rapidly than predicted. This is due to the excitation of spin waves, which reduce the magnetization[8] (see Example 11.2). The fractional reduction in the magnetization,

---

[7] W. Marshall and S. W. Lovesey, *Theory of Thermal Neutron Scattering: the Use of Neutrons for the Investigation of Condensed Matter* (Oxford University Press, Oxford, UK 1971); L. J. de Jongh and A. R. Miedema, "Experiments on simple magnetic model systems," *Advances in Physics* **23**, 1–260 (1974).

[8] Since spin waves involve the *correlated* motion of neighboring spins, they cannot exist in the Weiss model, which treats the spins as uncorrelated.

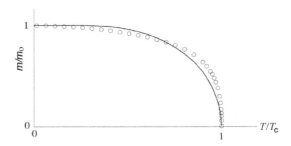

FIGURE 15.8   Full line: fractional magnetization of a spin 1/2 Weiss ferromagnet, as a function of reduced temperature T/T_c. Circles: data for Fe ($T_c = 1043$ K). From *Landolt-Bornstein Tables*, new series (gen. eds. K.-H. Hellwege and O. Madelung), volume 19a (ed. H. P. J. Wijn, compiled by M. B. Stearns), p. 37 (Springer, 1986).

$1 - m/m_0$, is proportional to the number of spin waves excited, which varies as $T^{3/2}$ at low T, whereas the Weiss model predicts that:

$$1 - \frac{m}{m_0} = 2e^{-2T_c/T}, \tag{15.20}$$

(see Problem 15.16).

The other difference is that m increases more rapidly than predicted as T drops below T_c. It is found that the dependence of m on T near T_c can be described by a power law: $m \propto |t|^\beta$, where β (not to be confused with $\beta = \frac{1}{k_B T}$) is called a critical exponent. The Weiss model predicts $\beta = \frac{1}{2}$, but the observed value is about $\frac{1}{3}$. A similar discrepancy is found in many other systems that exhibit second order phase transitions. We shall see that this discrepancy is not due to the details of the model, but is characteristic of all *mean field* models; that is, models in which the interaction between individual elementary units such as spins or molecules is replaced by an average.

The discrepancy between the magnetization predicted by mean field theory and observation may not look very pronounced, but the discrepancy in the heat capacity is much more dramatic. If we substitute B_{int} for $\mu_0 H$ in Equation (2.15), we find that the magnetic contribution to the internal energy of a ferromagnet of volume V in zero external field is

$$U_m = -\int_0^M M\, dB = -\int_0^m Vm\lambda_w\, dm = -\frac{1}{2}V\lambda_w m^2.$$

The magnetic contribution to the heat capacity of a ferromagnet per unit volume is thus, according to the Weiss model,

$$C_m \equiv \frac{1}{V}\frac{dU_m}{dT} = -\frac{1}{2}\lambda_w \frac{d(m^2)}{dT}.$$

Above T_c, $m = 0$, so that $C_m = 0$, while just below T_c, Equation (15.19) gives $m^2 \approx 3m_0|t|$.

Hence, just below T_c

$$C_m(t \to -0) = \frac{3\lambda_w m_0^2}{2T_c} = \frac{3}{2}k_B n, \tag{15.21}$$

while just above T_c, $C_m(t \to +0) = 0$. Thus, the model predicts a heat capacity which is finite, but has a large discontinuity at T_c.

These predictions completely disagree with experiment. While it remains true that $\int_{T_c-\delta T}^{T_c+\delta T} C_m \, dT$ vanishes as $\delta T \to 0$, unlike a first order transition where this integral converges to the latent heat, the heat capacity is found to diverge at T_c. Typically, $C_m \propto |t|^{-\alpha}$, where $\alpha \sim 0.1$. Furthermore, this divergence is observed not only when T_c is approached from below, but also when it is approached from above. In the region $t > 0$, the magnetic heat capacity in zero field, which is predicted by the model to be zero, is found experimentally to tend to infinity as $t \to 0$! For a thorough discussion of the data on a wide variety of magnetic materials, see the article by de Jongh and Miedema referred to in Footnote 7. This dramatic disagreement is due to fluctuations, as will be discussed at the end of this chapter.

Example 15.4: (An Order-Disorder Transition)

A system mathematically analogous to the spin 1/2 ferromagnet in zero external field, also showing a second order transition, is a 50:50 binary alloy which can exist in either an ordered or a disordered phase. As a simple example, consider β-brass, whose structure was described in Example 4.2. This alloy contains equal numbers of copper and zinc atoms, which we call A and B, and there are equal numbers of two types of sites, labeled a and b. Each a site is surrounded entirely by b sites, and vice versa (see Figure 4.6). The disordered phase, in which the a and b sites are occupied at random by A and B atoms, has higher entropy than the ordered phase, in which one type of site is occupied entirely by A atoms and the other site by B atoms. On the other hand, the ordered phase may have lower energy. If it does, we would expect it to be stable at low temperature, while the disordered phase should be stable at high temperature. However, the case is more complex than in the first order transitions considered in Section 15.1, since there can be varying degrees of disorder, and, as in the case of a ferromagnet, the transition is second order rather than first.

We use a simplified version of the Bragg-Williams mean field theory[9] of such a system. Let the fraction of a sites occupied by A atoms (which is equal to the fraction of b sites occupied by B atoms) be f. The quantity analogous to the magnetization in the Weiss model is the order parameter $\psi \equiv 2f - 1$, since $f = 1/2$ in the completely disordered phase and f can be either 0 or 1, corresponding to $\psi = \pm 1$, in the completely ordered phase.

Think of the line joining nearest neighbor atoms as a "bond" with a certain energy. The structure is such that all such lines connect an a to a b site. We assume for simplicity that all bonds between like atoms have the same energy, and take this as our zero of energy, while a bond between unlike atoms has energy ϵ. All interactions other than those between nearest neighbors are neglected. We shall see that if ordering is to occur ϵ must be negative; that is, the attractive force between unlike atoms must exceed that between like atoms. If the structure is body-centered cubic, as shown in Figure 4.6, each atom has eight bonds to its neighbors, and each bond is shared by two atoms, so that there $4N$ bonds per unit volume, where N is the total number of atoms.

1. Find the fraction of bonds that are between unlike atoms, and the total energy of these bonds in terms of ψ.

The combined probability of an a site being occupied by atom A and a b site by B is f^2. Similarly, the combined probability of an a site being occupied by atom B and a b site by A is $(1-f)^2$. Hence, the total fraction of AB bonds (bonds between unlike atoms) is $f^2 + (1-f)^2$. Similarly, the combined fraction of AA and BB bonds is $2f(1-f)$. These fractions add up to 1, as they must.

---

[9] See C. N. R. Rao and K. J. Rao, *Phase Transitions in Solids* (McGraw-Hill, New York, 1978), pp. 184–186, for a fuller account of this theory.

Since the energy of an AA bond and that of a BB bond is taken as zero, the total energy of $4N$ bonds is:

$$U = 4N\epsilon \left[f^2 + (1 - f)^2 \right]$$
$$= 2N\epsilon(1 + \psi^2),$$

where we have substituted $f = \frac{1+\psi}{2}$, $1 - f = \frac{1-\psi}{2}$.

2. Calculate the entropy as a function of ψ, and hence find the free energy $F(\psi)$. Minimize F to obtain an equation for the equilibrium value of ψ. Find the conditions for this value to be non-zero.

Substituting the multiplicity Ω from Equation (4.7) in $S = k_B \ln \Omega$, we find:

$$S = Nk_B \left\{ \ln 2 - \frac{1}{2} \left[(1 + \psi) \ln(1 + \psi) + (1 - \psi) \ln(1 - \psi) \right] \right\},$$

where we have dropped terms $\sim \ln N$, which are small relative to N when N is large.

$$F = U - TS$$
$$= N \left[2\epsilon(1 + \psi^2) \right.$$
$$\left. + k_B T \left(\frac{1}{2} \left[(1 + \psi) \ln(1 + \psi) + (1 - \psi) \ln(1 - \psi) \right] - \ln 2 \right) \right].$$

Note that F is an even function of ψ.

Minimizing F with respect to ψ:

$$\left(\frac{\partial F}{\partial \psi} \right)_T = N \left\{ 4\epsilon \psi + \frac{1}{2} k_B T \left[1 + \ln(1 + \psi) - 1 - \ln(1 - \psi) \right] \right\}$$
$$= N \left[4\epsilon \psi + \frac{1}{2} k_B T \ln \left(\frac{1 + \psi}{1 - \psi} \right) \right]$$
$$= 0.$$

Solving for ψ;

$$\psi = -\frac{k_B T}{8\epsilon} \ln \left(\frac{1 + \psi}{1 - \psi} \right),$$

which can be rearranged to give:

$$\psi = -\tanh(4\beta\epsilon\psi), \tag{15.22}$$

where $\beta = \frac{1}{k_B T}$.

Since $\tanh(x)$ has the same sign as x, and $\beta > 0$, Equation (15.22) has no non-zero solution for any T if $\epsilon > 0$. This makes physical sense: If the bonds between like atoms have lower energy than those between unlike atoms, ordering in the sense that we have defined it is energetically unfavorable (there can be another sort of ordering, called *disproportionation*, where like atoms cluster together and one no longer has an alloy).

If $\epsilon < 0$, Equation (15.22) has exactly the same mathematical form as Equation (15.14), with ψ substituting for m/m_0, and $4|\epsilon|$ for $\mu\lambda_w m_0$. It can have non-zero solutions if and only if $T \leq T_c$, where $T_c = \frac{4|\epsilon|}{k_B}$.

3. Examine the behavior of $\psi(T)$, and give an analytic expression for it just below T_c.

Since the equation for ψ is identical, when expressed in terms of T/T_c, to the equation for m/m_0 in the Weiss model, its temperature dependence is the same as that shown for m/m_0 in

Figure 15.8. Just below T_c, ψ is small, so that we can expand Equation (15.22) to third order in ψ and obtain:

$$\psi = \pm|3t|^{1/2}, \quad \text{where} \quad t \equiv \frac{T}{T_c} - 1,$$

just as in Equation (15.19).

As in the magnetic case, the observed behavior of ψ agrees qualitatively with the predictions of this model, but quantitatively, the data near the transition deviate considerably.[10] Furthermore, just as in the magnetic case, the observed heat capacity in the vicinity of T_c differs drastically from the prediction of the mean field model (see Problem 15.19).

15.6 The van der Waals Model of an Imperfect Gas

One might think that the discrepancies between the Weiss and Bragg-Williams theories and experiment are specific to the models, but we shall find that, in fact, they are unavoidable consequences of very general assumptions about the behavior of the free energy at a critical point. Before introducing the general theory, we will develop a model of an imperfect gas, known as the van der Waals model.[11] An *imperfect gas* is one in which interaction between the molecules significantly affects the thermodynamic properties. We shall see that in spite of the fact that the imperfect gas is physically a very different system from the ferromagnet or the binary alloy, near the critical point the van der Waals model makes predictions that are closely analogous to those of the Weiss and Bragg-Williams models (Weiss was in fact aware of the analogy when he introduced his molecular field hypothesis of ferromagnetism). Furthermore, the two models disagree with experiment in analogous ways.

A typical intermolecular interaction potential[12] is illustrated schematically by the full line in Figure 15.9. The interaction can be separated into two parts: a strong short-range repulsion and a weaker long-range attraction. The potential assumed in the van der Waals model is shown by the dashed line. The model makes two approximations. At short distances, the interaction is assumed to be a *hard sphere repulsion* like the interaction between colliding billiard balls. Two molecules cannot occupy the same volume, but it is assumed that if they do not touch, there is no repulsive force between them. The model thus approximates the effect of the short range interaction by reducing the volume available for molecular motion from V to $V - Nb$, where N is the number of molecules and b the *co-volume*; that is, the volume from which the center of any particular molecule is excluded because of the presence of all the other molecules (b was estimated by van der Waals to be four times the actual molecular volume). If this hard sphere repulsion were the only interaction the effect would simply be to replace the true particle density, $n = \frac{N}{V}$, by an effective density $\frac{N}{V-Nb}$.

The second approximation concerns the long range attractive potential, that is more difficult to treat. For this, the model makes an assumption which is analogous to the mean field approximation ($B_{int} = \lambda_w m$) of the Weiss theory: Each molecule is assumed to move in an average potential which

---

[10] See E. W. Elcock, *Order-Disorder Phenomena* (Methuen, London, UK, 1956), pp. 89–91.

[11] Johannes Diderik van der Waals (1837–1923), Dutch chemical physicist, winner of the Nobel Prize in 1910. Because van der Waals had no Latin, knowledge of which was at that time legally required of Leiden students, legislation was required before he could present his doctoral thesis *On the Continuity of the Gaseous and Liquid State* [University of Leiden, 1873; English translation edited and introduced by J. S. Rawlinson (North Holland, 1988)] which contained the theory outlined here. This thesis is probably the most influential Ph.D. thesis ever written, in the sense that its influence has lasted longest. In the 30 years *since its centenary*, well over 400 papers deriving directly from it have been published, and in spite of its limitations, it is still probably the most widely used model of an imperfect gas.

[12] The potential shown is the Lennard-Jones or "6-12" potential, $V(R) = V_0[(R/R_0)^{-12} - 2(R/R_0)^{-6}]$, where R_0 and $-V_0$ are the values of R and $V(R)$ at the minimum of V.

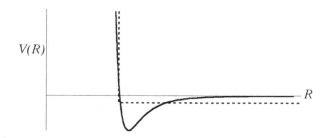

FIGURE 15.9   Full line: a typical intermolecular interaction potential $V(R)$, as a function of the separation R between the centers of two molecules. Dotted line: the model potential used in van der Waals' theory. The vertical section of the potential represents the minimum distance of approach of two molecules, assumed to be hard spheres, while the horizontal section represents a constant potential $-an$, where n is the density and a is a positive constant to be fitted to experiment.

is independent of position and is proportional to the density.[13] Since the force is attractive, the potential is negative. All the energy levels are assumed to be moved down by the same amount, so that their occupancies do not change and the entropy is unaffected by this interaction. With these assumptions, the contribution of the potential to the free energy per particle is $-an$, where a is a positive constant.

With these two approximations, the free energy of a van der Waals gas (that is, a gas conforming to the van der Waals model) is:

$$F = Nk_BT \left\{ \ln\left[\frac{N}{n_q(V - Nb)} \right] - 1 \right\} - a\frac{N^2}{V} \tag{15.23}$$

where the first term is simply the free energy of an ideal gas [Equation (10.9)] with n replaced by $\frac{N}{V-Nb}$, and the second term is the contribution of the attractive interparticle potential $-an = -a\frac{N}{V}$ for N particles.

van der Waals' equation of state is obtained by differentiating Equation (15.23) with respect to volume:

$$p = \left(\frac{\partial F}{\partial V} \right)_{T,N} = \frac{Nk_BT}{V - Nb} - a\frac{N^2}{V^2}$$

$$= \frac{nk_BT}{1 - bn} - an^2 \tag{15.24}$$

since $n = N/V$. At low density ($bn \ll 1$, $an \ll k_BT$), the equation of state reduces to $p = nk_BT$, the ideal gas law. At high density $p \to \infty$ as $n \to b^{-1}$, which is the density at which the molecules are packed as closely as they can be.

For a given temperature, Equation (15.24) is a cubic equation in n, which is monotonic at high T, but is S-shaped at low T, as illustrated by the isotherms plotted in Figure 15.10. The critical isotherm,

---

[13] This assumption can be shown to be rigorously valid in the limit of an infinitesimally weak potential with infinite range [J. L. Lebowitz, "Exact derivation of the van der Waals equation," *Physica* **73**, 48 (1974)]. See B. H. Flowers and E. Mendoza, *Properties of Matter* (John Wiley & Sons, New York, 1970), for a simple, but thorough discussion of the microscopic basis of the van der Waals model. The model is, of course, approximate, but is based on physics as good as any other mean field model, and the equation of state is derived rigorously from it. The disdain with which it is sometimes treated by theoretical physicists (for an example, see Problem 15.21) is quite unjustified. This disdain may be partly due to the bad arguments which are sometimes used to justify the model. For example, I have heard a distinguished professor of chemistry justify the attractive term in the equation of state Equation (15.24), which varies as V^{-2}, on the grounds that the long-range attractive intermolecular potential is known to vary as R^{-6} (where R is the interatomic separation) and $\langle R^{-6} \rangle \propto V^{-2}$ (see Footnote 12). In fact, the model assumes an interaction potential proportional to n, which varies as $\langle R^{-3} \rangle$. The V^{-2} dependence comes from the decrease in density as the volume increases, as can be seen from the derivation of Equation (15.24).

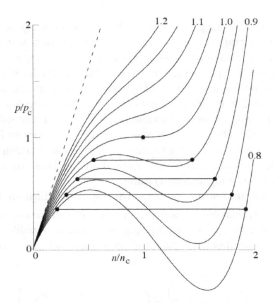

FIGURE 15.10 van der Waals' equation of state [Equation (15.26)] in terms of the reduced variables n/n_c and p/p_c (curves), with T/T_c as parameter. The horizontal straight lines show the true isotherms in the liquid-gas coexistence region (see text), the filled circles indicating the limits of this region. The dashed line shows the ideal gas isotherm (Boyle's law) for $T = 1.2T_c$.

marked "1.0," has zero slope at its point of inflection, marked with a filled circle. As we shall see, this is the critical point where the distinction between vapor and liquid vanishes. We denote the temperature, pressure, and density at the critical point as T_c, p_c, and n_c, respectively. Differentiating Equation (15.24) and finding the point of inflection (see Problem 15.7) gives:

$$T_c = \frac{8a}{27bk_B}, \qquad p_c = \frac{a}{27b^2}, \qquad n_c = \frac{1}{3b}. \tag{15.25}$$

It is convenient to write Equation (15.24) in terms of the dimensionless "reduced" variables, $\frac{p}{p_c}$, $\frac{n}{n_c}$, and $\frac{T}{T_c}$. In terms of these variables, Equation (15.24) takes the "universal" form:

$$\boxed{\frac{p}{p_c} = \frac{8}{3\left(\frac{n_c}{n}\right) - 1}\left(\frac{T}{T_c}\right) - 3\left(\frac{n}{n_c}\right)^2.} \tag{15.26}$$

The isotherms predicted by Equation (15.26) are plotted in Figure 15.10, with $\frac{T}{T_c}$ as parameter.

We see from Equation (15.26) that the equation of state of all van der Waals gases is the same when expressed in terms of the reduced variables $\frac{p}{p_c}$, $\frac{n}{n_c}$, and $\frac{T}{T_c}$. This is an example of the *law of corresponding states*. Even though the equations of state of real gases differ from Equation (15.24), it can be shown[14] that the equation of state obeys the law of corresponding states for all gases in which the intermolecular force law has the same mathematical form; that is, the potential is the same except for scale factors on energy and distance (this is true, e.g., for the Lennard-Jones potential given in Footnote 12, in which only the

---

[14] See D. ter Haar, *Elements of Statistical Mechanics*, 1st ed. (Rinehart, New York, 1954), pp. 180–182; G. D. J. Phillies, *Elementary Lectures in Statistical Mechanics* (Springer, New York, 2000), pp. 260–262.

parameters V_0 and R_0 vary from one gas to another). It is found to be a good approximation for a large number of gases over a wide range of parameters; that is, the equation of state is very closely the same when expressed in terms of the reduced variables. (Readers are encouraged to re-read the quote by Leo Kadanoff at the start of Chapter 1.) For example, according to the van der Waals model, $\frac{p_c}{n_c k_B T_c} = 0.375$. The observed value of this ratio is within $\pm 5\%$ of 0.294 in most monatomic and homonuclear diatomic gases, whose T_c ranges from 5 to 600 K. The limits of the coexistence region of different gases near the critical point, expressed in terms of reduced variables, show the same universality, as is discussed in more detail below (see Figure 15.13). The law of corresponding states was very useful in the early days of low temperature physics when Dewar, Kamerling-Onnes,[15] and others were attempting to liquefy gases, since it enabled them to predict the critical temperature and the Joule-Thomson inversion temperature (see Problem 5.10) from data at much higher temperatures.

While the physical significance of the isotherms with $\frac{T}{T_c} > 1$, which are monotonic, is clear, those for $\frac{T}{T_c} < 1$ need some interpretation. In particular, a real isotherm cannot have a negative slope, since this corresponds to a negative compressibility, making the substance unstable (see Problem 15.8).[16] What this means can be seen by considering the chemical potential, which can be obtained by differentiating Equation (15.23):

$$\zeta = \left(\frac{\partial F}{\partial N} \right)_{T,V} = k_B T \left[\ln \left(\frac{n}{n_q(1-bn)} \right) + \frac{bn}{1-bn} \right] - 2an, \tag{15.27}$$

or, in terms of reduced variables,[17]

$$\frac{\zeta}{k_B T_c} = \frac{T}{T_c} \left[\ln \left(\frac{3n_c}{n_q} \right) - \ln \left(\frac{3n_c}{n} - 1 \right) + \frac{1}{3n_c/n - 1} \right] - \frac{9n}{4n_c}. \tag{15.28}$$

Equation (15.28) is plotted for some values of $\frac{T}{T_c}$ between 0.8 and 1.2 in Figure 15.11(a). We see that for $T < T_c$ the $\zeta(n)$ curve is S-shaped, and that over a certain range, there are three values of n which have the same value of ζ. For the middle value of n, $\frac{dp}{dn} < 0$; that is, the compressibility is negative so that the system is unstable. The outer two values are both stable and correspond to two different phases which are in equilibrium with each other (since their chemical potentials are equal) and can coexist if the pressure is the same. Since one of the phases has a much higher density and lower compressibility than the other, we can identify the two phases as the liquid and gas. Thus, if for a given temperature we solve Equation (15.26) for $n(p)$, we can then use Equation (15.28) to obtain $\zeta(p)$. The results of such a calculation are shown in Figure 15.11b. The two lower branches correspond to the two phases, and the point where they cross gives the transition pressure at that temperature. This pressure is indicated by the horizontal line connecting the filled circles in Figure 15.10. Between these circles the two phases coexist and the true isotherm is the horizontal line.[18]

---

[15] Heike Kamerling-Onnes (1853–1926), Dutch physicist and head of physics at Leiden University for 42 years, was the first to succeed in liquefying helium, for which he received the Nobel Prize in 1913.

[16] The heat capacity is also negative in this region, so that the system is thermodynamically as well as mechanically unstable (see Footnote 16 of Chapter 5 and the paper by Lynden-Bell referred to therein).

[17] It might appear that Equation (15.28) is not universal in the sense that Equation (14.25) is, since it contains n_q, which depends on the spin and mass of the particles. However, n_q only appears in first term in the square bracket, which does not contain the density n, so that this term is independent of n (though not of T), and the *shape* of the $\zeta(n)$ isotherm (which is all that matters here) is independent of n_q.

[18] The curves in Figure 15.11 were obtained from calculations based on the unpublished work of R. Cahn and A. Manoliu (see Kittel and Kroemer, pp. 292–293).

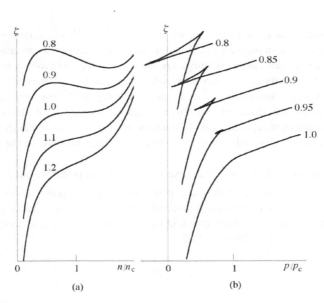

FIGURE 15.11 Chemical potential ζ of a van der Waals gas, for different values of T/T_c in the region of the critical point: (a) as a function of reduced density n/n_c and temperature; (b) as a function of reduced pressure p/p_c, at T_c and below. At a given temperature the gas-liquid transition occurs at the pressure where the two lower branches cross. Note that the vertical scale is expanded relative to (a).

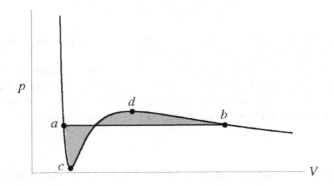

FIGURE 15.12 Isotherm in the (p, V) plane for a van der Waals gas at $T = 0.85T_c$, illustrating Maxwell's equal area rule and showing the limits of the superheated liquid (a to c) and the supercooled vapor (d to b). The intermediate region, c to d, is unstable.

Before the advent of the computer, it was tedious to calculate the coexistence curve this way, and Maxwell devised an ingenious method, known as the *equal area rule*, to avoid it.[19] To visualize the rule, we plot the isotherms in the (p, V) plane, as shown for an arbitrarily chosen temperature below T_c in Figure 15.12. The two points delimiting the coexistence region are labeled a and b, and the problem is to find the pressure at these points. Maxwell assumed that the chemical potential can be defined at any point on a van der Waals isotherm, including the unstable region, so that the integral of $d\zeta$ along the curve $acdb$ exists. Since $\zeta_a = \zeta_b$ this integral must be zero, that is, $\int_a^b d\zeta = 0$. This means that $\int_a^b dG = 0$, while from Equation (15.2) we see that for fixed N and T, $dG = V\,dp$. It follows that the true isotherm,

---

[19] J. Clerk Maxwell, "On the dynamical evidence of the molecular constitution of bodies," *Nature* **11**, 357 (1875).

that is, the one for which $\zeta_a = \zeta_b$, is the one for which $\int_a^b V\,dp = 0$. This integral is simply the difference between the two shaded areas in Figure 15.12, and it follows that these two areas must be equal.[20]

This argument is questionable, since the integral is taken over a region of the curve that is mechanically and thermodynamically unstable, and hence does not correspond to a physically realizable equilibrium state, so that it is not at all obvious that ζ can be defined in this region.[21] However, it is found that the rule gives results in agreement with those obtained by the rigorous procedure, and it has the advantage that it needs only the equation of state, which may be easier to obtain than the actual chemical potential. We shall see that the two methods agree in the van der Waals case, at least close to the critical point, where the algebra is simple.

To examine the behavior of the van der Waals gas near the critical point, we put $\frac{n}{n_c} = 1 + r$, $\frac{p}{p_c} = 1 + s$, $\frac{T}{T_c} = 1 + t$, where r, s, and t are all small, and expand Equation (15.26) as a Taylor series in r, s, and t. To get a non-trivial result, we need only go to first order in s and t, but we have to go to third order in r, just as in the Weiss model of a ferromagnet we had to go to third order in x to obtain Equation (15.19). The equation of state becomes (see Problem 15.9):

$$s \approx 4t + \frac{3}{2}(4tr + r^3),\tag{15.29}$$

where the term in $r^2 t$ has been dropped, since we shall find that for $t < 0$, $r \sim |t|^{1/2}$, so that this term is of order t^2.

In this approximation, the expression for ζ is:

$$\frac{\zeta - \zeta_c}{k_B T_c} \approx t\left[\ln\left(\frac{3n_c}{n_{qc}}\right) - \frac{3}{2}\right] + \frac{9}{16}(4tr + r^3),\tag{15.30}$$

where $\zeta_c \equiv \zeta(T_c)$ and $n_{qc} \equiv n_q(T_c)$. Only the r-dependent (that is, density dependent) terms matter for our purposes, since we are not concerned with the absolute value of ζ.

For a given t, the values of r which define the limits of the coexistence region, are those that simultaneously give the same ζ in Equation (15.30) and the same s in Equation (15.29), since the chemical potential and pressure (represented by s) are both constant throughout this region. Since the two values of r have opposite signs, it is impossible for ζ or s to be the same for both values of r unless:

$$r(4t + r^2) = 0.\tag{15.31}$$

For $t > 0$, Equation (15.31) has only one real root, $r = 0$. For $t < 0$, it has three roots, $r = 0$ and $r = \pm 2|t|^{1/2}$, and the corresponding value of s is $-4|t|$. The $r = 0$ solution is unstable, and we can identify the other two solutions as corresponding to the gas and liquid densities, n_g and n_ℓ, respectively. The difference between the two densities is $\Delta n \equiv n_\ell - n_g = 2r = 4n_c|t|^{1/2}$. The critical exponent β (that is, the exponent in the power law $\Delta n \propto |t|^\beta$) is $1/2$, as in the Weiss model.[22] The equal area rule gives the same result (see Problem 15.9).

Other properties are also analogous. For $t = 0$, corresponding to $T = T_c$, Equation (15.29) becomes $s \propto r^3$, so that the critical isotherm $p(\Delta n)$ is cubic with zero slope at the critical point, like the critical

---

[20] Another way to see this is to consider a hypothetical reversible heat engine (see Chapter 3) operating around a cycle, one side of which is the straight line ab and the other the van der Waals curve. Since the temperature is the same at all points on the cycle, the work done, which is just the integral $\oint_a^b p\,dV$, must be zero. This argument, which was Maxwell's, is open to the same objection as that in the text.

[21] Maxwell was well aware of this difficulty, but (characteristically) ignored it.

[22] In a ferromagnet, the spontaneous magnetization m is a measure of the difference between the phases (ferromagnetic and paramagnetic, respectively) below and above T_c; Δn is the analogous measure of difference for a condensible gas. These are both examples of an order parameter (see Section 15.7).

isotherm $B(m)$ in the Weiss model. Just above the critical point, the isothermal compressibility $\left(\frac{\partial n}{\partial p}\right)_T$ varies as t^{-1}, like the magnetic susceptibility in the Weiss model. From Equation (6.50), the divergent compressibility implies that density fluctuations are very large at the critical point. The corresponding fluctuations in the refractive index produce strong light scattering, a beautiful phenomenon known as critical opalescence.[23] Similarly, the divergent susceptibility of a ferromagnet leads to large fluctuations in the magnetization near T_c (see Problem 15.10). These fluctuations are very important, and are, in fact, large enough to put the whole basis of the theory in doubt, as we shall see in the final section of this chapter.

Like the Weiss model, the van der Waals model agrees qualitatively, but not quantitatively with experiment. Figure 15.13 shows the limits of the coexistence region for three gases near the critical point, in terms of reduced density and temperature. The dashed line shows the prediction of the van der Waals model. The experimental data for the three gases (and many others) lie on a single line, as predicted by the law of corresponding states, but this line differs considerably from the van der Waals prediction. In particular, near T_c, the critical exponent β is $\sim 1/3$, rather than 1/2 as predicted.[24]

As already pointed out, the re-entrant region (between the points c and d in Figure 15.12) of the van der Waals isotherm is not physically realizable. On the other hand, the regions between a and c and between b and d are mechanically stable, but the chemical potential is higher than on the equilibrium isotherm, which in this range of n is the straight horizontal line representing the gas-liquid mixture.

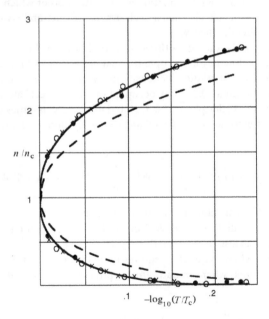

FIGURE 15.13 Coexistence region for argon $\bullet$, methane o, and xenon $\times$, in terms of reduced density n/n_c and temperature T/T_c. (Adapted from Pitzer, K.S., *J. Chem. Phys.*, 7, 583–590, 1937.) The fact that all the points lie on a single line shows that the law of corresponding states holds. The dashed line shows the limits of the coexistence region predicted by the van der Waals model. For a more extensive collection of data on fluids illustrating the law of corresponding states, see the references in Footnote 24.

---

[23] Equipment giving very elegant demonstration of critical opalescence in hexane is available from Leybold Didactic Gmbh.

[24] An early comprehensive analysis of the data then available showed that a large number of different fluids obey the law of corresponding states and have $\beta = 1/3$ within experimental error [D. A. Goldhammer, "Study of the theory of corresponding states," *Zeitschrift für physikalische Chemie* **71**, 577 (1910)]. For more recent and precise data confirming this result, see E. A. Guggenheim, *Thermodynamics*, 5th ed. (North Holland, Amsterdam, the Netherlands, 1967), pp. 135–140.

These are regions of metastability, a to c representing superheated liquid and d to b supercooled gas. They can be achieved if care is taken to prevent nucleation of the other phase (see Section 15.3). The pressure in a superheated liquid can even be negative (see, e.g., the $\frac{T}{T_c} = 0.8$ isotherm in Figure 15.10), because the intermolecular attraction gives the liquid some tensile strength; this is the case for sap in a tall tree (see Problem 8.3; in particular, the footnote to it), where the liquid is confined to narrow channels and is stabilized against evaporation by surface energy.

15.7 Landau's Theory of Phase Transitions

The Weiss, Bragg-Williams, and van der Waals models make very crude approximations and appear to be limited in their application. Nevertheless, they predict analogous behavior in systems with very different physics, suggesting that they can be generalized. They are, in fact, examples of a class of theories called *mean field theories*, which can in principle be adapted to any phase transition. This generalization was formalized by Landau, who introduced the concept of an *order parameter*, defined in the following paragraph. The *only* assumption of Landau's theory is that the free enthalpy is an analytic function of this parameter in the region of the critical point. The fact that the predictions of this general theory fail to agree with experiment in most cases shows that it is this basic assumption that the free enthalpy (or, equivalently, the free energy) is an analytic function at the critical point which is at fault, rather than the details of the models.[25] The formalism also helps to clarify the distinction between first and second order transitions. We will now outline the theory.

In all phase transitions, there is some quantity whose equilibrium value is zero at temperatures above the transition and non-zero below it; for example the magnetization m in a ferromagnet, the dielectric polarization p in a ferroelectric, the liquid-gas density difference Δn in an imperfect gas, the deviation from $1/2$ of the fraction of sites in a 50:50 alloy occupied by the "correct" atom (see Example 15.4), and so on. We call this quantity the order parameter ψ. Landau's theory assumes that the free enthalpy can be expressed as an analytic function $G(\psi, T)$, so that sufficiently close to T_c, where ψ is small, G can be expanded in powers of ψ:

$$G(\psi, T) = G_0(T) + G_1(T)\psi + \frac{1}{2}G_2(T)\psi^2 + \frac{1}{3}G_3(T)\psi^3 + \frac{1}{4}G_4(T)\psi^4 + \cdots, \qquad (15.32)$$

where the numerical coefficients are chosen for later convenience.

In equilibrium, $\frac{\partial G}{\partial \psi} = 0$. For all $T > T_c$, $\psi = 0$ in equilibrium, so that $G_1(T) = 0$ for all $T > T_c$. By analytic continuation, this must also be true for $T < T_c$.

We consider first the case where symmetry requires that G be an even function of ψ; this is the case, for example, for the ferromagnet, where m and $-m$ are physically equivalent. Then $G_3(T) = 0$ and we can write:

$$G(\psi, T) = G_0(T) + \frac{1}{2}G_2(T)\psi^2 + \frac{1}{4}G_4(T)\psi^4 + \cdots. \qquad (15.33)$$

If the system is to be stable for some ψ within the range of validity of Equation (15.33), $G_4(T)$ must be positive, but G_2 can have either sign. In Figure 15.14, Equation (15.33) is plotted for $G_2(T)$ positive, zero, and negative; ψ has its equilibrium value where G is a minimum. For $G_2 > 0$ there is one minimum, at $\psi = 0$,

---

[25] The belief that physical quantities must be analytic functions is deeply ingrained, and the possibility that this need not be the case near a critical point took a long time to enter the consciousness of theorists. For example, the eminent Dutch theoretical physicist J. J. van Laar, when faced with overwhelming data showing that $\beta \neq 1/2$ (see Footnote 24), insisted that the experiments must be in error, since β *had* to go to $1/2$ sufficiently close to T_c.

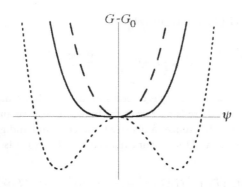

FIGURE 15.14 Equation (15.33) for three possible $G_2(T)$. Full line: $G_2 = 0$ ($T = T_c$); dashed line: $G_2 > 0$ ($T > T_c$); and dotted line: $G_2 < 0$ ($T < T_c$).

but for $G_2 < 0$ there are two equivalent minima with $\psi \neq 0$. The critical temperature T_c is defined as the temperature below which $\psi \neq 0$ in equilibrium. $G_2(T)$ changes sign at T_c; that is, $G_2(T_c) = 0$. Note that below T_c there is a maximum rather than a minimum at $\psi = 0$. Near T_c, we can expand G_2 as a Taylor series in t, where $t \equiv \frac{T}{T_c} - 1$. $G_2(T_c) = 0$, so that to lowest order in t, $G_2(T) \approx Kt$, where K is a constant which must be positive, since $G_2 > 0$ for $t > 0$.

To find the equilibrium value of ψ at a given temperature we differentiate Equation (15.33):

$$\frac{\partial G}{\partial \psi} = Kt\psi + G_4\psi^3 = 0, \tag{15.34}$$

where K and G_4 are both positive. If $t > 0$ (that is, $T > T_c$), the only solution is $\psi = 0$, but for $t < 0$ there are two stable solutions, which are physically equivalent:

$$\psi = \pm \left(\frac{K|t|}{G_4} \right)^{1/2}. \tag{15.35}$$

The critical exponent $\boldsymbol{\beta}$ (that is, the power of $|t|$ in the expression for ψ) is $\frac{1}{2}$. Thus, $\boldsymbol{\beta}$ is independent of the model and depends only on the basic assumption of all mean field theories, that $G(T, \psi)$ is an analytic function.

We can also obtain the heat capacity near T_c from the Landau theory. It is given by:

$$C_p = T \left(\frac{\partial S}{\partial T} \right)_p = -T \left(\frac{\partial^2 G}{\partial T^2} \right)_p, \tag{15.36}$$

since: $S = - \left(\frac{\partial G}{\partial T} \right)_p$.

Just below T_c, ψ is small, so that to lowest nontrivial order:

$$G - G_0(T) \approx -\frac{1}{2}K|t|\psi^2 = -\frac{K^2 t^2}{2G_4} = -\frac{K^2}{2G_4 T_c^2}T^2,$$

$$C_p(t \to -0) = -T\frac{\partial^2 G_0}{\partial T^2} + \frac{K^2}{G_4 T_c}. \tag{15.37}$$

Above T_c, $\psi = 0$, so that $G = G_0(T)$. Hence, just above T_c,

$$C_p(t \to +0) = -T \frac{\partial^2 G_0}{\partial T^2}.$$

Thus, as in the Weiss model, the heat capacity is discontinuous at T_c, but remains finite.

We now turn to the case where there is no symmetry between positive and negative ψ; for example, the gas-liquid case where ψ is the difference in density between liquid and gas. We can no longer drop the term in ψ^3 in Equation (15.32), and the Taylor expansion of G now reads:

$$G(\psi, T) = G_0(T) + \frac{1}{2}G_2(T)\psi^2 + \frac{1}{3}G_3(T)\psi^3 + \frac{1}{4}G_4(T)\psi^4 + \cdots . \qquad (15.38)$$

Note that the term linear in ψ still vanishes, since in equilibrium $\psi = 0$ above T_c, there being no distinction between gas and liquid in this region.

Taking the derivative of Equation (15.38) with respect to ψ, we find that if $G_3^2 > 4G_2G_4$, $G(\psi)$ has two inequivalent minima whose ordering may depend on T, as illustrated in Figure 15.15. The two minima correspond to two different phases, which can coexist in equilibrium when they have the same G; that is, when the two minima are at the same level, as in the full line in Figure 15.15. This corresponds to a first order transition, since the equilibrium value of ψ changes abruptly from zero to a finite value when the ordering of the minima changes. However, on both sides of the transition, two minima exist. Since there is a free enthalpy barrier between them, the higher minimum can be metastable, so that supercooling and superheating can occur. In a second order phase transition, on the other hand, there are no metastable states (see Figure 15.14), and supercooling and superheating are impossible.

In a gas, G_3 and G_2 are functions of pressure as well as of temperature. On the critical isobar (that is, when $p = p_c$ while T is varied) G_3 and G_2 simultaneously go to zero at T_c and a second order transition occurs. If $p < p_c$, the transition is first order, while if $p > p_c$, there is no transition at all since $G(\psi)$ has only one minimum.

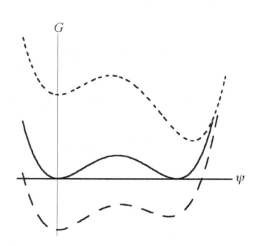

FIGURE 15.15 Possible free enthalpy functions when $G_3^2 > 4G_2G_4$. Full line: $T = T_t$; and dotted line: $T < T_t$; and dashed line: $T > T_t$.

Example 15.5: Superconductors

In 1911 Kamerling-Onnes and his student, Gilles Holst,[26] discovered that many metals, when cooled below a certain temperature T_c, lose all electrical resistance and enter the superconducting phase. Such a metal is called a superconductor, and its resistive phase is called the normal phase. The transition in zero magnetic field is second order. In the normal phase, the electronic heat capacity per unit volume is γT, where γ is a constant which depends on the metal (see Equation (13.7)). In the superconducting phase, the electronic heat capacity is a more complicated function of temperature, but it can be approximated by aT^3, where a is a constant, over most of the range of temperature in which the metal is superconducting.[27] The lattice heat capacity is the same in both phases. At T_c, there is an abrupt discontinuity in the heat capacity, as predicted by Landau theory; in fact, the second order transition of a Type I superconductor is one of the very rare cases where Landau theory is accurately obeyed. This is a consequence of the very long range of the interaction which is responsible for superconductivity (see Section 15.8).

1. Find the transition temperature in terms of γ and a. From Chapter 5, the entropy is:

$$S(T) = V \int_0^T \frac{C \, dT}{T},$$

where C is the heat capacity per unit volume, and we have used the fact that the third law of thermodynamics requires that the entropy be zero at 0K.

Hence, the difference between the entropies of the normal and superconducting states is:

$$\Delta S \equiv S_n - S_s = V \int_0^T \frac{\gamma T - aT^3}{T} \, dT = V \left(\gamma T - \frac{a}{3} T^3 \right). \tag{15.39}$$

The transition at T_c is second order, so that the latent heat (which is equal to $T_c \, \Delta S$) is zero, and $\Delta S = 0$ at T_c.

Hence,

$$\gamma T_c - \frac{a}{3} T_c^3 = 0;$$

$$T_c = \left(\frac{3\gamma}{a} \right)^{1/2}. \tag{15.40}$$

2. If a superconductor is cooled in a magnetic field B, the loss of resistance is accompanied in many cases by complete expulsion of the magnetic flux, so that $B = 0$ inside the metal, a phenomenon called the Meissner-Ochsenfeld effect.[28] A superconductor exhibiting this effect is said

---

[26] In 1914, Holst, whom Kamerling-Onnes had not deigned to acknowledge in his published papers on superconductivity, was appointed the first director of the new Philips Research Laboratories, the first and still perhaps the greatest European industrial research laboratory [see H. B. G. Casimir, *Haphazard Reality* (Harper & Row, 1983), pp. 158–167, 230–238].

[27] At temperatures well below T_c, the heat capacity varies as $e^{-\Delta/k_B T}$, where 2Δ is the superconducting energy gap: This is the energy required to break up a bound electron pair (Cooper pair). For a more thorough treatment of superconductivity, see M. Tinkham, *Introduction to Superconductivity*, 2nd ed. (McGraw Hill, New York, 1996), or J. R. Waldram, *Superconductivity of Metals and Cuprates* (Institute of Physics, London, UK, 1996).

[28] The field is expelled by shielding currents which circulate in a thin "penetration" layer at the surface, where $B \neq 0$. Since this layer is typically $\sim 10^{-7}$ m thick it can be neglected for ordinarily sized samples. The fact that $B = 0$ in the superconducting phase, regardless of its history, shows that it is a true equilibrium phase. If this were not the case, we could not apply thermodynamics to it.

to be Type I. Since $B = \mu_0(H + m)$, where $m = \frac{M}{V}$, this implies that the magnetic moment in the superconducting state is $M_s = -HV$. Relative to this, the normal phase, which has only Pauli susceptibility (see Example 13.2), has negligible magnetization. In general, the moment is modified by a shape-dependent demagnetizing field due to the large magnetization; here we confine our attention to a long thin cylinder with its axis parallel to the field, for which the demagnetizing field is zero.

Show that the entropy S is independent of H in both phases.

For a magnetic material,

$$dF = -S\,dT - \mu_0 M\,dH,$$

(we cannot put $B = \mu_0 H$ here, as we did in Chapter 2, since m is large).

Hence, there is a Maxwell relation (see Appendix E):

$$\left(\frac{\partial S}{\partial H}\right)_T = \mu_0\left(\frac{\partial M}{\partial T}\right)_H.$$

In the superconducting state, $M_s = -HV$, so that $\left(\frac{\partial M_s}{\partial T}\right)_H = 0$.

In the normal state, the magnetization is due only to the Pauli susceptiblity, so that it is small and independent of T.

Hence, $\left(\frac{\partial S}{\partial H}\right)_T = 0$ in both phases

3. Find the free energy in the superconducting state at a given temperature as a function of H.

$$F(T, H) = F(T, 0) + \int_0^H \left(\frac{\partial F}{\partial H}\right)_T dH,$$

$$\left(\frac{\partial F}{\partial H}\right)_T = -\mu_0 M = \mu_0 VH,$$

in the superconducting state, so that:

$$F_s(T, H) = F_s(T, 0) + \frac{1}{2}\mu_0 VH^2, \tag{15.41}$$

where $F_s(T, 0)$ is the free energy in zero field.

4. If the field is increased from zero, at a certain critical field H_c, there is a first order transition to the normal state. As $T \to T_c$, $H_c \to 0$. Find a relation between H_c and the entropy difference between the normal and superconducting phases in zero field, and hence derive the critical field as a function of temperature.

At $H_c(T)$, where $T < T_c$, there is a first order transition, so that $F_s(T, H_c) = F_n(T, H_c)$.

Hence,

$$F_s(T, 0) = F_n(T, H_c) - \frac{1}{2}\mu_0 VH_c^2.$$

Since $S = -\left(\frac{\partial F}{\partial T}\right)_H$, and F_n is independent of H,

$$S_n - S_s = -\frac{1}{2}\mu_0 V\frac{d(H_c^2)}{dT}. \tag{15.42}$$

Note that H_c decreases with temperature, so that the derivative is negative and $S_n > S_s$, showing that the superconducting phase is more ordered than the normal phase.

From Equation (15.39),

$$S_n - S_s = V\left(\gamma T - \frac{a}{3}T^3\right) = V\gamma\left(T - \frac{T^3}{T_c^2}\right),$$

so that:

$$\frac{d\left(H_c^2\right)}{dT} = -\frac{2\gamma}{\mu_0}\left(T - \frac{T^3}{T_c^2}\right).$$

Integration gives:

$$H_c^2 = \frac{\gamma}{\mu_0}\left(\frac{T^4}{2T_c^2} - T^2\right) + A,$$

where A is a constant of integration.

When $T = T_c$, $H_c = 0$, so that $A = \frac{\gamma T_c^2}{2\mu_0}$, and:

$$H_c^2 = \frac{\gamma T_c^2}{2\mu_0}\left(1 - 2\frac{T^2}{T_c^2} + \frac{T^4}{T_c^4}\right).$$

Hence,

$$H_c(T) = H_0\left(1 - \frac{T^2}{T_c^2}\right), \tag{15.43}$$

where $H_0 = T_c\left(\frac{\gamma}{2\mu_0}\right)^{1/2}$ is the critical field at 0K. The predicted parabolic dependence of H_c on T, and the general tendency of H_0 to increase with T_c, is found to hold experimentally with fair accuracy. In particular, the fact that Equation (15.43) holds right up to T_c shows that Landau theory is valid in this case, since this equation was derived from the assumption that the free energy is analytic at T_c.

15.8 Envoi: Critical Fluctuations and the Failure of Mean Field Theory

Except in superconductors and a few other systems where the interaction which is responsible for the phase transition has a very long range, Landau theory disagrees with experiment in the same way as do mean field theories such as the Weiss model. This shows that their failure is not due to the specific assumptions of the model. It is, rather, a necessary consequence of the assumption that an analytic free energy function exists at the critical point, since this is the *only* assumption that Landau theory makes. This assumption is true *only if one can safely average over fluctuations*. However, we have seen that near the critical point, fluctuations become very large, and they involve large numbers of atoms. When the length scale of the fluctuations (which can be quantitatively expressed in terms of a correlation length ξ) becomes large compared with the range of the interactions, averaging is impermissible; molecules (or spins) in different regions of the sample experience different environments. It can be shown, by methods beyond the scope of this book,[29] that $\xi \to \infty$ as $|t| \to 0$. There exist systems where the interaction

---

[29] See, for example, H. E. Stanley, *Introduction to Phase Transitions and Critical Phenomena* (Oxford University Press, Oxford University Press, Oxford, UK, 1971), Chapter 7.

range is very long, for example Type I superconductors (see Example 15.5, and the books referred to in Footnote 27); for these the predictions of Landau theory are well obeyed. However, this is a rare case; for example in a ferromagnet, spins interact predominantly with their immediate neighbors, so that mean field theory breaks down when ξ significantly exceeds the interatomic spacing.

In Chapter 4, we found that in a simple noninteracting system the relative fluctuations of any macroscopic quantity about its mean are extremely small, of order $N^{-1/2}$ where N is the number of particles. We then developed thermodynamics and its statistical underpinning on the assumption that the systems with which we are dealing have well-defined average properties, with small fluctuations about the mean. In this chapter, we have found that this assumption ceases to hold at a critical point, where fluctuations become so large that one can no longer talk in terms of average properties. While some thermodynamic quantities remain well defined (e.g., temperature, since it is defined by the heat bath and not by the system under study), other quantities such as the density do not, and a new approach is needed. We have come full circle, to the point where a reexamination of the very basis of the theory is required. This has been the subject of active research over the past thirty years, and there is now a well established theory of critical phenomena based on the renormalization group. This theory is beyond the scope of this book.[30]

15.9 Problems

15.1   (a)   Show that in a first order phase transition:

$$\frac{d}{dT}\left(\frac{L}{T}\right) = \frac{C_1 - C_2}{T},\tag{15.44}$$

where L is the latent heat and C_1 and C_2 are the heat capacities of the higher and lower temperature phases, respectively, *measured along the coexistence curve*. One way to prove this is to consider 1 kmole of the substance going round the following closed cycle. The substance starts in phase 1 on the coexistence curve at temperature T. It is transformed to phase 2, and is then heated from T to $T + dT$, the pressure changing to remain on the coexistence curve. At this temperature, it is transformed back to phase 2, and cooled along the coexistence curve to its starting point. Since the cycle is reversible, the overall change of entropy round the cycle must be zero

   (b)   Show that if phase 1 is an ideal gas, while phase 2 is a liquid or solid of negligible volume, Equation (15.44) becomes:

$$\frac{dL}{dT} = C_{p1} - C_{p2},\tag{15.45}$$

where C_{p1} and C_{p2} are the respective heat capacities at constant pressure

   (c)   Under the same assumptions as (b), show that $C_1 < 0$ if $L \gg R_0 T$ (as is usually the case for a gas-liquid transition except very close to the critical point). Explain why this is the case, and show that it does not contradict the second law of thermodynamics, which requires that C_V and C_p be positive for a system in thermodynamic equilibrium.

15.2   Prove the statement in Section 15.4 to the effect that if $L \gg R_0 T$, raising the pressure raises the density of the gas on the coexistence curve much more rapidly than the corresponding rise in temperature lowers it. Treat the gas as ideal.

---

[30] For an introduction to modern methods of handling critical phenomena, see M. E. Fisher, "The renormalization group in the theory of critical behavior," *Reviews of Modern Physics* **46**, 597 (1974), or L. E. Reichl, *A Modern Course in Statistical Physics*, 2nd ed. (John Wiley & Sons, New York, 1998), Chapter 8. For a more advanced treatment, see N. Goldenfeld, *Lectures on Phase Transitions and the Renormalization Group* (Addison-Wesley, Boston, MA, 1992).

15.3 (a) Show that in the Weiss model, when $T = T_c$ (T_c is the Curie temperature), $m \propto B^{1/3}$ for small B, where m is the magnetization, and B is the applied magnetic field

     (b) Show that in the Weiss model, $\frac{m}{m_0} \approx 1 - 2e^{-2T_c/T}$ as $T \to 0$

15.4 If the pressure cooker in Example 15.4 is found to shorten the cooking time by a factor 6, estimate the average activation energy E_a (in eV) for the rate of cooking

15.5 For an Einstein solid in equilibrium with its vapor, considered in Example 15.1, find the enthalpies of the two phases and hence find the latent heat L. Note that L depends on temperature. Confirm by direct calculation from Equation (15.5) that the Clausius-Clapeyron relation [Equation (15.11)] is obeyed

15.6 (The dilution refrigerator.) The helium dilution refrigerator is the most commonly used means of achieving temperatures in the millikelvin range. It depends on the fact that at very low temperatures, a mixture of the two isotopes of helium, ^{3}He and ^{4}He, separates into a phase A containing almost pure ^{3}He and another phase B containing 6% ^{3}He dissolved in ^{4}He. Because ^{4}He is a Bose-Einstein condensate at this temperature (see Chapter 12), its entropy is zero. The ^{3}He in phase B acts as an almost perfect Fermi gas with $T_F \approx 0.2$ K, while phase A is a pure liquid, whose entropy can be neglected. In this problem, we assume that U/N is the same in both phases, which is the case if the potential energy of a ^{3}He atom is the same[31] when it is surrounded by ^{3}He as when it is surrounded by ^{4}He. Conversion of phase A to phase B can be approximated as a reversible process

     (a) Treating the ^{3}He in phase B as an ideal gas in equilibrium with phase A, find the entropy per kmole of phase B at a temperature $T > T_F$. In fact, the assumption of ideality is not correct and the result is only approximate (see the reference in Footnote 31)

     (b) If 10^{-4} g of phase A is converted to phase B per second, what is the rate of heat extraction in watts at 0.3 K?

15.7 By differentiating Equation (15.24) twice with respect to n, show that in the van der Waals model, the critical temperature, pressure, and density have the values given in Equation (15.25)

15.8 By considering stability against local fluctuations in density, show that a substance with negative compressibility (that is, with $\frac{dn}{dp} < 0$), is mechanically unstable, and spontaneously separates into two phases which both have positive compressibility

15.9 (a) Derive Equation (15.29) from van der Waals' equation of state

     (b) Show that Equation (15.29) can be written, to the same order of accuracy, as:

$$s = 4t - 6tv - \frac{3}{2}v^3 \tag{15.46}$$

     where $v \equiv \frac{V}{V_c} - 1$, V_c being the critical volume

     (c) Hence, show that at given t and s, the coexistence region is bounded by $v_a, v_b = \pm 2|t|^{1/2}$

     (d) Show that Maxwell's equal area rule is obeyed.

15.10 Show from Equation (6.34) that the mean square fluctuation in magnetization $\langle \Delta m^2 \rangle$ of a Landau ferromagnet diverges at the Curie temperature T_c

15.11 (a) Use Equation (4.4) to find the entropy of N spins with net moment M, where $M \ll M_0 = N\mu$. Hence, show that in the Weiss model the free energy of a ferromagnet has the Landau form near T_c, with F substituted for G and m for ψ, up to second order in m

     (b) Using the more accurate expression [Equation (4.3)] and Stirling's formula [Equation (D.6)], show that the free energy has the Landau form up to fourth order in m

15.12 (a) If the liquid drop in Example 15.2 is charged, it acquires electrostatic energy $-U_e$, where U_e is positive. To simplify the algebra, U_e may be taken for the purpose of this calculation to be independent of the drop radius R. Show that the difference in free enthalpies can now be

---

[31] While this is in accordance with our usual view of isotopes, it is not quite true in this case. For a more accurate treatment, see A. V. Lounasmaa, *Experimental Principles and Methods below 1 Kelvin* (Academic Press, New York, 1974), Chapter 3.

written $\Delta G = 4\pi\sigma\left(\frac{R^3}{R_c} - R^2\right) + U_e$, where R_c is given by Equation (15.6). Hence, show that if $U_e < \frac{16\pi}{27}\sigma R_c^2$, there is range of R (of order R_c) in which ΔG is negative, so that the drop does not grow to a visible size. Estimate the minimum value of U_e for the drop to become visible, using the parameters for water vapor given in Example 15.2

(b) In fact, the electrostatic energy of a single electron charge q on a sphere of radius R with a very large dielectric constant is[32] $-\frac{q^2}{8\pi\epsilon_0 R}$. Give R its value at which ΔG has its minimum value and hence estimate the degree of supercooling ΔT necessary in order for the drop to grow.

15.13 (Gas-liquid-solid equilibrium.)

(a) The vapor pressure line in Figure 15.3 separates the region of the (p, T) plane where the gas is stable from that where the condensed phase (liquid water, or ice) is stable. Copy Figure 15.3 and sketch on it the coexistence line for ice and liquid water, using the following data for the ice-water transition: $L = 6 \times 10^6$ J/kmole, $\rho(\text{water}) = 922$ kg/m^3, $\rho(\text{ice})$ 916 kg/m^3, $M_w = 18$. Neglect the pressure dependence of the ice-water volume difference and of the latent heat of melting, and don't forget that 1 bar = 10^5 Pa. Label the regions of the diagram where (respectively) gas, liquid, and solid are stable.

(b) Explain in terms of Landau theory why a solid-liquid coexistence line, unlike a liquid-gas line, has no critical point.

15.14 (a) Extend the free energy expression for a monatomic van der Waals gas [Equation (15.23)] to the case where the molecules have internal degrees of freedom (assume that equipartition applies to these). Hence, show that C_V is the same for a van der Waals gas as for an ideal gas. Explain in physical terms why this should be, given the assumptions of the van der Waals model

(b) Use Equation (5.43) to find an expression for C_p near T_c. What happens to C_p at T_c?

(c) Find the latent heat of a van der Waals gas near T_c as a function of T

(d) A fixed quantity of a van der Waals gas is enclosed in a box of fixed volume, and cooled from above just above T_c to just below it. Show that the heat capacity of this box and its contents (regarded as a single system of unknown composition) shows the discontinuity at T_c predicted by Landau theory. Find an expression for the magnitude of the discontinuity per kmole in terms of the critical parameters.

15.15 (Solid-solid phase transitions.) Many solids can exist in more than one crystalline form. While one of these has the lowest internal energy, and is therefore the stable phase at 0 K, there may be another form with higher entropy, which becomes the stable phase above some transition temperature. In the Debye model, the higher entropy can be attributed to a lower Debye temperature, so that at a given temperature there are more phonons excited. For a given atomic mass, a lower Debye temperature implies a softer material, and it is generally true that the high temperature phase is softer than the low temperature one. The transition is usually first order

(a) Consider a simple model of a solid which can exist in two phases, A and B. At 0 K, phase A has $U = 0$, while phase B has $U = U_0$ per kmole ($U_0 > 0$). Assume that they are both Debye solids with Debye temperatures θ_A and θ_B, where $\theta_A > \theta_B$. Assume also that the temperature of interest is well below θ_A and θ_B, so that the entropy is given by Equation (7.61), and assume that the pressure is zero so that $H = U$ and $F = G = N\zeta$. Find the transition temperature T_t, and the latent heat L.

Repeat the calculation assuming that T is well above θ_A and θ_B.

(b) The softer phase (that is, the one with the lower Debye temperature) usually has lower density than the harder phase. If the respective molar volumes are V_A and V_B, with $V_A < V_B$, how does T_t vary with pressure?

(c) Diamond and graphite, two crystalline forms of carbon, have almost the same internal energy at 0 K. Diamond and graphite have, respectively, Debye temperatures of about 2000 and 400 K

---

[32] D. J. Griffiths, *Introduction to Electrodynamics*, 3rd ed. (Simon & Schuster, New York, 1999), p. 94.

and mass densities of 3500 and 2250 kg/m^3. Assuming both substances to be incompressible, what is the minimum pressure necessary at 600 K to equalize the free enthalpies of the two phases so that conversion from graphite to diamond can take place? (In practice, higher pressure and temperature are needed to get the conversion to proceed at a reasonable rate.) Assume that Equation (7.61) holds for the entropy of diamond and Equation (7.62) for that of graphite (these are not very good approximations, but they give the right order of magnitude). How much heat is produced per kmole converted? Where does this energy come from?

15.16 Derive Equation (15.19) for the magnetization as a function of T near $T = 0$ of a spin 1/2 ferromagnet in the Weiss model, and show that this equation applies equally well to the order parameter ψ in the Bragg-Williams model of an ordered binary alloy

15.17 (a) A magnetic field H is applied to a thermally isolated superconductor at an initial temperature $T_i < T_c$, where T_c is the transition temperature in zero field. What temperature will the superconductor reach when it goes into the normal state? Neglect the lattice heat capacity, and assume that the field is high enough to keep the metal in the normal phase as it cools

Here cooling is produced by an *increasing* magnetic field, while in adiabatic demagnetization (see Problem 6.4) cooling is produced by a *decreasing* field; why the difference?

(b) Show that the latent heat absorbed in the transition from the superconducting to the normal state at H_c has a maximum when $T = T_c/\sqrt{2}$, and find this maximum value

15.18 If a man weighing 80 kg is walking on ice, and the contact area of the sole of his shoe is 0.01 m^2, how cold does the ice have to be in order that he not slip, assuming that slipping occurs when and only when liquid water is formed under the sole? The densities of ice and water, and the latent heat of melting, are given in Problem 15.13a

15.19 Show directly (that is, without invoking Landau theory) that the Bragg-Williams model of a binary alloy undergoing an order-disorder transition at $T = T_c$ predicts that the heat capacity C will have a discontinuity ΔC at T_c, and find ΔC

15.20 Use the Weiss model to find a transcendental equation for $m(T)$ of a classical ferromagnet, in which the elementary magnets of moment μ can be oriented in any direction, rather than being parallel or antiparallel to the field. It is shown in Chapter 7 that the response of these magnets to an applied field B is given by

$$\langle \mu \rangle = \mu \mathfrak{L}(\beta \mu B),$$

where $\mathfrak{L}(x)$ is the Langevin function [Equation (7.58a)]. Show that the magnetization near T_c is still given by Equation (15.18)

15.21 A widely used textbook of statistical mechanics[33] has this to say about the van der Waals equation of state:

"The repulsive forces between the atoms...lead to a smaller free volume for the gas. We must thus replace the factor V (in the equation $pV = RT$) by a slightly smaller factor, say $V - b$. If repulsive forces decrease the factor V, we should expect attractive forces to increase it; we must thus ...replace V by $(V - b + a)$ .... We now have the following equation of state ...

$$p(V - b + a) = RT, \tag{a}$$

(where) b and a are small corrections to V ...as long as the temperature or the density of the gas are sufficiently low. We can then rewrite (a) as follows:

$$\left(p - \frac{a'}{V^2} \right)(V - b) = RT, \tag{b}$$

[33] D. ter Haar, *Elements of Statistical Mechanics*, 3rd ed. (Butterworth-Heinemann, Oxford, UK, 1995), pp. 8–9.

where ... equations (a) and (b) differ only by terms which are of second order in $\frac{a}{V}$ or $\frac{b}{V}$, provided (that) ... $a' = aRT$. The reason for rewriting (a) in the form (b) is purely a historic one. It is done in order to show that to a first approximation the changes introduced by the ... forces between the atoms ... lead to the famous equation introduced by van der Waals ... (whose) importance lies mainly in the very great number of successful applications."

Critique this passage

15.22   Making the substitutions described in the text, and the values for the critical pressure, temperature, and density (See Problem 15.7), show how one obtains Equation (15.26) from Equation 15.24.)

16

Transport Processes

Who... is not familiar with Maxwell's memoir on his dynamical theory of gases? ... from one side enter the equations of state; from the other side, the equations of motion in a central field. Ever higher soars the chaos of formulae. Suddenly we hear, as from kettle drums, the four beats "put $n = 5$." The evil spirit **v** vanishes; and... that which had seemed insuperable has been overcome as if by a stroke of magic... One result after another follows in quick succession till at last.... we arrive at the conditions for thermal equilibrium together with expressions for the transport coefficients.

Ludwig Boltzmann

16.1 Gedanken

In introductory physics, we spend a good bit of time building up the concepts of energy and momentum conservation from Newton's laws and before that, kinematics, simply to describe how rigid macroscopic objects get from point A to point B. The breakthrough with energy and momentum conservation is that everything that happens in between is in a very real sense inconsequential. We apply energy and momentum conservation only by considering the initial and final states. That paradigm shift is only fully appreciated now when we realize that through energy (and momentum) conservation by the first law, and entropy maximization from the second and third laws, we don't have to contemplate the Deity-like task of contemplating 6.022×10^{23} applications of Newton's laws. Transport (electric, acoustic, thermal) comes naturally out of the energy/entropy picture of substances regardless of their state of matter.

16.2 Quasi-Ideal Gases

The perfect gas is an idealization of reality. As we have seen, this idealization gives rather accurate results for equilibrium properties of most gases, but there are some properties of these gases on which it can shed no light. In particular, if we are to understand transport processes such as diffusion and conduction in real gases, collisions must be taken into account (the photon gas is an exception; photons in a vacuum do not interact with each other). A gas which is ideal in its thermodynamic properties, but in which the molecules undergo collisions, is called *quasi-ideal*. We confine our attention to elastic collisions; that is, encounters between two particles in which the total kinetic energy is conserved, but the direction of motion of one or both particles is altered. Such a process is called *scattering*.

16.3 Scattering Processes: Mean Free Path

The simplest system to treat is an electron gas in a solid containing impurities. We saw in Chapter 14 that carriers (electrons or holes) in a semiconductor behave at low density like an ideal gas. Carriers move freely through the solid, but thermal vibrations, impurity atoms, and defects scatter them (that is, change their direction of motion) and limit the conductivity. We assume a temperature low enough that thermal vibrations can be neglected and concentrate on impurity scattering. If the impurity atom is uncharged, its interaction with the particle usually has a short range. If the particle gets close to the impurity it is deflected through a large angle, while if it does not get close it is unaffected. Such scattering can be described by a *collision cross-section* σ_c, which is the cross-sectional area of an imaginary sphere surrounding the impurity. If the trajectory of the center of the particle passes through this sphere, the particle is scattered elastically, and we assume that the new direction of motion is entirely random, bearing no relation to the original direction. It is customary to say, somewhat anthropomorphically, that the particle "has no memory" of its previous direction. If the trajectory does not pass through the sphere, it is assumed to be unaffected. In the simplest case, σ_c is independent of the velocity of the particle. Two trajectories, one involving such a "hard" collision, the other not, are illustrated in Figure 16.1a.

If the scattering impurity is charged, the Coulomb interaction deflects the carrier even at large distances, the deflection increasing as the distance gets smaller. Scattering by such a long-range interaction is illustrated in Figure 16.1b, for the case of an electron scattering from a negatively charged impurity. The process cannot be described by a unique collision cross-section since there is no limit to the distance at which the trajectory is affected; furthermore, the effect of scattering on the trajectory is strongly velocity dependent. We do not specifically treat such "soft" scattering here, although most of our general results can in fact be applied to it,[1] and Example 16.1 of this chapter gives a qualitative analysis.

Suppose there are n_I impurities per unit volume in the solid, with collision cross-section σ_c. Then in a layer of thickness dx, area A, perpendicular to the motion of a particle moving in the x-direction, there are $n_I A\, dx$ spheres, with a total cross-sectional area, as seen by the approaching particle, of $\sigma_c n_I A\, dx$. Since dx is infinitesimal, the spheres do not overlap. The impurity atoms are located at random positions relative to the point at which the particle's trajectory crosses the plane, so that the probability of collision with an impurity is the ratio of this cross-sectional area to the total area A; that is, $\sigma_c n_I\, dx$. Thus, the probability that the particle passes through unscattered is $(1 - \sigma_c n_I\, dx)$. Suppose that the particle starts from $x = 0$, and let the probability that it has reached a distance x without being scattered be $P(x)$. Since the two probabilities (the probability that the particle reaches x and the probability that it then passes through the layer dx unscattered) are independent, the probability that it reaches $x + dx$ is the product of these:

$$P(x + dx) = P(x)(1 - \sigma_c n_I\, dx).$$

Hence,

$$\frac{dP}{dx} = -\sigma_c n_I P(x), \tag{16.1}$$

with the initial condition $P(0) = 1$.

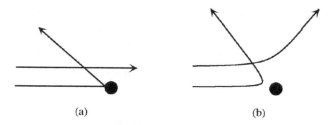

(a)   (b)

FIGURE 16.1   (a) "Hard" and (b) "soft" collisions of a particle with a fixed object.

[1] See, for example, D. L. Rode, *Low field electron transport* in *Semiconductors and Semimetals*, eds. R. K. Willardson and A. C. Beer, **10** (Academic Press, New York, 1975), pp. 1–189; in particular, Section 6.

The solution to Equation (16.1) is:

$$P(x) = e^{-x/\ell},$$ (16.2)

where:

$$\ell = (\sigma_c n_I)^{-1}.$$ (16.3)

The *mean free path* ℓ is defined as the average (mean) distance traveled by a randomly chosen particle before it undergoes a collision. If the particle has velocity v, the mean time of flight before a collision, or *scattering time*, is $\tau_s = \frac{\ell}{v}$. An analogous argument with time reversed shows that the mean distance gone since the last collision[2] is also ℓ and the mean time, τ_s. A mean free path or scattering time can often be defined even when σ_c cannot; however, in general, both ℓ and τ_s may be velocity dependent and they need not be strictly proportional to n_I^{-1}.

16.4 Mobility and Diffusivity: The Einstein Relation

We are now in a position to calculate the electrical conductivity of a quasi-ideal gas of particles with charge q and mass m. Suppose that we apply an electric field E_z in the z-direction. Choose a particle at random. The mean time since it underwent a collision is τ_s. Immediately after the last collision, its velocity was random in direction, so that on average its velocity parallel to the field was zero at that time. The effect of the electric field is to give it an acceleration $\frac{qE_z}{m}$ in the z-direction, so that at a time τ_s after a collision, its mean velocity is $v_z = \frac{qE_z}{m}\tau_s$. We call the mean velocity parallel to the field the *drift velocity* v_d. As long as v_d is much less than the root mean square thermal velocity,[3] it is proportional to E, the constant of proportionality being the mobility μ_e, where in our simple model:

$$\mu_e = \frac{q\tau_s}{m}.$$ (16.4)

The current density (the net charge crossing unit area per unit time) is $\mathbf{J}_e = nq\mathbf{v}_d$, where n is the electron density, so that the conductivity is:

$$\sigma_e \equiv \frac{J_e}{E} = nq\mu_e = nq^2\frac{\tau_s}{m}.$$ (16.5)

Even when τ_s cannot be defined in terms of microscopic processes, as in the case of Coulomb scattering, the mobility can still be defined as $\frac{v_d}{E}$, and Equation (16.4) then serves to define an effective mean scattering time.

Even if the particles are not charged, they diffuse if their chemical potential ζ is not spatially uniform. We will now show that if the temperature is uniform, there is an effective force $\mathbf{f} = -\nabla\zeta$ acting on the particle. Suppose that we have a particle with charge q, and an electric field $\mathbf{E}$ is applied in opposition to the chemical potential gradient, of such a magnitude that, on average, the particle does not move. This

---

[2] An apparent paradox (a special case of Poisson's paradox) arises here. If a particle has just undergone a collision, the mean time until the next collision is τ_s, by the argument in the text. However, the time reversal argument seems to imply that a randomly chosen particle spends a mean time $2\tau_s$ between collisions. This discrepancy is a consequence of the selection procedure, since a particle chosen at random is likely to have a longer than average free path. We can make this quantitative as follows. If τ_s is the mean time between collisions, the probability of a given particle spending a time t between collisions is $\tau_s^{-1}e^{-t/\tau_s}$. The probability of a randomly selected particle having a given value of t is enhanced by a factor t/τ_s. Hence, the probability of selecting a particle with a given value of t is $\tau_s^{-2}te^{-t/\tau_s}$, and the mean value of t for particles selected in this way is $\tau_s^{-2}\int_0^\infty t^2 e^{-t/\tau_s}\,dt = 2\tau_s$.

[3] For a degenerate Fermi gas, such as the electrons in a metal, the relevant velocity is the Fermi velocity v_F.

means that the net effective force on the particle is zero. The electrostatic force $q\mathbf{E}$, where $\mathbf{E} = -\nabla V$, is cancelled out by an equal and opposite effective force $\mathbf{f}$ due to the chemical potential gradient, so that $\mathbf{f} = -q\mathbf{E}$. In this situation, the electrochemical potential is $\zeta + qV$ and is uniform; that is, $\nabla(\zeta + qV) = 0$. Hence, $\nabla\zeta = -q\nabla V = q\mathbf{E} = -\mathbf{f}$. Since the chemical potential does not in itself depend on the presence of electric charge, the effective force on any particle, charged or not, due to a chemical potential gradient is $\mathbf{f} = -\nabla\zeta$. The acceleration due to this force is $\mathbf{f}/m$, producing a drift velocity $\mathbf{v}_d = -\tau_s \nabla\zeta/m$. If the particle density is n, the particle flux (the net number of particles crossing unit area per unit time) is:

$$\mathbf{J}_n = n\mathbf{v}_d = -n\frac{\tau_s}{m}\nabla\zeta.$$

The *diffusivity* D (generally called the *diffusion constant*; a misleading name since it is often by no means constant) is defined by $D \equiv -\frac{\mathbf{J}_n}{\nabla n}$, so that:

$$D = n\frac{\tau_s}{m}\left(\frac{\partial\zeta}{\partial n}\right)_T. \tag{16.6}$$

In an ideal gas, $\zeta = k_B T \ln(n/n_q)$, so that:

$$\left(\frac{\partial\zeta}{\partial n}\right)_T = \frac{k_B T}{n},$$

giving:

$$D = k_B T \frac{\tau_s}{m}. \tag{16.7}$$

Equations (16.4) and (16.7) together relate diffusivity to mobility for a quasi-ideal gas:

$$\boxed{\frac{D}{\mu_e} = \frac{k_B T}{q}.} \tag{16.8}$$

Equation (16.8) is known as the Einstein relation.

In a degenerate Fermi gas, only the electrons near the Fermi energy can respond to a field, and it might appear that this treatment will not work. However, it is shown in texts on solid state physics[4] that the effect of an electric field is simply to displace the entire distribution of electrons in k-space (see Chapter 6) by an amount $\Delta\mathbf{k} = m\mathbf{v}_d/\hbar$, where $\mathbf{v}_d = \mu_e\mathbf{E}$. This means that the distribution acquires a mean velocity $\mathbf{v}_d$, as in the ideal gas, so that Equation (16.5) still applies. However, the relation between ζ and n is different and the Einstein relation for a degenerate Fermi gas is (see Problem 16.1):

$$\frac{D}{\mu_e} = \frac{2}{3}\frac{E_F}{q}, \tag{16.9}$$

where E_F is the Fermi energy.

If the mean free path ℓ is independent of velocity, there is another way of writing D, using the fact that for an ideal gas, $\langle v^2 \rangle = \frac{3k_B T}{m}$. If we put $\tau_s = \langle v^{-1} \rangle \ell$, Equation (16.7) becomes:

$$\boxed{D = \frac{1}{3}v_m\ell,} \tag{16.10}$$

where $\langle v_m \rangle \equiv \langle v^2 \rangle \langle v^{-1} \rangle$. Note that in a quasi-ideal gas $v_m \propto T^{1/2}$, so that Equation (16.10) implies that $D \propto T^{1/2}$ if ℓ is independent of temperature. In Example 16.1 and Problem 16.3, we discuss the more

---

[4] See, for example, C. Kittel, *Introduction to Solid State Physics*, 7th ed. (John Wiley & Sons, New York, 1996), p. 156.

general case where σ_c depends on velocity, and hence on temperature. Equation (16.10) also applies to a degenerate Fermi gas, with v_m replaced by v_F, the Fermi velocity (see Problem 16.1). The Fermi velocity is independent of T if $T \ll T_F$, the Fermi temperature, so that in this case, D and μ_e have the temperature dependence (if any) of ℓ.

This model is applicable not only to scattering by fixed impurities, but also to interparticle scattering in a gas. It is reasonable to assume, at least for hard scattering (see Figure 16.1a), that when two particles collide, the velocities after the collision are random in direction and average to zero. For example, in a typical measurement of diffusivity, "labelled" molecules of one species diffuse through a "host" gas of another species. The labelled molecule is often a radioactive isotope for convenience of identification. The labelled molecule is (on average) brought to rest after each collision, just as if it had struck a stationary impurity. Thus, the mean free path is determined by the density n and collision cross-section of the host gas; Equation (16.3) holds, but with n_I replaced by n. This argument, although clearly shaky, since it ignores the fact that ℓ must depend on velocity, was used by Maxwell to estimate the size of molecules.[5] In air at 300 K and atmospheric pressure, $n \approx 3 \times 10^{25}$ m^{-3}, $v_m \approx 500$ m/s, $D \approx 2 \times 10^{-5}$ m^2/s, so that $\ell \sim 10^{-7}$ m and $\sigma_c \sim 3 \times 10^{-19}$ m^2. Two hard spheres of diameter d collide if the trajectories of their centers approach to within a distance d, so that $\sigma_c \sim \pi d^2$. By this means, Maxwell obtained a value of $d \sim 3 \times 10^{-10}$ m ≈ 3 Å, a remarkably accurate estimate not improved upon until the advent of X-ray crystallography.

Example 16.1: Dimensional Treatment of the Scattering Cross-Section

Transport theory is an area where dimensional methods[6] can be very useful, and are often more illuminating than exact mathematical analysis. Here we consider the effect of the intermolecular force law on the effective average cross-section σ_c, the mean scattering time τ_s, and the transport coefficients.

We start by assuming that the molecules of a quasi-ideal gas interact with each other through a radial force $f = KR^{-n}$, where R is the intermolecular separation, n is a positive integer ≥ 2, and K is a constant.

1. If σ_c depends only on K, v, and the molecular mass M, how do σ_c and τ_s depend on v, the average speed? Explain the dependence of σ_c on v in physical terms for the case of small n ("soft" collision) and large n ("hard" collision).

We put $\sigma_c = K^a v^b M^c$, no other form being dimensionally possible. We first have to determine the dimensions of the constant K. Since the dimensions of force are MLT^{-2}, those of K are $ML^{1+n}T^{-2}$.

The dimensions of σ_c are L^2 so that $L^2 = (ML^{1+n}T^{-2})^a (LT^{-1})^b M^c$

Equating dimension of M: $\qquad a + c = 0$,
Equating dimension of L: $\qquad (1 + n)a + b = 2$,
Equating dimension of T: $\qquad -2a - b = 0$.
Solving these three equations gives:

$$a = \frac{2}{n-1}, \quad b = \frac{-4}{n-1}, \quad c = \frac{-2}{n-1},$$

so that:

$$\sigma_c \propto v^{-4/(n-1)}.$$

---

[5] Before the discovery of radioactivity, which made possible the tracking of a radioactive isotope as it diffuses, diffusivity was difficult to measure directly. Maxwell actually measured the viscosity, which is simply related to diffusivity (see Equation 16.26 and Footnote 13).

[6] For an excellent introduction to dimensional analysis, see the article, "Scale and Dimension," by G. B. West in Cooper and West.

Soft collisions have small n and strong velocity dependence, since the faster the molecule, the less it is deflected by the intermolecular force. On the other hand, when $n \to \infty$, corresponding to hard ("billiard ball") collisions, σ_c is independent of v, being purely geometric.

2. Assuming that the relation $\tau_s = n^{-1} \langle \frac{1}{v\sigma_c} \rangle$ holds, where n is the particle density and τ_s the mean scattering time, find the dependence of τ_s on v.

$$\tau_s \propto n^{-1} \left\langle v^{[4/(n-1)-1]} \right\rangle = n^{-1} \left\langle v^{(5-n)/(n-1)} \right\rangle.$$

Note that if the force law exponent $n = 5$, the scattering time τ_s is independent of velocity, which greatly simplifies the mathematical analysis.[7]

16.5 General Treatment of Transport Processes in Gases

Other transport properties are also of interest. For example, if there is a temperature gradient, there is a flow of energy in the form of heat. The ratio of the energy flux[8] to the temperature gradient is called the thermal conductivity K_{th}. Similarly, in a fluid flowing past a fixed surface, there is a velocity gradient perpendicular to the direction of flow. Particles moving down the velocity gradient carry momentum with them, so that there is momentum flow down the gradient, as illustrated in Figure 16.2. By Newton's second law, this implies a shear stress in the fluid, opposing the flow. The ratio of the shear stress to velocity gradient is called the viscosity η.

We now develop a general theory,[9] based on the simple kinetic model described in the previous section, relating the flux of some quantity to its driving force. The four transport processes that we consider are:

electrical conduction	$\mathbf{J}_e = -\sigma_e \nabla V;$	(16.11)
diffusion of particles	$\mathbf{J}_n = -D \nabla n;$	(16.12)
thermal conduction	$\mathbf{J}_u = -K_{th} \nabla T;$	(16.13)
viscosity	$\mathbf{J}_p = -\eta \nabla v_f$	(16.14)

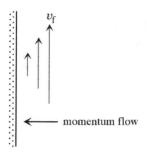

FIGURE 16.2 Viscous flow of a fluid near a boundary. The flow velocity v_f goes to zero at the boundary.

---

[7] Maxwell claimed in his 1866 memoir [reprinted in S. G. Brush (ed.) *Kinetic Theory*, Vol. 2 (Pergamon, New York, 1966)] that he had deduced the r^{-5} law of force from his measurements of gas viscosity. We now know that the long range force (which is what matters in a dilute gas) usually varies as r^{-7} (see Footnote 12 of Chapter 15). One suspects that Maxwell's data were not sufficiently extensive or precise to distinguish between these different power laws, and that he settled on the fifth power law for the excellent reason that it makes the calculations tractable, rather than because the data required it (see Boltzmann's description of Maxwell's style at the beginning of this chapter).

[8] *Flux* is the rate of flow of something (in this case energy) per unit area (see Glossary).

[9] This development is based on that of Kittel and Kroemer, pp. 397–404.

Here, J_e is the electric current density (charge flux), driven by the gradient of electrostatic potential V; J_n is the net particle flux, driven by the gradient of particle density n; J_u is the heat flux, driven by the gradient of temperature T; and J_p is the momentum flux (shear stress[10]), driven by the gradient of fluid velocity v_f. The electrical conductivity σ_e, the diffusivity D, the thermal conductivity K_{th}, and the viscosity η are collectively referred to as *transport coefficients*. We shall not go into the detailed calculations necessary to derive accurate values of these coefficients, but confine ourselves to showing how a unified approach relates them all to a single quantity, which we call the *universal mobility* μ_u. For example, we can write the electrical conductivity and the diffusivity:

$$\sigma_e = nq^2 \mu_u, \tag{16.15}$$

$$D = n \left(\frac{\partial \zeta}{\partial n} \right)_T \mu_u, \tag{16.16}$$

where:

$$\mu_u \equiv \frac{\tau_s}{m} = \frac{\mu_e}{q}. \tag{16.17}$$

As before, Equation (16.17) serves to define an effective scattering time. It must be remembered, however, that only in certain simple cases (such as scattering by hard spheres, by uncharged impurities, or by phonons at a temperature well above the Debye temperature), is τ_s a well-defined quantity. We confine our attention to such cases,[11] and derive relations corresponding to Equations (16.15) and (16.16) for the thermal conductivity and the viscosity.

16.6 Thermal Conductivity

If there is a temperature gradient, there is a flow of energy, since particles moving from a hotter to a colder region will, on average, have more thermal energy than those moving in the other direction. In the presence of a temperature gradient, ζ cannot be treated as a potential (see Section 8.5 and Problem 10.6d), and we must use a different approach. In a diffusion experiment, the diffusing particle is labelled (for instance, by making it radioactive). In principle, the particle can be labelled in another way (for instance, by its energy). Consider a gas in a temperature gradient, assume that the particle density is uniform and focus attention on particles with energy between E and $E + dE$. At a distance x from some arbitrary reference plane, the temperature is $T(x)$. The local density of particles with energy E at this point is:

$$n_\epsilon(E, x)\, dE \equiv g(E) f[E, T(x)]\, dE,$$

where $g(E)$ is the density of states and $f(E, T)$ is the distribution function. If T varies with x, so does n_ϵ, even though the total particle density $n = \int_0^\infty n_\epsilon(E, x)\, dE$ is constant. The flux of these particles is:

$$J_{n\epsilon}\, dE = -D(E) \frac{\partial n_\epsilon}{\partial x}\, dE$$

$$= -D(E)g(E) \frac{\partial f(E, T)}{\partial T} \frac{dT}{dx}\, dE, \tag{16.18}$$

---

[10] Stress is actually a second order tensor. In order to keep our notation uniform, we consider one particular component of the flow velocity v_f. ∇v_f and J_p are perpendicular to this component.

[11] The approach breaks down if different transport processes are affected by different types of scattering, since then the effective τ_s depends on the process considered. This is the case, for example, in metals at moderate temperatures; see N. W. Ashcroft and N. D. Mermin, *Solid State Physics* (Saunders, Philadelphia, PA, 1975), Chapter 13.

where $D(E)$ is the diffusivity of particles with energy E. Each particle carries an energy E, so that the contribution of these particles to the energy flux is $EJ_{n\epsilon}\,dE$. If we assume that D is independent of E, we can take it outside the integral over E, and the total particle flux due to the temperature gradient is:

$$J_n = \int_0^\infty J_{n\epsilon}\,dE = -D\frac{dT}{dx}\int_0^\infty g(E)\frac{\partial f(E, T)}{\partial T}\,dE$$

$$= -D\frac{dT}{dx}\frac{d}{dT}\int_0^\infty g(E)f(E, T)\,dE$$

$$= 0, \tag{16.19}$$

since $\int_0^\infty g(E)f(E, T)\,dE = n$, which is independent of T. However, in general, D is a function of E and the integral $\int_0^\infty J_{n\epsilon}\,dE \neq 0$. Since the net particle flux must vanish, a pressure difference builds up, causing a hydrodynamic counterflow. An extreme example of this is the fountain effect in superfluid helium (see Chapter 12), but in normal fluids it is a small effect.

To avoid this complication, we confine our attention to the case where D is independent of energy, so that we can write for the energy flux (heat current):

$$J_u = \int_0^\infty EJ_{n\epsilon}\,dE = -D\frac{dT}{dx}\int_0^\infty Eg(E)\frac{\partial f(E, T)}{\partial T}\,dE. \tag{16.20}$$

The energy density is $u = \int_0^\infty Eg(E)f(E, T)\,dE$, so that the heat capacity at constant volume *per unit volume* is:

$$C = \frac{du}{dT} = \int_0^\infty Eg(E)\frac{\partial f(E, T)}{dT}\,dE$$

whence:

$$J_u = -K_{\text{th}}\frac{dT}{dx},$$

where the thermal conductivity is:

$$\boxed{K_{\text{th}} = DC.} \tag{16.21}$$

For an ideal gas at a temperature where the equipartition theorem holds, $C = \frac{1}{2}nn_f k_B$, where n_f is the number of degrees of freedom per molecule (five for a diatomic gas: see Sections 7.9 and 10.6), so that if Equation (16.10) holds, the thermal conductivity is:

$$K_{\text{th}} = \frac{1}{6}nn_f k_B v_m \ell.$$

However, the assumption concerning $D(E)$ is not correct for an ideal gas, and the accurate calculation is considerably more complicated.[12]

---

[12] See, for example, S. Chapman and T. G. Cowling, *The Mathematical Theory of Non-uniform Gases: An Account of the Kinetic Theory of Viscosity, Thermal Conduction and Diffusion in Gases*, 3rd ed. (Cambridge University Press, Cambridge, UK, 1990), and S. G. Brush, "The Chapman-Enskog solution of the transport equation for moderately dense gases," in *Kinetic Theory*, vol. 3 (Pergamon, Press, Oxford, UK, 1972).

The case of a degenerate Fermi gas is much simpler, since $\frac{\partial f(E,T)}{\partial T} = 0$ except very close to the Fermi energy, so that we can take $D = D(E_F)$. From Equation (13.7), the electronic heat capacity per unit volume is:

$$C_e = \frac{\pi^2}{3} k_B^2 g(E_F) T$$

$$= \frac{\pi^2}{2} \frac{n}{E_F} k_B^2 T \text{ if the electrons are free.}$$

From Equations (16.9) and (16.17),

$$D = \frac{2}{3} E_F \mu_u.$$

Hence, the thermal conductivity of a degenerate free electron gas is:

$$K_{\text{th}} = \frac{\pi^2}{3} \mu_u n k_B^2 T. \tag{16.22}$$

Taking the electrical conductivity σ_e from Equation (16.5), we find that the thermal and electrical conductivities are related by a universal constant called the *Lorentz number* L_0:

$$\frac{K_{\text{th}}}{\sigma_e} = L_0 T, \tag{16.23}$$

where for a degenerate electron gas, $L_0 = \pi^2 k_B^2/3q^2 = 2.45 \times 10^{-8} \text{ W}\,\Omega\,\text{K}^{-2}$. Equation (16.23) is called the Wiedemann-Franz law. The above calculation assumes free electrons, but it can be shown that it applies, with the same value of L_0, to any degenerate electron gas so long as the dominant scattering mechanism is the same for both electrical and thermal conductivity. This is true in metals at low temperatures, where impurity scattering is the only mechanism, and usually at room temperature and above, where phonon scattering dominates; in between, it breaks down (see Footnote 11). In an ideal conducting gas, the Wiedemann-Franz law also holds, with a different expression for L_0, so long as D is independent of E (see Problem 16.2).

In a metal, as we saw in Chapter 13, the lattice contribution to the heat capacity is much greater than the electronic contribution, except at extremely low temperatures. In view of this, one might at first sight expect from Equation (16.21) that the thermal conductivity would be primarily due to phonons rather than to electrons, so that a metal would have a similar thermal conductivity to an electrical insulator. Common experience shows that this is not the case. The reason is that at ordinary temperatures $(T > \theta_D)$, the phonon mean free path is extremely short because of phonon-phonon and phonon-electron scattering, so that the electrons dominate the thermal conductivity of metals. At low temperatures, on the other hand, phonon-phonon scattering becomes weak and the thermal conductivity of a good quality single crystal insulator is comparable to that of a metal (see Problem 16.5).

16.7 Viscosity

The calculation of the viscosity follows analogous lines and makes similar approximations. We take the z-direction in Figure 16.2 to be the direction of fluid flow and the $x = 0$ plane to be the boundary of the fluid. Whereas in the calculation of the thermal conductivity the particles were labelled by their energy, here they are labelled by the z-component of their momentum, mv_z, where m is the mass of a particle. Consider the density $n_v(v_z)\,dv_z$ of particles whose z-component of velocity is between v_z and $v_z + dz$, in a layer at a distance x from the boundary. If the gas in this layer is flowing with velocity $\mathbf{v}_f$, the entire distribution is displaced in velocity space by $\mathbf{v}_f$; that is, a particle which would have velocity $\mathbf{v}_{\text{th}}$

(the thermal velocity) if the gas were at rest, now has a velocity $\mathbf{v}_{\text{th}} + \mathbf{v}_f$. If we take $\mathbf{v}_f$ to be in the z-direction, the distribution over the velocity components, v_x and v_y, is unchanged; we need only consider the distribution over v_z. The z-component of the velocity is changed from $v_{z\text{th}}$ to $v_{z\text{th}} + v_f$, where $v_{z\text{th}}$ is the z-component of the thermal velocity. Note that v_f is the same for all particles in the layer; that is, at any given distance from the boundary v_f is constant over the thermal distribution. Let the probability that the thermal velocity lies between $v_{z\text{th}}$ and $v_{z\text{th}} + dv_{z\text{th}}$ be $P(v_{z\text{th}}) \, dv_{z\text{th}}$. Then the density of particles in this layer with actual velocity between v_z and $v_z + dv_z$ is:

$$n_v(v_z) \, dv_z = nP(v_{z\text{th}}) \, dv_{z\text{th}} = nP(v_z - v_f) \, dv_z, \tag{16.24}$$

since $dv_z = dv_{z\text{th}}$, v_f being constant over the distribution.

If we assume $v_f \ll v_{z\text{th}}$, the right-hand side of Equation (16.24) can be expanded in a Taylor series, giving:

$$n_v(v_z) \approx n \left[P(v_z) - v_f \frac{dP(v_z)}{dv_z} \right],$$

where $P(v_z) \, dv_z$ is the probability that the z-component of velocity is between v_z and $v_z + dv_z$ when the gas is at rest. Note that $P(v_z)$ is a property of the gas at a given temperature and does not depend on x.

Since v_f is a function of x, $n(v_z)$ is also a function of x through v_f. Particles "labelled" with velocity v_z diffuse with a flux:

$$J(v_z) \, dv_z = -D(v_z) \frac{dn_v(v_z)}{dx} \, dv_z$$
$$= nD(v_z) \frac{dv_f}{dx} \frac{dP(v_z)}{dv_z} \, dv_z,$$

where $D(v_z)$ is the diffusivity of a particle with velocity v_z.

Each particle carries momentum mv_z, so that the momentum flux is:

$$J_p = mn \frac{dv_f}{dx} \int_0^\infty D(v_z) v_z \frac{dP(v_z)}{dv_z} \, dv_z. \tag{16.25}$$

If D is independent of v_z, this becomes

$$J_p = mnD \frac{dv_f}{dx} \int_{-\infty}^\infty v_z \frac{dP(v_z)}{dv_z} \, dv_z.$$

Since $v_z P(v_z) \to 0$ as $v_z \to \pm\infty$, integration by parts gives $-\int_{-\infty}^\infty P(v_z) \, dv_z = -1$ for the integral. Hence,

$$J_p = -\eta \frac{dv_f}{dx},$$

where the viscosity is:

$$\boxed{\eta = mnD = \rho D,} \tag{16.26}$$

$\rho \equiv mn$ being the mass density.

For a quasi-ideal gas in which $D = k_B T \mu_u = \frac{1}{3} v_m \ell$,

$$\eta = \rho k_B T \mu_u = \frac{1}{3} \rho v_m \ell. \tag{16.27}$$

To summarize, we have shown, using plausible assumptions, that all four transport coefficients can be related to a single microscopic quantity, the universal mobility μ_u, or, equivalently, to the mean free path ℓ. More accurate calculations (see the books referred to in Footnotes 11 and 12) introduce numerical coefficients of order 1, but do not alter the overall picture.

16.8 Effect of Density on Transport Coefficients: The Low Density (Knudsen) Regime

From Equation (16.10), $D \propto \ell$ at a given temperature, and from Equation (16.3), $\ell \propto n^{-1}$. Thus, D varies inversely as the density n. On the other hand, C in Equation (16.21) is the heat capacity per unit volume, so that it is proportional to n. Since the thermal conductivity K_{th} of a quasi-ideal gas is the product DC, it follows that K_{th} is independent of the density, so long as $\ell \propto n^{-1}$. The same applies to viscosity, as can be seen from Equation (16.27). However, when the pressure and hence the density is very low, ℓ is limited by scattering from the walls of the container rather than by intermolecular collisions, and it is no longer true that $\ell \propto n^{-1}$. This low pressure regime is called the *Knudsen regime*. In this regime, it is still a reasonable approximation to use the results obtained above, but with ℓ taken to be the smallest dimension of the container, so that it is independent of density. Thus in the Knudsen regime, $K_{th} \propto C \propto n$, and $\eta \propto n$. It follows that in order to reduce the thermal conductivity or viscosity of a gas by reducing its pressure, it is necessary to have a pressure low enough that the mean free path due to collisions exceeds the smallest dimension of the container. We found earlier that in air at atmospheric pressure (1 bar) and room temperature, $\ell \sim 10^{-7}$ m. If the smallest dimension of the container is ~ 1 mm $= 10^{-3}$ m, we need a pressure below 10^{-4} bar (~ 0.1 torr) to achieve the Knudsen regime, since $n \propto p$ for a quasi-ideal gas. The pressure in a typical commercial vacuum flask is $\sim 10^{-6}$ bar, so that the thermal conductivity of the air in it is reduced to about 1% of its atmospheric value. Similar reasoning applies to the viscosity.

The fact that viscosity is independent of pressure[13] except at very low pressures explains why Boyle, to his surprise and chagrin, was unable to reduce the damping of a swinging pendulum by reducing the air pressure. The pumps available in the seventeenth century could not reach the low pressure needed to reduce the viscosity significantly.

Example 16.2: Knudsen Flow

Flow of a gas in the Knudsen regime is quite different from the hydrodynamic fluid flow (viscous at low velocity, turbulent at high) that is characteristic of a gas in which the mean free path is small compared with the dimensions of the container.[14]

Suppose that a chamber of volume V, containing N molecules of a quasi-ideal gas, is connected by a tube of diameter d and length L ($L \gg d$) to a perfect vacuum. The pressure is so low that the mean free path due to intermolecular collisions is small relative to the diameter of the tube. Flow under such low pressure conditions is known as *Knudsen flow*. If scattering at the wall of the tube is diffuse, so that a molecule loses all memory of its previous velocity when it collides with it, the effective mean free path ℓ is approximately d.

---

[13] Maxwell deduced this result from kinetic theory in 1860. Having been told by George G. Stokes that the experimental data showed that, as naive common sense suggests, the viscosity of a gas depends on density, he concluded that kinetic theory must be wrong. Later, however, he made his own measurements, which showed that the viscosity is indeed independent of density over the accessible pressure range, and by 1866 he had concluded that kinetic theory accounts accurately for gas viscosity. The previous experiments had in fact been interpreted by Stokes on the assumption that the viscosity *does* decrease as the pressure is reduced; a vivid illustration of Einstein's remark that our theories determine what we can observe [see S. G. Brush, *The Kind of Motion We Call Heat* (Elsevier, Amsterdam, the Netherlands, 1976, 1992), pp. 189–91, 435–6].

[14] See N. W. Robinson, *The Physical Principles of Ultra-High Vacuum Systems and Equipment* (Chapman and Hall, London, UK, 1968), Chapter 8, for a thorough treatment of transport in the Knudsen regime.

1. Treating the flow of gas through the tube as a diffusive process, show that the number of gas molecules that pass through the tube in unit time is Sn, where $n = \frac{N}{V}$ is the density at the entrance to the tube and S is independent of n. S is called the conductance (or speed) of the tube (do not confuse it with entropy!). Find S, assuming that $\ell = d$. Find the volume of gas flowing per unit time, the volume being measured at the inlet pressure.

From Equation (16.10), putting $\ell = d$,

$$D = \frac{1}{3} v_m d.$$

Let the particle flux in the tube be J_n. This is constant along the length of the tube, and since D is also constant, the density gradient is constant and equal to n/L. Hence,

$$J_n = \frac{Dn}{L}.$$

The rate of flow of molecules through the tube is:

$$-\frac{dN}{dt} = \frac{\pi}{4} d^2 J_n = Sn,$$

where:

$$S = \frac{\pi}{12} v_m \frac{d^3}{L}. \tag{16.28}$$

At the inlet pressure N molecules occupy a volume $V = N/n$. Hence, $\frac{dV}{dt} = n^{-1} \frac{dN}{dt} = -S$, so that S is the volume of gas extracted per unit time, independent of inlet pressure so long as it is low enough for Knudsen conditions to obtain. As the pressure in the vessel being evacuated goes down, a given number of molecules occupy a larger and larger volume, so that fewer and fewer are removed.

A more precise calculation of S, properly averaging over the velocity distribution, gives Equation (16.28) with a slightly different numerical coefficient (see the reference in Footnote 14)

2. Find the conductance of a hole of area A, under conditions of Knudsen flow.

Every molecule that reaches the hole goes through it, so that the flow is simply the number of such molecules per unit time. This number is AJ_n, where J_n is the particle flux striking the surface, which is, from Equation (H.4),

$$J_n = \frac{1}{4} n \langle v \rangle.$$

Hence,

$$S = \frac{AJ_n}{n} = \frac{1}{4} \langle v \rangle A.$$

16.9 The Diffusion Equation

We now revert to the regime where the mean free path is small, and ask what happens when the particle flux J_n varies with position. For simplicity, we assume that the variation is in the x-direction only. The total number of particles is conserved, so that if $\frac{\partial J_n}{\partial x} \neq 0$, the net flow of particles out of the region between x and $x + dx$ must produce a corresponding change in density (see Figure 16.3). The rate of increase in

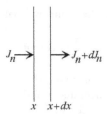

FIGURE 16.3 The number of particles per unit area of the layer between x and $x + dx$ is $n\,dx$, and changes at a rate $-dJ_n$ per unit time.

the number of particles per unit area between x and $x + dx$ is the difference between the incoming and outgoing flux; that is:

$$\frac{\partial n}{\partial t}\,dx = J_n - (J_n + dJ_n),$$

where $dJ_n = \frac{\partial J_n}{\partial x}\,dx$. Hence,

$$\frac{\partial n}{\partial t} = -\frac{\partial J_n}{\partial x}. \tag{16.29}$$

Similarly, energy is conserved, so that the rate of change of energy density is related to the energy flux in the same way:

$$\frac{\partial u}{\partial t} = -\frac{\partial J_u}{\partial x}. \tag{16.30}$$

In one dimension, Equation (16.12) becomes:

$$J_n = -D\frac{\partial n}{\partial x}. \tag{16.31}$$

Combining Equation (16.31) with the conservation relation (Equation 16.29) gives:

$$\boxed{\frac{\partial n}{\partial t} = D\frac{\partial^2 n}{\partial x^2}.} \tag{16.32}$$

This is the one-dimensional *diffusion equation* for the density n.

Similarly, Equation (16.13) can be combined with Equation (16.30) to obtain the *equation of heat conduction*:

$$\frac{\partial T}{\partial t} = -\frac{1}{C}\frac{\partial J_u}{\partial x} = \frac{K_{\text{th}}}{C}\frac{\partial^2 T}{\partial x^2}, \tag{16.33}$$

where C is the heat capacity per unit volume and we have put $\frac{du}{dT} = C$. Equations (16.32) and (16.33) are mathematically identical, and $\frac{K_{\text{th}}}{C}$ is called the *thermal diffusivity*. The thermal diffusivity is equal to D in the simple model discussed above, but in general, it need not be. In an insulator, where the thermal conductivity and the heat capacity are both due to phonons, the thermal diffusivity is roughly equal to the phonon diffusivity. In a metal at ordinary temperatures, this is not the case, since, as pointed out at the end of Section 16.6, the heat capacity is due to phonons, while the thermal conductivity is primarily due to electrons, which have a much higher diffusivity.

The same equation governs the flow of a viscous fluid at low velocities, the analog of the diffusivity being the kinematic viscosity $\frac{\eta}{\rho}$, which is also equal to D in the simple model discussed in Section 16.5.

The solution of Equation (16.32), and its three-dimensional analog:

$$\frac{\partial n}{\partial t} = D \nabla^2 n, \qquad (16.34)$$

is the subject of many massive tomes[15] and only two very simple cases will be considered here.

One case, of great importance in the semiconductor industry, is the semi-infinite slab in which a fixed number of particles diffuse in from the surface. N_0 impurity atoms per unit area are deposited on the surface of the slab (taken as the $x = 0$ plane) and subsequently diffuse into it. The density profile at a time t is then given by:[16]

$$n(x, t) = \frac{N_0}{2(\pi D t)^{1/2}} \exp\left(-\frac{x^2}{4Dt}\right), \qquad (16.35)$$

as can be seen by substituting Equation (16.35) into Equation (16.32) and integrating over x to show that $\int_0^\infty n(x, t)\, dx = N_0$ for all t. Equation (16.35) is plotted in Figure 16.4 (the curve marked n_D).

In a typical semiconductor, which is solid, diffusion is usually extremely slow at room temperature, but D increases to a usable value ($\sim 10^{-15}$ m^2/s) at a few hundred °C. The standard way of making a $p - n$ junction (see Section 14.6) uses this solid state diffusion. One starts with a slab of semiconductor containing a uniform distribution of one type of impurity, (say, an acceptor with density n_A), and deposits a thin layer of the other type on the surface. If the material is then heated for a few minutes so that the donors diffuse into the material, $n_D(x, t)$ is given by Equation (16.35), and a junction forms where $n_D = n_A$, as illustrated in Figure 16.4. The distance (typically a few microns) from the surface to the junction can be controlled by adjusting the temperature (which determines D) and the diffusion time t. This is one of the basic processes in the manufacture of integrated circuits.

Because of the precise analogy between the equations describing different transport processes, the same mathematical results apply, with appropriate redefinition of symbols, to any transport process. For example, Equation (16.35) describes the temperature distribution in a semi-infinite slab after a given quantity of energy has been deposited at the surface (e.g. by a laser flash).

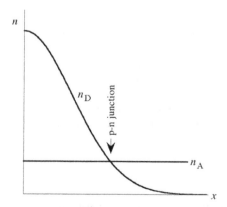

FIGURE 16.4 Impurity concentrations as a function of distance below the surface of a semiconductor, in which a $p - n$ junction is made by diffusion of donors into a slab of initially p-type material, whose acceptor concentration is n_A. The profile of the donor concentration n_D follows Equation (16.35). The carrier concentration is $|n_D(x) - n_A|$, so that the semiconductor is n-type to the left of the junction and p-type to the right.

---

[15] For example, J. Crank, *Mathematics of Diffusion* (Oxford University Press, Oxford, UK, 1956); H. S. Carslaw and J. C. Jaeger, *Conduction of Heat in Solids*, 2nd ed. (Oxford University Press, Oxford, UK, 1986).
[16] ibid.

Another important application of Equation (16.33) is to heat flow in a sound wave, which was mentioned in Section 10.8. In a sound wave, air is periodically compressed and expanded. If heat does not have time to flow from the warm compressed regions to the cool expanded regions, the compression and expansion are isentropic. We now show that this is always the case at sufficiently low frequencies.

Suppose that at some instant, a plane wave of frequency ν propagating in the x-direction produces (by isentropic compression) a temperature deviation from the average given as a function of position by:

$$\Delta T(x) = A \sin(kx), \tag{16.36}$$

where $k \equiv \frac{2\pi\nu}{c_s}$, c_s being the velocity of sound. Rather than solving the full combined wave and diffusion equation, we estimate how rapidly thermal conduction makes this temperature deviation decay, by solving Equation (16.32) with Equation (16.36) as the initial condition. The solution, readily obtained by separation of variables, is:

$$\Delta T(x, t) = A e^{-t/\tau} \sin kx \tag{16.37}$$

where $\tau = \frac{C}{K_{th}k^2}$, as can be proved by substitution.

If $2\pi\nu\tau > 1$, thermal diffusion does not have time to occur, so that the compression is isentropic. Substituting for k, we find that $\tau \propto \nu^{-2}$, and the condition $2\pi\nu\tau > 1$ gives an *upper* limit to ν:

$$\nu < \nu_c = \frac{c_s^2 C}{K_{th}}. \tag{16.38}$$

At frequencies in the vicinity of ν_c, there is significant heat flow in the wave. Since heat flow is an irreversible process, entropy is generated; that is, mechanical energy is converted to heat and the sound is attenuated.

Let us put in some numbers. In a gas, $\frac{K_{th}}{C} \approx D = \frac{1}{3}v_m\ell$ from Equation (16.10), while from Equation (10.35), $c_s^2 = \frac{\gamma k_B T}{m}$, where γ is the ratio of heat capacities and m is the mass of a molecule. Hence,

$$\nu_c \approx \frac{3\gamma k_B T}{2\pi m v_m \ell}.$$

Since $\ell \propto n^{-1}$, $\nu_c \propto n$. For air at 300 K and 1 bar, $\ell \sim 10^{-7}$ m, $D \sim 2 \times 10^{-5}$ m^2/s, $c_s \approx 350$ m/s, so that Equation (16.38) gives $\nu_c \sim 500$ MHz, well above the audible range. However, if the pressure is reduced to the point where ℓ is determined by the size of the container, say 10^{-1} m, acoustic waves with frequencies of order kHz are strongly attenuated (viscosity also contributes to the attenuation under these circumstances). It is this attenuation, rather than the low pressure in itself, that is responsible for the fact that a bell ringing inside an evacuated container cannot be heard.

16.10 Envoi

Historically, kinetic theory preceded statistical mechanics, and the kinetic theory of transport in gases was well advanced before the foundations of the equilibrium statistical theory were laid by Boltzmann and Gibbs. For example, the first version of the memoir by Maxwell[17] referred to in the epigraph to this chapter was published in 1858, while Boltzmann's recognition of the connection between probability and entropy, expressed in Equation (5.4), dates from 1877, and it was another 25 years before classical statistical mechanics was put on a firm logical basis by Gibbs.[18] However, while the discovery of quantum theory

---

[17] Reprinted in S. G. Brush (ed.) *Kinetic Theory*, vol.1 [Pergamon Press, Oxford, UK, 1965], pp. 148–171.

[18] J. W. Gibbs, *The Elementary Principles of Statistical Mechanics* (1902) [reprinted in 1981 by the Oxbow Press, Woodbridge, CT]; Bailyn, pp. 459–467.

in the first half of the 20th century revolutionized statistical mechanics, its contributions to transport theory were restricted to some rather special cases. It is remarkable that the Boltzmann equation,[19] which describes the time evolution of the distribution of molecular velocities in a gas out of equilibrium, and dates from 1872, is still the basis of most advanced treatments of transport processes. Recent advances in this field have owed more to new mathematical techniques, and to the enormous increase in computing power, than to new physical insights.

16.11 Problems

16.1   Assuming that Equations (16.5) and (16.6) hold for a Fermi gas at 0 K, prove Equation (16.9), and show that Equation (16.10) holds, with v_F substituted for v_m

16.2   Show that if the diffusivity of a particle is independent of its energy, the Wiedemann-Franz law $\frac{K_{th}}{\sigma_e} = L_0 T$ (Equation (16.23)) holds for a classical (quasi-ideal) gas as well as for a Fermi gas, but with a different value of the Lorentz number L_0. Find L_0

16.3   Using Equation (16.7) and the dimensional analysis of Example 16.1, find the temperature dependence at constant pressure of the diffusivity and viscosity of a quasi-ideal gas in which the intermolecular force is proportional to R^{-n}

16.4   In experiments at very low temperatures, it is often necessary to run electrical connections from a point at a relatively high temperature (typically about 1.2 K, the lowest temperature readily accessible with liquid helium) to one at much lower temperature, which may be at a few millikelvin. While it is desirable to keep the resistance R of the wire as low as possible, it is essential to minimize the heat flow down the wire. At these low temperatures, the heat capacity is primarily electronic, and the mean free path of electrons in most "normal" (that is, non-superconducting) metals is independent of temperature, being determined by impurity scattering. If the higher temperature is T_1 and the lower temperature is T_2, show that the heat flow is proportional to $\frac{T_1^2 - T_2^2}{R}$, and find the constant of proportionality

16.5   In an insulator at a temperature well below its Debye temperature, phonon-phonon scattering conserves phonon momentum and hence does not affect the thermal conductivity. In a single crystal of good quality, the only mechanism limiting the phonon mean free path is scattering at the surfaces. Thus phonons at such a temperature can be regarded as a gas in the Knudsen regime. Show that the thermal conductivity K_{th} varies as T^3. Estimate K_{th} at 100 K in a diamond crystal 1 mm across, and compare it with that of pure copper, for which $\sigma_e \sim 5 \times 10^6$ mho-cm at this temperature. The Debye temperature of diamond is approximately 2000 K

16.6   (a) If the cross-section for scattering of an electron by an air molecule is $\sim 10^{-19}$ m^2, at approximately what air pressure (at 300 K) will 90% of the electrons emitted from the cathode of a TV tube reach the screen 0.3 m away? Assume that scattered electrons do not reach the screen, that the electrons take the shortest possible path from cathode to screen, and that the air is an ideal gas. Give your answer in bar (1 bar $\equiv 10^5$ Pa)

       (b) In the tube described in (a), assume that the air pressure is 10^{-9} bar. The electron current from the cathode to the screen is 10 μA. If every collision of an electron with an air molecule produces one positive ion, how many positive ions are produced per second? The presence of these positive ions, which are accelerated towards the cathode and do damage when they strike it, is a major cause of failure of TV tubes

16.7   Consider two chambers that are thermally insulated from each other except for a hole whose diameter is small compared with the mean free path. The chambers contain the same quasi-ideal gas, but are maintained at different temperatures, T_1 and T_2

---

[19] See Bailyn, pp. 457–459.

(a) Show that diffusive equilibrium is achieved (so that the net particle flux through the hole is zero) when the pressures in the two chambers are in the ratio:

$$\frac{p_1}{p_2} = \left(\frac{T_1}{T_2}\right)^{1/2}. \tag{16.39}$$

Explain why Equation (16.39) only applies at pressures such that the mean free path is large relative to the diameter of the hole

(b) If the area of the hole is A, what is the rate of heat flow through it?

16.8 A vessel of volume V and initial pressure p_0 is to be evacuated through a tube of conductance S. Assuming that the pump maintains zero pressure at the outlet of the tube, show that the pressure in the vessel varies as $e^{-t/\tau}$, where t is the time, and find τ

If $V = 0.1 \text{ m}^3$, $p_0 = 10^{-4}$ bar, $d = 5$ cm, $L = 1$ m, and the gas is nitrogen, roughly how long will it take to reach a pressure of 10^{-12} bar (about 10^{-9} torr)? Take $v_m = v_{\text{th}}$ and $T = 300$ K. This calculation gives a lower limit to the pumping time; in practice, it usually takes much longer because of *outgassing* (the desorption of gas from surfaces), and because gas trapped in cracks and crevices takes a long time to escape. These processes can be accelerated by heating the entire vessel to a high temperature while under vacuum, and, whenever possible, apparatus for use at very high vacuum is designed to be baked at a few hundred °C under vacuum before use

16.9 In the technique of molecular beam epitaxy (MBE), semiconductor or metal layers of very accurately controlled thickness (epilayers) are laid down in a high vacuum by directing beams of the constituent atoms onto a surface called the substrate. It is this technique, or variations of it, which is generally used to create the two-dimensional structures mentioned in Section 7.4. The atoms come from ovens, each containing a solid or liquid element in equilibrium with its vapor, connected to the vacuum chamber by an aperture that can be closed by a shutter. Figure 16.5 shows schematically a typical geometry. The vacuum is so good that the probability of the atom undergoing a collision during its journey from the furnace to the substrate is negligible [see part (b) of this problem]. The fraction of the atoms hitting the surface which remain on the substrate is called the *sticking coefficient*, which for simplicity we take to be 1

(a) Suppose that a 10 nm layer of iron (atomic weight 56, mass density $8 \times 10^3 \text{ kg m}^{-3}$) is to be grown. The aperture has area 10^{-3} m^2, is at a distance 0.3 m from the substrate, and is located so that the atoms strike the substrate normally. The oven temperature is 1600 K, at

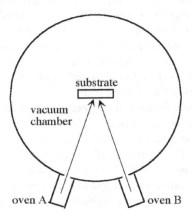

FIGURE 16.5 Schematic of MBE apparatus (omitting diagnostic tools). Atoms of elements A and B exit the hot ovens through small apertures, controlled by shutters, and strike the substrate, forming the compound AB. A typical MBE system has five or more ovens for different elements.

which the vapor pressure is 0.1 Pa. Assuming a sticking coefficient of 1, how long should the shutter be kept open? How much error in thickness will be introduced by an error of 10 K in the oven temperature? Treat iron vapor as an ideal gas with a latent heat of vaporization of 350 MJ/kmole

(b) Gas molecules in the vacuum chamber can strike the substrate and be incorporated into the growing layer. If no more than 10^{24} impurities per m^3 ($\sim$ 10 parts per million) can be tolerated, and the growth rate is 1 μm per hour, what is the maximum pressure of air permissible in the vacuum chamber (assumed to be at 300 K)? Again, assume a sticking coefficient of 1. Show that at this pressure the probability that an atom from the furnace collides with a gas molecule before it reaches the substrate is indeed negligible, as was assumed in (a)

16.10 (Wind chill.) If a liquid or solid is in equilibrium with its vapor, the rate of evaporation, defined as the mean number of atoms or molecules evaporating per unit time and area, must equal the rate of condensation from the vapor. This is analogous to Kirchhoff's law of radiation (see Section 11.4)

(a) Assuming that water vapor is a quasi-ideal gas and that the sticking coefficient (defined in Problem 16.9) for a water molecule on liquid water is 1, find the rate of evaporation of water from a free surface into dry air at 300 K, given that the saturation vapor pressure (that is, the pressure at which the vapor and the liquid are in equilibrium) at this temperature is 4×10^3 Pa. Assume that the air in contact with the surface is kept perfectly dry by a continuous draft. If the latent heat of evaporation is 4 MJ/kmole, at what rate must heat be supplied if the remaining water is not to cool down? Compare the rate of heat loss from a wet human body in a strong wind with the average heat output of the body when resting, which is $\sim$ 100 W

(b) Find the diffusivity of water vapor in air at 300 K and 1 bar pressure, if the cross-section for collision of a water molecule with a nitrogen or oxygen molecule is 3×10^{-19} m^2. The molecular weight of H_2O is 18

(c) Now suppose that the air just above the water surface in (a) is stationary, so that its water vapor content can build up, but that 1 mm away the air is perfectly dry. Show that the evaporation is now diffusion-limited, so that the relative humidity at the surface is almost 100%; that is, the water vapor density there is close to its saturation value. By what factor is the rate of evaporation reduced, relative to that in (a)?

16.11 The aurora borealis is produced by electrons entering the atmosphere from above. The electrons collide with the nitrogen and oxygen molecules in the atmosphere, exciting the molecules so that they emit visible light. Consider the following simple model of the process. The temperature is assumed to be uniform so that the atmospheric pressure p varies exponentially with height h; that is, $p(h) = e^{-h/h_0}$ (see Example 8.2). The electrons are assumed to enter the atmosphere vertically from outer space. The cross-section σ_{ex} of a molecule for excitation by an electron is taken as constant, and this is assumed to be the only process by which electrons interact with molecules. Assume that once an electron has collided with a molecule, it no longer has enough energy to excite another. Hence, for every molecule excited, an electron is "lost." Note that since the flux of electrons J is coming from above, $\frac{dJ}{dh}$ is positive

(a) Show that the rate at which molecules are excited (and hence, light is emitted) per unit volume is a maximum at a height given by $h_{max} = h_0 \ln(h_0/\ell)$, where ℓ_0 is the mean free path of an electron in air at ground level ($h = 0$) calculated assuming that $\sigma_{ex} = \sigma_c$, the collision cross-section. What is the mean free path at height h_{max}?

(b) Find h_{max}, given that $p_0 = 10^5$ Pa, $T = 300$ K, the acceleration due to gravity $g = 10$ m/s^2, $\sigma_c = 10^{-21}$ m^{-2}. Treat air as an ideal gas with a mean molecular weight of 30

16.12 Make an argument (semi-classical) for an upper bound on transport properties in any system

Non-equilibrium Thermodynamics

Philosophy gets on my nerves. If we analyze the ultimate ground of everything, then everything finally falls into nothingness.

Ludwig Boltzmann

17.1 Gedanken

Let us return briefly to one of our first examples by which we began to construct the physical concepts of heat and temperature. You go to the doctor's office, and they place a thermometer in your mouth to determine your temperature. The thermometer is left in your mouth for some period of time so that an amount of heat is transferred from your body to the thermometer until the thermometer is in thermal equilibrium with your body. This is then your measured temperature.

However, while your body and the thermometer are in thermal equilibrium, your body is not in thermal equilibrium with the room. Indeed, as we have mentioned before, there is a specific term for your body when it is in thermal equilibrium with the room — the state of death. Your body is a biochemical engine that is constantly converting food into energy and expelling waste. While it may be tempting to consider your body to be a reservoir thermodynamically, it is more proper to think of your body as being in a steady state out of equilibrium with one's environment.

17.2 Time-Dependent States

Broadly, one can argue that thermodynamics is simply the consequence of any set of energies (with a cardinality of 2 or more) that a system can possess. This is why one of the simplest thermodynamic systems we have considered is the spin 1/2 states in an external magnetic field. The applicability of the thermo-statistical picture then comes from the large number of N spins in the ensemble. For the moment, let's take another look at our quantum mechanical underpinning of statistical mechanics. The idea of a state here is the solution of Schrödinger's equation, which is really just the Hamiltonian for microscopic systems. Hence, we can generalize what we have been doing in our development of statistical mechanics in a briefer, cleaner form:

$$\hat{Z}\Psi = \left(\sum_i e^{-\beta\hat{H}_i}\right)\Psi = \sum_i \sum_{n=0}^{\infty} \frac{\left(-\beta\hat{H}_i\right)^n}{n!}\Psi$$

$$= \sum_i \sum_{n=0}^{\infty} \frac{(-\beta E_i)^n}{n!}\Psi = \sum_i e^{-\beta E_i}\Psi, \tag{17.1}$$

where the value E_i is the eigenvalue corresponding to the eigenstate Ψ of Schrödinger's equation. In quantum mechanics, the time-dependent states are then found by the propagator acting on the time-independent states of the system. We wish to bring this to mind for two reasons. First, as we consider relaxing our equilibrium temperature condition, keep in mind that unless the system itself undergoes a physical change, the underlying equilibrium states of the system are not altered. Second, the concept of the total energy of the system, even out of equilibrium, is a valid quantity. Indeed, in the next chapter on biological systems, it is the energy of the system that is modified by the presence of catalysts and enzymes that allow certain changes to occur to take such systems from one state to another.

To understand this more clearly, let us review the occupation numbers for a set of time-independent states. For simplicity sake, let us imagine we have a three-level system with energies E_0, E_1, and E_2. As we've seen previously (see Problem 13.12), the free energy is ambiguously defined up to the energy of the lowest level. That is, we can, with just as much generality, describe our three states as having energies, 0, $E_1 - E_0$, and $E_2 - E_1$ by subtracting off the energy of the lowest (ground) state. We make note of this to remind the reader that it is the energy differences between states that are the potential barriers to particle populations.

The free energy of the system is then:

$$F = -k_B T \ln(Z), \tag{17.2}$$

where Z of course is the partition function which we can explicitly write in this case as:

$$Z = 1 + e^{-\beta(E_1 - E_0)} + e^{-\beta(E_2 - E_1)}. \tag{17.3}$$

If we imagine allowing N non-interacting particles to populate these three states, then at an equilibrium temperature T the population of each state, n_0, n_1, and n_2 is simply $n_i = p_i N$, where p_i is determined by the Boltzmann factor normalized by Z. Instead of this absolute measure, we instead look at the relative number in each of the three states, which is given by the ratio of the Boltzmann factors. So, the relative occupation numbers $n_1/n_0 = p_1/p_0$ for the number of particles in the first excited state to the ground state is:

$$n_1/n_0 = p_1/p_0 = e^{-\beta(E_1 - E_0)}/[Z \times (Z)^{-1}] = e^{-\beta(E_1 - E_0)}/1$$
$$= e^{-\beta(E_1 - E_0)} = e^{-\beta(\delta E_1)}. \tag{17.4}$$

We now rearrange this expression to solve for $T(\beta = (k_B T)^{-1})$:

$$\ln(n_1/n_0) = -\beta \delta E_1, \tag{17.5}$$

from which substituting for β gives:

$$T = \frac{-\delta E_1}{k_B \ln(n_1/n_0)}. \tag{17.6}$$

In equilibrium at finite temperatures, the occupation number n_0 will always be larger than any of the excited states, so we see from the above relationship that the term $k_B \ln(n_1/n_0)$ can be thought of naively as $-dS_1$ between the first excited state and the ground state, where the negative sign is added explicitly to indicate a positive definite temperature. We lose no generality in then asserting that this is the same temperature one would find from looking at the relative occupation numbers between the second excited state relative to the ground state or the relative occupation numbers between the second and first excited states. Indeed, we can further generalize our specific example to any number of excited states in a system to say that the equilibrium temperature is simply the case where the relative occupation numbers between all of the time-independent energy levels in the system yield the same value for the temperature.

Non-equilibrium temperatures, to invoke a misnomer as temperature was formally defined in equilibrium, would then be a condition under which the occupation numbers of the microscopic states are kept from reaching (or remaining in) their equilibrium configuration. One can then think of two immediate non-equilibrium situations: one in which the occupation numbers result in a temperature gradient between multiple states or one in which the occupation numbers vary in time to a degree that is larger than the fluctuations one expects in equilibrium.

Now, we can imagine a process where heat is suddenly, but in a non-equilibrium fashion, dumped into this system (see Appendix I concerning the correspondence principle). The energy levels remain the same (under an assumption that $đW = 0$). So, we understand the system moves to a new equilibrium temperature T' by the addition of the large amount of heat rearranging the individual quantum state occupation numbers. The equilibrium temperature is then reached when the occupation numbers for the states reach those predicted by the Boltzmann probability distribution. The result above reminds us of our discussion of Arrhenius' law from Chapter 6. It is no coincidence that, such as we discussed then, the relaxation to the new equilibrium temperature will be dominated by an exponential term that can give us important physical insight into the thermal transport properties of the system (see Chapter 16).

17.3 Relaxation to Equilibrium: Newton's Law of Cooling

Newton's law of cooling states that the rate of heat loss by an object to its surroundings is proportional to the temperature difference between the object and those surroundings. This version of Newton's law of cooling assumes a constant heat transfer coefficient (so independent of the object's temperature). However, similar to the form of the Stefan-Boltzmann law, the surface area of the object (the area of heat transfer between the object and its surrounding environment) plays a role in the heat flux, so that:

$$\frac{đQ}{dt} = c_h A[T(t) - T_a], \tag{17.7}$$

where $đQ$ is the heat flow out of (or into) the object as the temperature of the object changes in time from its value $T(t)$ to the ambient temperature of the surrounding environment, T_a, A is the surface area of the body through which the heat flow occurs, and c_h is the heat transfer coefficient[1] in units of W m^{-2} K^{-1}.

Obviously, this relationship as stated here assumes that the temperature of the object is uniform throughout, and that the surrounding environment can be assumed to be a reservoir at constant temperature T_a. The heat transfer coefficient varies from one material to another and is often found empirically. The value of the heat transfer coefficient also depends upon whether the radiative transfer to the surroundings (air) occurs in a conductive or convective regime. While a convective heat transfer (and its coefficient) is in general larger, over short time periods the method of measurement remains the same regardless of whether the temperature difference results in a laminar, convective, or turbulent flow of the medium (air, liquid) surrounding the object. The only necessary assumption for the equation to hold is that c_h does not change (the method of heat transport remains the same) during the time interval in question.

Assuming that the total heat capacity of the object relates the heat radiated (lost by the object, so negative) and the temperature of the object as a function of time, then:

$$C = -\frac{đQ}{dT(t)} = -\frac{đQ}{dt}\frac{dt}{dT(t)}, \tag{17.8}$$

---

[1] In real experiments, the heat transfer coefficient is dependent, not only upon the material losing the heat, but the adjoining material to which the heat is being lost. In the treatment here, we are considering an idealized situation and ignoring other effects, such as in fluids and gases that the motion of the material itself can enhance the heat transfer coefficient for the material.

so that:

$$\frac{dT(t)}{dt} = -\frac{c_H A}{C}[T(t) - T_a].\tag{17.9}$$

Examining the units indicates that the quantity $C/(c_h A)$ is a characteristic time constant of the system, t_0, and the differential form of the relationship where the rate of temperature change is proportional to the temperature implies an exponential approach to the ambient temperature of the environment. It can be shown (see Problem 17.1) that the form of the solution for $T(t)$ is therefore:

$$T(t) = T_a + [T(0) - T_a]e^{-\frac{t}{t_0}},\tag{17.10}$$

where $T(0)$ is the initial temperature of the object at $t = 0$.

17.4 Time Dependence and the Laws of Thermodynamics

With over a century of success applying the axioms of equilibrium thermostatistics to a variety of physical systems, physicists have turned their eye to a host of other problems that are implicitly non-equilibrium or for which the assumptions of equilibrium thermostatistics do not produce accurate predictions. With this in mind, we turn our attention back to the four laws of thermodynamics to see the implications of re-introducing time as a degree of freedom (time was a quantity we removed by considering only state variables and state functions in our treatment of equilibrium thermodynamics). Our purpose here is to ask what we lose by examining a time-dependent version of the four laws. For the moment, given that the zeroth law concerns the definition of equilibrium that we are explicitly relaxing, we will take up the zeroth law last to motivate the rest of our work in this chapter.

Our original version of the First Law:

$$dU = đQ + đW,\tag{17.11}$$

lends itself quite well to a time-dependent version:

$$\frac{dU}{dt} = \frac{đQ}{dt} + \frac{đW}{dt},\tag{17.12}$$

as energy balance in each moment in time accurately predicts that any change in energy per unit time of the system would necessarily need to be balanced moment by moment. That is, since the first law is essentially conservation of energy, and energy is always conserved, then the first law translates fairly well (for non-relativistic speeds in a system, anyway). However, as moment to moment there can be both work done by the system as well as work done on the system (most often necessary to drive the system away from equilibrium), we make a purposeful change in that we consider $đW/dt$ to be the work done by the system as an energy cost and an additional term dD/dt to be the external driving of the system. Hence the first law becomes:

$$Power = \frac{dU}{dt} = \frac{đQ}{dt} + \frac{đW}{dt} + \frac{dD}{dt}.\tag{17.13}$$

The second law may be stated in any number of ways. In our treatment of equilibrium thermodynamics, we could simply state that:

$$dS \geq 0.\tag{17.14}$$

Hence, for any reversible process $dS = 0$ over a closed loop, and for irreversible processes, $dS > 0$ over any closed path. From this, we obtained both our macroscopic and, eventually, microscopic understanding that the universe is always tending to a state of increasing entropy. Naively, we could simply assert that entropy is always increasing and that a time-dependent version of our second law is simply:

$$\frac{dS}{dt} \geq 0. \tag{17.15}$$

A far more rigorous investigation of a non-equilibrium version of the second (and third) law relies on the idea of *entropy production*, which we will examine in the next section.

Our transitive version of the zeroth law is the only direct casualty of relaxing the equilibrium condition, precisely because it is the equilibrium condition. While the absolute temperature of any two objects in equilibrium implies the equivalence of any other object with the same condition, the time dependence alone of these quantities does not. If we imagine the same amount of energy added to three systems with three different heat capacities, the changes in temperature for each will be different. Conversely, establishing that the rate of temperature change in two objects is the same tells us nothing of their thermodynamic state at time t later without having a detailed knowledge of the history of both objects. Not having to detail the history of an object was the very reason for removing time as a thermodynamic quantity in favor of the state variables of equilibrium thermodynamics. Having noted that the states of the system are still present under the approach to equilibrium (or conversely being driven from it) is there some manner in which to recapture a picture of non-equilibrium thermodynamics in terms of the equilibrium states?

17.5 Entropy Production

In this section, we will consider the construction of a macroscopic calculation of a time-dependent entropy. Since the second law implies that entropy must increase over time,[2] we wish to develop a manner in which to determine the entropy produced in any process. First, we will do this for a macroscopic treatment, and then see that a microscopic treatment develops in the same manner.

We begin by considering any macroscopic composite system that can be thought of as composed of two components (two volumes, two groups of particles, etc.) The system need only conform to an extensive parameter that has values for each of the components and a fixed total such that:

$$X_A + X_B = X_{total}, \tag{17.16}$$

and therefore $dX_{total} = 0$ necessarily implying that $dX_A = -dX_B$, so that if we now consider the entropy production by changing the values of any of the extensive parameters of the thermodynamic system then:

$$\frac{\partial S}{\partial X} = \mathsf{F}, \tag{17.17}$$

where $\mathsf{F} = 0$ in equilibrium. So, for the case of two systems separated by a diathermal wall, the extensive parameter is the energy U, and the two component system evolves such that:

$$\frac{\partial S_A}{\partial X_A} + \frac{\partial S_B}{\partial X_A} = \frac{\partial S_A}{\partial X_A} - \frac{\partial S_B}{\partial X_B} = \frac{\partial S_A}{\partial U_A} - \frac{\partial S_B}{\partial U_B}$$
$$= T_A^{-1} - T_B^{-1} = \mathsf{F}, \tag{17.18}$$

so, the system evolves until there is no temperature difference between the two subsections of the overall system. We have looked at this before, but now wish to consider it from a slightly different perspective. We can consider the difference in inverse temperature between the two subsections to be motivator of change driving the two subsections into equilibrium when $\mathsf{F} \neq 0$ and obtaining the value of zero when the two subsections reach equilibrium.

---

[2] Unless one removes every irreversible process from their universe, then entropy will increase over time.

Hence, S, the entropy, is produced (maximized) as long as there is a difference in the intensive parameter (temperature) that drives the extensive parameter energy to flow (a flux) between the two subsections. The generalized motivators F_i are called *affinities* for any general X_i that satisfies the behavior above. For any choice of X_i there is a corresponding flux that may be defined as:

$$J_i = \frac{\partial X_i}{\partial t}. \tag{17.19}$$

So, the entropy of the system achieves a local maximum when the particular affinity reaches zero, at which point the flux simultaneously reaches zero. In general, there is an affinity for each extensive parameter that can contribute to the production of entropy S. Hence, we can think of the entropy of the system evolving in time as:

$$\frac{dS}{dt} = \sum_i \frac{\partial S}{\partial X_i} \frac{\partial X_i}{\partial t}, \tag{17.20}$$

or simply:

$$\frac{dS}{dt} = \sum_i F_i J_i. \tag{17.21}$$

Instead of simply considering this relationship for two discrete subsections, we extend the result to a large number of infinitesimally small elements to generate a continuum relationship. Namely, we can define a local value for the entropy as a function of the local values of the extensive parameters in the system such that:

$$dS = \sum_i F_i dX_i. \tag{17.22}$$

Now, by this we have been able to define an entropy that is useful in both continuous and non-equilibrium systems by adopting the equilibrium entropy as the local entropy for the set of local extensive parameters X_i, implying that each affinity F_i is the corresponding local intensive condition identically defined from the equilibrium case. For our example above where the extensive parameter is the energy, we can now properly consider situations such as a metal bar whose temperature varies continuously from one end to another as a thermal gradient, whereas we initially only defined the temperature as an equilibrium quantity with the zeroth law.

If we consider one of the extensive parameters, for instance the energy U, and its behavior in a time-dependent and continuous treatment of the entropy production, it is beneficial to think in terms of the densities of both the entropy and energy as our local quantities. So our relationship above becomes:

$$ds = \sum_i F_i dx_i, \tag{17.23}$$

for any of the extensive quantities (energy, particle number) where $s = S/V$ and $x_i = X_i/V$.[3]

---

[3] By this choice, the volume is no longer one of the extensive quantities in the list. Instead of an equation for particle number we have an expression for particle density, etc.

The entropy production now depends upon the entropy density, which itself can change locally due to the flux of entropy into or out of a region in the system in addition to any explicit time dependence of the entropy density. Therefore, we can easy argue that there is a continuity equation:

$$\frac{ds}{dt} = \frac{\partial s}{\partial t} + \nabla \cdot \mathbf{J}_s, \tag{17.24}$$

where $\mathbf{J}_s$ is the entropy current density and is defined by:

$$\mathbf{J}_s = \sum_i \mathsf{F}_i \mathbf{J}_i, \tag{17.25}$$

which is simply a vector sum of any of the extensive parameters (such as the energy or particle number) as the corresponding flux density flows in any spatial direction. Any or all of these parameters may be conserved, in which case:

$$\frac{\partial x_i}{\partial t} + \nabla \cdot \mathbf{J}_i = 0, \tag{17.26}$$

however, each of the extensive parameters can be conserved or not conserved, such as above where we considered a time rate of change of the second law and specifically incorporated an external driving term $\partial D / \partial t$ that would be injecting energy into the system.

The first term in Equation (17.24) is easily obtained from Equation (17.23):

$$\frac{\partial s}{\partial t} = \sum_i \mathsf{F}_i \frac{\partial x_i}{\partial t}, \tag{17.27}$$

whereas the second term in Equation (17.24) is determined directly by the divergence of the entropy current density:

$$\nabla \cdot \mathbf{J}_s = \nabla \cdot \left(\sum_i \mathsf{F}_i \mathbf{J}_i \right), \tag{17.28}$$

which yields two terms:

$$\nabla \cdot \left(\sum_i \mathsf{F}_i \mathbf{J}_i \right) = \sum_i (\nabla \mathsf{F}_i \cdot \mathbf{J}_i + \mathsf{F}_i \nabla \cdot \mathbf{J}_i). \tag{17.29}$$

Combining all of this allows us to rewrite Equation (17.24) as:

$$\frac{ds}{dt} = \sum_i \mathsf{F}_i \frac{\partial x_i}{\partial t} + \sum_i (\nabla \mathsf{F}_i \cdot \mathbf{J}_i + \mathsf{F}_i \nabla \cdot \mathbf{J}_i), \tag{17.30}$$

where all of the conserved extensive quantities in the first and last terms above cancel by the condition in Equation (17.26) to leave:

$$\frac{ds}{dt} = \sum_i (\nabla \mathsf{F}_i \cdot \mathbf{J}_i), \tag{17.31}$$

so the time dependence of the entropy density is proportional to the gradient of each affinity along the direction of the associated current density for the corresponding extensive parameter in the continuous system.

17.6 A Microscopic View

A microscopic consideration of entropy production (and a time-dependent or non-equilibrium temperature) could proceed directly from Equation (17.6), in that a time-dependent or non-equilibrium temperature simply reflects a time-dependent or non-equilibrium occupancy of the underlying equilibrium quantum states.

$$T(t) = \frac{-\delta E_1}{k_B \ln(n_1(t)/n_0(t))}. \tag{17.32}$$

Considering the time-dependence connected to the occupancies now allows us to retain a naturally time-dependent measure of the energy of the thermodynamic system:

$$< E(t) > = \sum_i p_i(t)E_i, \tag{17.33}$$

where the probability of finding a particle in the ith state can be thought of from its statistical measure, namely:

$$p_i(t) = \frac{n_i(t)}{N}, \tag{17.34}$$

for a fixed number of particles N in the system. From Chapter 6, let us consider the most fundamental expression we have for the entropy:

$$S = -k_B \sum_i p_i \ln(p_i). \tag{6.31}$$

If we simply take a time derivative of this expression of the entropy then:

$$\frac{dS}{dt} = -k_B \sum_i \left(\frac{dp_i}{dt} \ln(p_i) + p_i \frac{1}{p_i} \frac{dp_i}{dt} \right), \tag{17.35}$$

which reduces to:

$$\frac{dS}{dt} = -k_B \sum_i \frac{dp_i}{dt} \ln(p_i) + \sum_i \frac{dp_i}{dt}, \tag{17.36}$$

and as the probability of finding the system in any state is properly normalized, then the second sum is zero and we obtain:

$$\frac{dS}{dt} = -k_B \sum_i \frac{dp_i}{dt} \ln(p_i). \tag{17.37}$$

If we now think of the probabilities p_i in terms of their statistical measure, namely, $p_i = n_i/N$ for a closed system of N particles, we can imagine a multi-level system in equilibrium, the largest occupation number for finite temperature being in the ground state, n_0. If we now drive that system away from equilibrium, the ground state initially and locally obeys the third law in that $dS/dt > 0$ for particles being driven out of ground state $dn_0/dt < 0$. In order to examine what occurs for the entire system, we first examine our stalwart example of the two-level system (Figure 17.1).

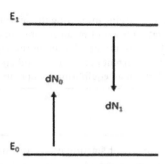

FIGURE 17.1 A two-level system is driven from equilibrium, changing the occupation numbers of both of the underlying quantum states.

Example 17.1: The Two-Level System Revisited

We return to our two-level paradigm to gain a clearer picture what is happening as we take our system, initially in equilibrium, and drive it away from equilibrium. Since there are only two states, a ground state with energy E_0 and an excited state with energy E_1 let us rewrite our last equation above for the two terms:

$$\frac{dS}{dt} = -k_B \left[\frac{dp_0}{dt} \ln(p_0) + \frac{dp_1}{dt} \ln(p_1) \right]. \tag{17.38}$$

Since the sum of the probabilities is equal to unity, $p_0 + p_1 = 1$ (any individual particle must be in either state if we consider our measurement to be faster than the transition time between states) then:

$$\frac{dp_0}{dt} = -\frac{dp_1}{dt}, \tag{17.39}$$

and we can formulate an expression for either the lower state or the excited state:

$$\begin{aligned}
\frac{dS}{dt} &= -k_B \frac{dp_0}{dt} \left(\ln(p_0) - \ln(p_1) \right) = -k_B \frac{dp_1}{dt} \left(\ln(p_1) - \ln(p_0) \right) \\
&= -k_B \frac{dp_0}{dt} \left[\ln(p_0) - \ln(1 - p_0) \right] = -k_B \frac{dp_1}{dt} \left[\ln(p_1) - \ln(1 - p_1) \right] \\
&= -k_B \frac{dp_0}{dt} \ln\left(\frac{p_0}{1 - p_0} \right) = -k_B \frac{dp_1}{dt} \ln\left(\frac{p_1}{1 - p_1} \right),
\end{aligned} \tag{17.40}$$

and realizing the relative sizes of p_0 and p_1 for any finite temperatures (imagine a low temperature where $p_0 \gg p_1$), then dS/dt is positive for either expression, meaning there is an increase in entropy in both of the levels, and thus, the entire system. (This is left as an exercise at in Problem 17.3.) The process cannot go on without limit, as both expressions indicate that $dS/dt = 0$ when the occupancy of each state reaches a value of $1/2$. In such a configuration, the microstates of the system have been driven to the maximum entropy for the entire system. Individual particles may continue to fluctuate between the two levels, but the system has been driven as far from its equilibrium condition as possible. We could think of this system as now being in a steady state configuration, with equal numbers of particles in the two levels and a balance being achieved between migration of particles out of each level to maintain that balance.

Recall in our discussion on gases the relationship between the temperature of a gas and the root mean square kinetic energy. Now consider a process that drives such a gas system away from its equilibrium condition as predicted by equilibrium thermodynamics. The particles in the gas are still moving with a distribution that has a root mean square value, just one that is not predicated on equilibrium statistics. One could imagine measuring the velocities of all the particles in the gas at one instant in time and calculating the root mean square value. The quantity thus calculated would still be the temperature of the gas, but not the equilibrium temperature, rather the value of the temperature as calculated by the current occupation numbers of the kinetic energy quantum states of the gas: A temperature, but not an equilibrium temperature.[4]

17.7 Envoi

Thermodynamic equilibrium was a convenient starting point for our treatment of thermostatistical systems. It allowed us to ignore the history of an object or a system and simply consider the current state of the physical system. For over 100 years, physicists have been developing an increasingly more sophisticated picture of equilibrium thermostatistics since the watershed assumption by Boltzmann that $S = k_B ln(\Omega)$. Now, with over a century of success at examining systems under the fundamental constraint of equilibrium, contemporary physicists have begun the daunting task to expand our knowledge of a plethora of physical systems that are not well described as occurring under equilibrium conditions. If one imagines every equilibrium state with a state variable X as being approachable from above or below to that value of X, then one can understand that there are (at least) twice as many transient or non-equilibrium processes as there are equilibrium conditions in the universe. Although we began our work in this textbook considering the situation of measuring the equilibrium temperature of the human body, it is clear that the human body is more accurately thought of as a non-equilibrium system running in the steady state, achieving a balance that is not a state of equilibrium with its surrounding environment. As such, non-equilibrium thermodynamics is not only at the forefront of our understanding of contemporary physics, but also our understanding of the mechanisms of life itself.

17.8 Problems

17.1 Complete the steps from Equation (17.9) to Equation (17.10)

17.2 Using published values of heat transfer coefficients, determine characteristic time scales for common substances such as water and metals. As the measured values vary greatly with the conditions under which they are measured, use an order of magnitude calculation to get a sense of the characteristic time scales

17.3 From Equation (17.31) determine the affinities associated with (a) a single component energy current density J_k and (b) a single component particle number density J_k. (Neglect consideration of any gravity stratification in density that occurs for the vertical component in real systems.)

17.4 Beginning from a situation where p_0 is large and p_1 is small, as occurs in a ground state equilibrium situation for the two-level system, demonstrate that the entropy increases for both the ground state as well as the excited state in Equation (17.40)

17.5 For the two-level system in non-equilibrium, Example 17.1, discuss the validity of the following statement: $dS/dt \rightarrow 0$ necessarily implies that as this steady state is achieved, $dp_0/dt = dp_1/dt = 0$

17.6 Determine the version of Equation (17.37) in terms of the number of particles in each state

17.7 Using the partition function for the two-level system, Equation (6.5) where $n = 2$, take the derivative with respect to time of these probabilities and demonstrate that $dp_0/dt = -dp_1/dt$ as found in Equation (17.39). Discuss the form of the solution and how it reconciles with Section 17.3 and Section 6.11

---

[4] In the granular physics community, this root mean square kinetic energy of a driven, dissipative collection of macroscopic particles undergoing collisions is called the granular temperature where $k_B = 1$.

18

Biological Systems

All science is either physics or stamp collecting.

Lord Ernest Rutherford

18.1 Gedanken

Under what conditions may we treat a biological system under equilibrium conditions? Under what conditions must we treat a biological system under non-equilibrium conditions? Are such biological systems limited to microscopic entities? Are larger biological structures (herds, flocks, schools) describable in either equilibrium or non-equilibrium thermodynamics?

A proper textbook should not end with answers, but with questions.

18.2 Biology versus Physics

It may be unsettling to some to apply the principles of statistical thermodynamics to biological systems. Who wants to think of their locomotion as having anything in common with a random walk? After all, don't living things work by a different set of rules than 10^{23} identical particles? Psychology would tell us that despite individuality, a statistical measure of a significant population will still predict what one of its constituent members is most likely to do. Whether the population of interest is cells, a virus swarm, a colony of ants, or a community of people, statistical data can be collected and then fit to models. It is the ability to make and apply models that gives physics its power: One does not go around touting a model in physics which does not work. So we will begin with an overview that discusses the differences between inanimate objects (physics problems) and living processes (biology problems), and remember always that any model that is successful is dependent upon the assumptions that may sometimes be quite narrow.

As we developed our statistical picture of thermodynamics, we gave a lot of attention to the absolute zero of the temperature scale as a necessary point at which to anchor our axioms. This was a result of needing a particular temperature scale where the accessible energy went to zero in the limit of zero absolute temperature. At the opposite end of the temperature scale was something we considered to be infinity, which is almost always ludicrous as it would take an infinite amount of energy to reach an infinite temperature. A more appropriate manner in which to think about it would be the high temperature limit. In our simple spin systems, the high temperature limit is then on the order of the energy $k_B T$, where if the energy to flip a single spin is ϵ and there are N spins, the high temperature limit is where $T = N\epsilon/k_B$. So, in reality, finite systems have a finite energy and thereby a finite temperature range over which statistical models are appropriate.

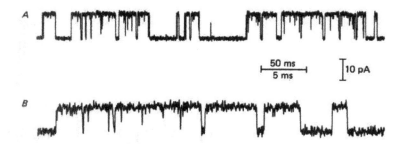

FIGURE 18.1   Two-state activity of large-conductance Ca^{2+}-activated K^+ channel. Single channel ion currents displayed at two different time bases in A and B. The upper current level indicates when the channel is in the open state and the lower current level indicates when the channel is in the closed state. (Reproduced from McManus, O.B. and Magleby, K.L., *J. Physiol.*, 402, 79–120, 1988.)

So, as we now turn to consider biological systems, it is important to realize that certain biological processes do not occur over an infinite temperature range. Consider any biotic that lives in water. Broadly speaking, if the biotic lives in such an environment there are already two natural extremes to consider: 273 K where water freezes and 373 K where water boils. This overly simplistic picture has already confined our concern to an energy range of 100 K, a small fraction of any of the Debye temperatures we discussed in describing solids. However, we are going to take our cue from the effectiveness of the Debye model (a small collection of representative states that accurately predicts the accessible degrees of freedom of the solid over an appropriate temperature range) to forge ahead to apply a similar approach to biotics.[1] Moreover, any process that demonstrates fluctuations to nearby accessible states can be thought to be equilibrium in nature if separated significantly from other states over an appropriately long time scale (see Figure 18.1[2]).

18.3 A Bit of Vocabulary

Just as the first day of a first physics course requires students to assimilate a bit of specific vocabulary, our treatment of biochemistry here will require the same investment. We will not put aside everything we have built up to this point, quite the opposite: when possible, we will use our idealized models from physical systems to understand the processes we are trying to learn in biochemistry.[3]

Amino Acids are the building blocks of proteins. There are essentially twenty amino acids that comprise all proteins, differing in size and shape, charge and chemical reactivity, characteristics of hydrophobia, and hydrogen bonding. We accentuate the last two items in the list on purpose: the role of water (H_2O). Where there is water, there is life. Whereas in our physics examples we examined sequences of different spins that belonged to ensembles based upon the number in the up or down state, amino acids play the role of an alphabet for the words that are proteins. This is a useful analogy, just as vocabulary can be built of successively longer words, even incorporating new words as we create new inventions (video cassette recorders), so all of life depends upon proteins constructed from this alphabet over billions of years.

---

[1]   The term "biotic" is a nice generic term that is equally useful to evoke that the system in question is part of a living organism that can be either microscopic or macroscopic in scale.

[2]   O. B. McManus and K. L. Magleby, Kinetic states and modes of single large-conductance calcium-activated potassium channels in cultured rat skeletal muscle, *Journal of Physiology*, 402, pp. 79–120 (1988).

[3]   The treatment in this chapter is by no means exhaustive. I do suggest an extremely thorough treatment available in a book that a colleague suggested to me that has inspired much of what is presented here: J. M. Berg, J. L. Tymoczko and L. Stryer, *Biochemistry (6th Ed.)* (2007) [W. H. Freeman Co., New York].

Proteins are linear polymers (structures) built from amino acids. They are macromolecules in biological systems that provide for the very functions that allow life to maintain itself. The functions of proteins and how these functions are accomplished is the essence of biochemistry. While it is tempting to think of a protein much as we would our linear spin polymer comprised of left and right pointing segments, proteins fold up into three-dimensional structures that are determined by the amino acids comprising them. Hence, flexibility and rigidity of that structure is often an important characteristic of a protein.

DNA and **RNA**, in comparison, are nucleic acids that form long linear polymers. The sequences of bases that comprise the chain of nucleic acids are where genetic information is carried. The macromolecule of DNA has a well-known double helix structure and can unbind and bind its two helical strands much like a zipper. It is by this structural process that DNA replicates and transfers genetic information from one generation to the next. DNA is responsible for propagating the information by which proteins are synthesized, but it is a special class of RNA, a messenger, that actually copies the information during a process called *transcription*.

Enzymes are another macromolecular structure that plays the role of catalyst in biological processes. If the seemingly esoteric and brief descriptions above seem to be summarizing what is a far more complex sequence of processes, enzymes play the important role of making it all work by taking that large, complicated process and breaking it down into manageable pieces. As we will show later in this chapter, enzymes may be thought of as intermediate energy states whose presence enhance reaction rates in biochemical systems by effectively lowering the free energy barrier that must be overcome in the process. In the absence of enzymes, many processes cannot occur. Enzymes are specific and are able to cut long DNA strands into smaller pieces.

Vectors are particular DNA molecules used as a platform for artificially transporting foreign DNA strands into another cell where the foreign DNA will genetically express itself and/or be replicated by the host. The vector is the transport vehicle and once it contains the foreign DNA is considered *recombinant DNA*. The foreign DNA are pieces of DNA strands that have been artificially cut at a specific site along the strand with a special enzyme called a *restriction enzyme*. Vectors are chosen because they can be suitable hosts to the foreign DNA fragment. Proteins with new functions can be created via recombinant DNA.

Plasmids are the most common type of vectors and are found in bacteria as circles of DNA that act as alternative chromosomes. Plasmids are located within a bacterium cell, but can replicated independently within a suitable host and can be moved from one bacterium to another. This mobility and independent replication ability of the plasmid are what make them such a viable vector.

Bacteriophage lambda is another choice vector for the transport of recombinant DNA. Denoted more simply as λ phage, this vector is actually a bacterial virus that transports the recombinant DNA by infecting its host, the bacterial species *E. coli*. What makes the λ phage particularly useful is its ability to either destroy its host bacteria or become integrated within it.

Hemoglobin is a particular protein of interest. Along with myoglobin, it was one of the first proteins to have its structure determined to high-resolution.[4] Hemoglobin is a transport vehicle for oxygen in the bloodstream of vertebrates, delivering the gas from the lungs to the tissues. Located in red blood cells, hemoglobin also transports carbon dioxide from the tissues back to the lungs.

Myoglobin is another protein of interest, albeit as a reservoir of oxygen in the muscles as opposed to a transport vehicle in the bloodstream. Both hemoglobin and myoglobin share a similar structure, but hemoglobin contains four cooperative sites for oxygen binding, whereas myoglobin contains only one. As such hemoglobin can be thought of as having four myoglobin-like sub-structures. However, myglobin is not simply four times less effective at oxygen storage. The cooperative binding of oxygen in hemoglobin makes the effective ratio more on the order of one tenth for myoglobin to hemoglobin.

---

[4] In 1958 by John Kendrew and Max Perutz for which they shared the 1962 Nobel Prize in Biochemistry.

18.4 Biochemical Reactions

One of the most fundamental processes you can imagine happening in your body is the movement of oxygen (O_2) from one point to another. Your lungs draw in air and expel carbon dioxide, and your blood (hemoglobin) transports that oxygen from where it is available (the lungs) to where it is needed (pretty much everywhere else). In physics textbooks, one often only focuses on one process:

$$A \rightharpoonup B, \tag{18.1}$$

but if one thinks about hemoglobin as the transporter of oxygen, what changes when the hemoglobin arrives at the muscle tissue, for instance, that motivates it to give up the oxygen where it is needed? The answer is the conditions present in the muscle tissue trigger the release of oxygen, namely, the presence of large amounts of carbon dioxide and hydrogen ions.

In biochemistry this process is called the *Bohr effect* for Christian Bohr (not Neils) who proposed this process in 1904. The presence of large amounts of carbon dioxide and hydrogen atoms leads to a lowering of pH in hemoglobin that motivates the release of the oxygen through a chemical reaction that is enhanced by the presence of an enzyme, *carbonic anhydrase* present in the red blood cell. Once returned to the lungs, the red blood cells undergo a reversal of this process at a different pH. While this certainly sounds more complicated than flipping a spin from one orientation to another, all of the same thermostatistical physics we have developed thus far is still at play. We simply need to understand the physical pictures we have been using in the biochemical environment in which they occur.

First, we are dealing with a process that is reversible; it is the environment in which the hemoglobin finds itself that alters or biases the situation to motivate the reversible process to proceed in one direction or another. Second, the process is a chemical reaction which can occur with or without the presence of an enzyme which acts as a catalyst. We will examine the physics of the chemical processes first, and then examine how enzymes enhance chemical reaction rates.

In order to understand the reactions, we consider a steady supply of the reactants and products rather than a closed system where the process consumes the materials within a certain amount of time. We consider processes of two reactants and two products and in equilibrium:

$$A + B \rightleftharpoons C + D, \tag{18.2}$$

so that in equilibrium the process (in the presence of equal amounts or molar concentrations) of the materials may proceed in either direction. In order to describe the chemical process for a general ratio of reactants/products, we consider the change in the Gibbs free energy from one side of the reaction to the other and can write:

$$\Delta G = \Delta G_0 + k_B T \ln(\kappa_{conv}), \tag{18.3}$$

where κ_{conv} is the conversion constant for reaction defined by the molar concentrations of the reactants present:

$$\kappa_{conv} = \frac{[C][D]}{[A][B]}, \tag{18.4}$$

and ΔG_0 is the standard free energy change of the reaction (usually defined at a standard pH of 7). So that we can see how this relationship works, let us imagine a reaction that is in equilibrium when the ratio of $[C][D]$ to $[A][B]$ is 1:2 ($\kappa_{conv}^{eq} = 0.5$) at a temperature of 300 K. In equilibrium, $\Delta G = 0$ so the relationship becomes:

$$0 = \Delta G_0^{eq} + k_B T \ln(\kappa_{conv}), \tag{18.5}$$

which then yields a standard free energy of:

$$\Delta G_0^{eq} = -k_B T \ln(0.5) = +2.87 \times 10^{-22} \text{J/particle}, \tag{18.6}$$

and upon converting this from particles to moles, $\Delta G_0^{eq} = 1.728 \text{ kJ/mole}$. If we consider the reaction in equal concentrations of $[A], [B], [C]$, and $[D]$, we find that:

$$\Delta G_m = 1.728 \text{ kJ/mole} + N_A k_B T \ln(1) = 1.728 \text{ kJ/mole}. \tag{18.7}$$

That is, for the situation we would consider to be "natural" for the reaction (equal amounts of the reactants/products), there is a free energy barrier to the reaction proceeding. The environment must change to different proportions to lower the free energy barrier so that the reaction $A + B \rightharpoonup C + D$ can occur spontaneously, that is, to be energetically favorable.

Alternately, let us suppose that the reactants and products change so that their relative concentrations produce a $\kappa_{conv} = 0.1$, that is the ratio of $[C][D]$ to $[A][B] = 1{:}10$. Then our free energy change for the reaction is now biased in favor of the reaction, because:

$$\Delta G_m = 1.728 \text{ kJ/mole} - N_A k_B T \ln(0.1) \tag{18.8}$$
$$= 1.728 \text{ kJ/mole} - 5.740 \text{ kJ/mole}$$
$$= -4.012 \text{ kJ/mole},$$

makes the reaction energetically favorable.

We should pause and clarify that the equilibrium we are speaking of is a chemical equilibrium. If we return to our reaction above and consider the return reaction $C + D \rightharpoonup A + B$ we find exactly the opposite situation, and want to highlight how this makes physical sense.

Returning now to our discussion of the transport of oxygen in the red blood cells, the environment in the lungs and the environment in the muscle are what change the ratios of κ_{conv} to make the release of oxygen preferential in the muscle tissue, and then the release of carbon dioxide preferential in the lungs, each gas being replaced with the one that makes the return trip in the other half of the cycle. Environment biases the free energy barrier.

18.5 Enzymes as Catalysts: Intermediate States

Now, we wish to examine the role that enzymes play as catalysts to chemical reactions. As we saw in the prior section, free energy barriers inhibit reactions. Enzymes catalyze reactions by providing an intermediate state that has a lower free energy potential (Figure 18.2). Here, we generalize the reactions simply by the pathway from reactants to products:

$$R \rightleftharpoons P. \tag{18.9}$$

Instead, consider an intermediate state, X, between R and P, such that:

$$R \rightleftharpoons X \rightleftharpoons P, \tag{18.10}$$

where the energy of the intermediate state is between the energies of R and P.

For our purposes, let us put aside the chemistry and return to our spin 1/2 system one more time. Imagine that instead of just the two states, spin "up" and spin "down" (as defined by the external magnetic field), we consider an intermediate state of the system where the spin points orthogonal to the field. Furthermore, let us examine the Boltzmann statistics by a convenient choice of parameters. For the strict two-state system, we imagine energies of 0 and $k_B T$ so that:

$$Z = 1 + e^{-\beta \epsilon} = 1 + e^{-1} = 1.367, \tag{18.11}$$

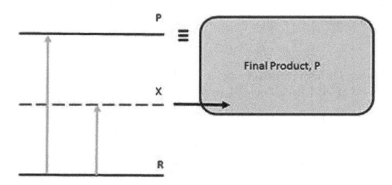

FIGURE 18.2 A catalyst, X, may be modeled as an intermediate energy state that delivers the system from R to P at a lower energy cost.

so that the populations of the two states are 0.731 for the ground state and 0.269 for the upper state. So, at a temperature where the energy of the state equals $k_B T$ only 26.9% of the particles are excited from the ground state.

Now, to demonstrate the effect of an enzyme, let us model the intermediate state at $\epsilon/2$ so that now the Boltzmann statistics are:

$$Z = 1 + e^{-\beta\epsilon/2} + e^{-\beta\epsilon} = 1 + e^{-0.5} + e^{-1} = 1.974, \qquad (18.12)$$

so that the populations of the three states are 0.507 for the ground state, 0.307 for the intermediate state, and 0.186 for the excited state. At first glance, one might think this has been counterproductive as only 18.6% of the particles have reached the excited state. However, we are instead interested in examining the population of the ground state. In the two-state configuration, 73.1% of the particles have remained in the ground state, whereas in the presence of the intermediate state, only 50.7% of the particles remain the ground state (Figure 18.3). Now, we improve our simple model with the assumption that any particle that makes it to the intermediate state is carried through the two-state barrier to the final state P above. Hence, in the presence of the enzyme as modeled here, nearly twice as many particles make the transition

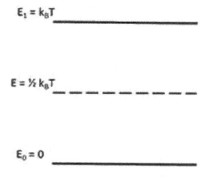

FIGURE 18.3 The modified spin 1/2 system demonstrates enhanced statistics for vacating the ground state at temperature T by the introduction of a metastable no-spin state at half the activation energy. The example is used to demonstrate the enhancement of a catalytic state for our ubiquitous spin 1/2 system under the assumption that all particles in the metastable state flip to the higher energy state. For a real metastable state, the fluctuations would in fact deliver the system to either the spin up or spin down state.

from R to P. Hence, enzymes not only make biochemical processes more efficient, their presence or absence controls where and when such processes can or cannot take place in a larger structure such as the human body.

18.6 Ligand-Receptor Binding

To better understand the binding process (oxygen binding to hemoglobin or myglobin, or DNA fragments to vectors, or a drug to a particular target in the body), let us work with a simple model of ligand-receptor binding. Imagine a solution mixed with our ligand, or target molecule, as a lattice of $N + L$ sites, and a single receptor that can sit on any of the N sites containing solution (unbound) and L sites that contain a ligand (bound).

We have chosen this model to once again take advantage of our two-state system. Rather than assign an energy to each state let us learn from our extensive use of the two-state system and say that the energy for the receptor to sit in solution is zero, and the energy to be bound to a ligand is simply $\delta\epsilon$. Then we can immediately write down the canonical ensemble expression from the partition function:

$$Z = N + Le^{-\beta\delta\epsilon}, \tag{18.13}$$

where N is the degeneracy of the unbound states of receptor in solution, and L is the degeneracy (number) of ligands in solution. The probability of the receptor being bound to the ligand, P_{bound}, is then:

$$P_{bound} = \frac{Le^{-\beta\delta\epsilon}}{N + Le^{-\beta\delta\epsilon}}. \tag{18.14}$$

We can divide by N, the number of unbound receptors to obtain:

$$P_{bound} = \frac{\frac{L}{N}e^{-\beta\delta\epsilon}}{1 + \frac{L}{N}e^{-\beta\delta\epsilon}}. \tag{18.15}$$

Rather than considering the number of ligands and solution, let's divide both quantities by the volume of our lattice to produce a result in terms of the ligand concentration c_L to a reference concentration c_s. Then:

$$P_{bound} = \frac{\frac{c_L}{c_s}e^{-\beta\delta\epsilon}}{1 + \frac{c_L}{c_s}e^{-\beta\delta\epsilon}}. \tag{18.16}$$

The result above is just as applicable to the binding of oxygen to the single heme site of myoglobin. To solidify the connection, we consider the stoichiometry of the binding, namely:

$$L + R \rightleftharpoons LR. \tag{18.17}$$

The dissociation constant K_d is defined by the mass action as:

$$K_d = \frac{[L][R]}{[LR]}, \tag{18.18}$$

which allows us to consider the amount of bound ligand-receptors to be given by:

$$[LR] = \frac{[L][R]}{K_d}, \tag{18.19}$$

so that we may generalize this to a comparable situation for our canonical ensemble result above in the case of more than one receptor. Statistically, if we have multiple receptors and multiple ligand-receptor bindings, then:

$$P_{bound} = \frac{[LR]}{[R] + [LR]}, \tag{18.20}$$

for which our prior relationship of $[LR]$ allows us to write as:

$$P_{bound} = \frac{\frac{[L][R]}{K_d}}{[R] + \frac{[L][R]}{K_d}}, \tag{18.21}$$

which can be reduced to:

$$P_{bound} = \frac{\frac{[L]}{K_d}}{1 + \frac{[L]}{K_d}}. \tag{18.22}$$

We have seen the behavior of $x/(1 + x)$ in our spin 1/2 systems and know that it tends between the limits of 0 and 1 for small and large x, respectively. K_d is commonly chosen to define where the probability of binding is 1/2.

Our two descriptions of ligand-receptor binding can then be reconciled by noting that Equations 18.15 and 18.21 imply that:

$$\frac{[L]}{K_d} = \frac{c_L}{c_s} e^{-\beta \delta \epsilon}. \tag{18.23}$$

Typically, the concentration c_s is of order 1 molar concentration. However, in drug development, where the potency of a particular drug may necessitate proving it can demonstrate larger binding affinity, a different concentration may be necessary to prove the drug's effectiveness. Such considerations are also affected by potential side effects caused by binding to other molecules besides the target. This reflects the challenges of modern drug development.

Example 18.1: Binding of Oxygen to Myoglobin; Carbon Monoxide Poisoning

Oxygen (O_2) is carried in the blood by hemoglobin, which is bright red when the oxygen is attached and darker when the oxygen has been removed; this is the reason for the difference in color between arterial and venous blood. In the lungs, venous blood is exposed to oxygen at close to its atmospheric chemical potential, and absorbs it; when the arterial blood reaches cells that need oxygen, it encounters a lower chemical potential and gives up the oxygen. We consider how the process of taking up oxygen in the lungs is affected by the presence of carbon monoxide (CO). Since hemoglobin consists of four myoglobin molecules, which interact in a rather complicated way, it is simpler to consider the binding to a single myoglobin molecule. This is the molecule that binds oxygen in muscle, and for the purpose of this calculation we assume that it, rather than hemoglobin, is the carrier of oxygen in blood.

O_2 binds to myoglobin with an energy E_O; that is, if we take the energy of deoxygenated myoglobin as zero, the oxygenated myoglobin molecule has an energy $-E_O$. CO also binds to myoglobin, with an energy $-E_C$; E_C is greater than E_O. There is only one binding site per myoglobin molecule, so that O_2 and CO cannot both be bound at the same time, and CO tends to displace O_2. To make this quantitative, we find the grand partition function for the O_2 and

n_O	n_C	E
0	0	0
1	0	$-E_O$
0	1	$-E_C$

FIGURE 18.4 Energy and occupancy for myoglobin in the presence of oxygen and carbon monoxide.

CO occupancies.[5] We begin by drawing the energy level diagram shown in Figure 18.4. We define n_O as the O_2 occupancy and n_C as the CO occupancy; these can each be 0 or 1, but not 1 simultaneously, since there is only one binding site. There are thus three possibilities, shown in Figure 18.4. The Gibbs factor is $e^{\beta[n_O \zeta_O + n_C \zeta_C - E]}$, where ζ_O and ζ_C are the respective chemical potentials of O_2 and CO in the lungs. Hence, the grand partition function, in terms of the respective oxygen and carbon monoxide activities, $\alpha_O = e^{\beta \zeta_O}$ and $\alpha_C = e^{\beta \zeta_C}$, is:

$$\Xi = 1 + \alpha_O e^{\beta E_O} + \alpha_C e^{\beta E_C}, \tag{18.24}$$

where the first term corresponds to deoxygenated myoglobin ($n_O = n_C = 0$), the second to oxygenated myoglobin ($n_O = 1, n_C = 0$), and the third to the case where CO instead of O_2 is bound to the myoglobin ($n_O = 0, n_C = 1$).

If there is no CO present, $\alpha_C = 0$. Then the probability of the myoglobin being oxygenated when it leaves the lungs is, from Equation (8.30) in Chapter 8,

$$\langle n_O \rangle = \frac{\alpha_O}{\Xi} \frac{\partial \Xi}{\partial \alpha_O} = \frac{\alpha_O e^{\beta E_O}}{1 + \alpha_O e^{\beta E_O}}$$
$$= \frac{1}{\alpha_O^{-1} e^{-\beta E_O} + 1}. \tag{18.25}$$

For an ideal gas at a given temperature, the density n (and therefore α) is proportional to pressure p, so that Equation (18.24) can be written:

$$\langle n_O \rangle = \frac{p}{p_{1/2} + p}, \tag{18.26}$$

where $p_{1/2}$ is the pressure at which half the binding sites are occupied at that temperature. Equation (18.25) also describes the adsorption of a gas onto a surface as a function of gas pressure and is known as the Langmuir adsorption isotherm.

Oxygen is a very good approximation to an ideal gas for which $\alpha \propto n$, and in the atmosphere, its activity is $\alpha_O \sim 10^{-8}$. From the temperature dependence of $p_{1/2}$, it has been found[6] that $E_O \approx 0.58$ eV. At the human blood temperature of 310 K, for which $k_B T = 0.027$ eV,

$$\langle n_O \rangle = \frac{1}{10^8 e^{-0.58/0.027} + 1} \approx 0.95.$$

Thus, arterial blood is almost fully oxygenated.

---

[5] This model is essentially that of Kittel and Kroemer (pp. 140–146). It is a greatly simplified picture of the true situation. See Frauenfelder, Sligar, and Wolynes, "The energy landscape and motions of proteins," *Science* **254**, 1598 (1991), for a readable discussion of the binding of CO to myoglobin.

[6] The available data are sparse and not altogether consistent with our simple model (see Footnote 4), so the figures used here are only illustrative. We follow the custom in physics of using the electron volt (eV) as the unit of energy on an atomic scale. In chemistry, it is more usual to express energies in calories or kilocalories per mole. 1 eV per molecule is 23 kcal per mole.

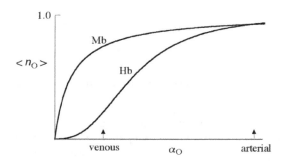

FIGURE 18.5   Oxygen uptake curves for myoglobin (Mb) and hemoglobin (Hb). The arrows indicate typical values of the oxygen activity α_O for arterial blood and for venous blood after heavy exercise. The Mb curve is Equation (18.24); the Hb curve was obtained by fitting an empirical formula (the Hill equation) to the data on human Hb.

Now suppose that there is some carbon monoxide in the air, perhaps due to a leaky tailpipe. Now we must include the third term in the sum in Equation (18.23) and obtain:

$$\langle n_O \rangle = \frac{\alpha_O}{\Xi} \frac{\partial \Xi}{\partial \alpha_O} = \frac{\alpha_O e^{\beta E_O}}{1 + \alpha_O e^{\beta E_O} + \alpha_C e^{\beta E_C}}$$

$$= \frac{1}{\alpha_O^{-1} e^{-\beta E_O} + 1 + \frac{\alpha_C}{\alpha_O} e^{\beta(E_C - E_O)}}. \tag{18.27}$$

If the concentration of CO in the air is 0.2%, $\frac{\alpha_C}{\alpha_O} \approx 10^{-2}$. Data on the temperature dependence of $\langle n_O \rangle$ in the presence of CO give $E_C - E_O \approx 0.16$ eV. Putting in these numbers, we find:

$$\langle n_O \rangle = \frac{1}{10^8 e^{-0.58/0.027} + 1 + 10^{-2} e^{0.16/0.027}} \approx 0.2,$$

which is too low to sustain life.

In myoglobin, the oxygen binding energy is independent of the mean occupancy $\langle n_O \rangle$, which is given as a function of O_2 activity α_O by Equation (18.24). This uptake curve is the upper curve (marked "Mb") in Figure 18.5. The variation of $\langle n_O \rangle$ is not great over the relevant range of oxygen pressure, indicated by the vertical arrows, so that myoglobin is an inefficient oxygen carrier. Hemoglobin consists of four myoglobin-like molecules weakly bound together. When one of these myoglobin-like molecules binds an oxygen molecule, it distorts in such a way as to increase the binding energy of oxygen to the other three molecules. Thus, E_O is a function of $\langle n_O \rangle$, and the uptake curve takes the "sigmoid" form shown by the lower curve (marked "Hb").[7] The uptake is much more sensitive to α_O, so that hemoglobin can transfer more oxygen per binding site than can myoglobin. This is an example of a cooperative effect. We saw such effects when we discussed phase transitions in Chapter 15.

18.7 Envoi

This chapter has not meant to be exhaustive in its treatment of biochemical systems in terms of thermodynamics. Rather, what we present here should be the starting point for another course or one's interest in pursuing research in biochemical physical systems from a thermodynamic point of view. It has been said that the twentieth century was a century of physics and the twenty-first century will be a century

---

[7] Adapted from Figure 41 of L. Garby and J. Meldon, *The Respiratory Functions of Blood* (Plenum, New York, 1977).

of biochemistry.[8] It would be more appropriate to say that the twentieth century was a modern time of great discovery in physics, biology, and chemistry as separate disciplines, and the great advances in the twenty-first century will be where scientists in each of these fields attack open questions at the interface of these disciplines. Physics, biology, and chemistry will move forward the furthest where there are scientists willing to step outside of their specific disciplines and work collaboratively together. Where that works the best will always be with the involvement of a physicist to explain and expand on the fundamentals of thermodynamics and thermostatistics. The laws of both hold court from the smallest cell all the way up the scale of the universe itself. A physicist who is willing to collaborate with colleagues in biology and chemistry will never lack interesting problems to solve.

18.8 Problems

18.1 Calculate the energy range corresponding to water based biotics whose processes take place between (at most) the freezing and boiling points of water at STP

18.2 There are two orthogonal states in the enzyme example, so the degeneracy is appropriately 2 for the intermediate state. Rework the statistics to show the enhancement in this situation by examining the population of the ground state to show further enhancement for the enzyme

18.3 We considered the canonical and grand canonical ensembles in our treatment of binding in Section 18.5 and Example 18.1. Show that one obtains the same result as Equation (18.14) by developing the microcanonical ensemble of states for a single receptor for N lattice sites and L ligands. Hint: Here it is helpful to calculate Ω from considering N sites in which L have a ligand and N-L have solution

18.4 Calculate the free energy barrier to the reverse process from Equations (18.2) to (18.8) if the ratio of $\kappa_{conv}^{eq} = 2$ instead of 0.5 as in the forward process

---

[8] Craig Venter and Daniel Cohen, "The Century of Biology," *New Perspectives Quarterly, 21, 73–77 (2004).*

Appendix A: Expansion in Series

A.1 The Use of Infinite Series in Physics

One of the most powerful mathematical tools in the physicist's armory is expansion of an analytic ("smooth") function in a series of ascending powers of a small parameter. While infinite series form an important part of any undergraduate mathematical curriculum, a physicist looks at series from a point of view somewhat different than that of most mathematicians.

Suppose that we have analyzed some physical phenomenon and have found a function that describes it. Initially, we will assume that it is a function of a single variable which we call $f(x)$ (Section A.5 extends the analysis to functions of multiple variables). Often this function is inconveniently complicated in general, but is simple for a certain special value of its argument. Series expansion provides a simple expression for its behavior close that value. As we shall see, the question of convergence, which dominates the discussion of infinite series in most textbooks of mathematics, is rarely important in physics.

A.2 The Taylor-McLaurin and Binomial Series

By the right choice of variable, we can take the special value of x to be zero with no loss of generality, and we assume that $f(x)$ is analytic in the region $x = 0$; a function is *analytic* in a certain domain if all its derivatives exist in that domain. This is not true of all functions; for example x^a is not analytic at $x = 0$ unless a is a non-negative integer, and the analyticity must always be checked.[1] If the function itself is not analytic, it can often be made so by substitution; for example, if $f(x) \sim x^{-n}$ for small x, where n is an integer, $1/f(x)$ is analytic.

If $f(x)$ is analytic, it can be expanded as a Taylor-McLaurin series:

$$f(x) = \sum_{n=0}^{\infty} a_n x^n. \tag{A.1}$$

The coefficients can be found, by taking successive derivatives of $f(x)$ and then putting $x = 0$, to be:

$$a_n = \frac{1}{n!} \frac{d^n f}{dx^n}, \tag{A.2}$$

where the derivative is evaluated at $x = 0$.

---

[1] An important example of non-analyticity is given in Chapter 3, where it is shown that the assumption that the free energy is analytic at a critical point leads to results that are in disagreement with experiment. Physics sometimes uses functions (e.g., fractals), which are not analytic for *any* value of their argument.

Even though the series [Equation (A.1)] may not be convergent for all x, if $f(x)$ is analytic at $x = 0$, the series still converges for *sufficiently small x*. In physics, we are often only interested in the lowest non-zero x-dependent term, although we often evaluate the next term to get an idea of how large an error we have introduced by cutting off the series.[2]

We very commonly find an expression that contains exponentials, and the function $f(x) = e^x$ provides one of the simplest examples of series expansion. Since $\frac{d^n f}{dx^n} = e^x$ and $e^0 = 1$, the series is:

$$e^x = 1 + x + \frac{1}{2}x^2 + \frac{1}{6}x^3 + \cdots,$$ (A.3)

where the dots indicate that the series continues. If we are interested in what happens in the vicinity of some other value of x, say $x = a$, we can substitute $\xi \equiv x - a$ and write:

$$e^x = e^{(a+\xi)} = e^a \left(1 + \xi + \frac{1}{2}\xi^2 + \frac{1}{6}\xi^3 + \cdots\right).$$

In the linear approximation, we drop all terms non-linear in x (or ξ); for example, in Equation (A.3), we would put $e^x \approx 1 + x$. For $x = 0.1$, this gives 1.1, compared with the correct value 1.10517, a precision that is quite sufficient for many purposes.

The expansions of the hyperbolic functions $\sinh(x) \equiv \frac{1}{2}\left(e^x - e^{-x}\right)$, $\cosh(x) \equiv \frac{1}{2}\left(e^x + e^{-x}\right)$, follow at once:

$$\sinh(x) = x + \frac{1}{6}x^3 + \cdots, \quad \cosh(x) = 1 + \frac{1}{2}x^2 + \cdots.$$ (A.4)

Note that while $\sinh(x)$ is linear for $x \ll 1$, $\cosh(x)$ is quadratic.

The inverse of the exponential function is the natural logarithm. Since $\ln(x)$ is not analytic at $x = 0$, we put $\xi = x - 1$ and expand about $\xi = 0$, obtaining:

$$\ln(1 + \xi) = \xi - \frac{1}{2}\xi^2 + \frac{1}{3}\xi^3 + \cdots.$$ (A.5)

The binomial series is a special case of Equation (A.1), in which $f(x) = (1 + x)^a$, so that:

$$\frac{d^n f}{dx^n} = a(a - 1)\ldots(a - n + 1)(1 + x)^{a-n},$$

and the Taylor expansion is:

$$(1 + x)^a = 1 + ax + \frac{a(a - 1)}{2!}x^2 + \frac{a(a - 1)(a - 2)}{3!}x^3 + \cdots.$$ (A.6)

If a is a positive integer, the series terminates, and we have the well-known Pascal triangle or binomial distribution (see Appendix D). However, Equation (A.6) is valid for any a, fractional or negative, although it is then only convergent for $|x| < 1$. For example, if $a = -1$, we have:

$$(1 + x)^{-1} \approx 1 - x + x^2 - x^3 + \cdots,$$ (A.7)

which is a useful formula for getting rid of awkward denominators.

---

[2] For slowly convergent non-alternating series, this is only a rough guide and may seriously underestimate the error. For example, the Riemann zeta function $\zeta(n) \equiv \sum_{r=1}^{\infty} r^{-n} \approx 1.645$ for $n = 2$. Cutting the series off at $r = 2$ gives 1.25, an error of 0.395, but the $r = 3$ term is $1/3^2 = 0.111$.

Other important functions are trigonometric:

$$\sin(x) = x - \frac{1}{6}x^3 + \cdots, \quad \cos(x) = 1 - \frac{1}{2}x^2 + \cdots. \tag{A.8}$$

The series for $\tan(x)$ can be obtained directly from Equation (A.1), but it is instructive to find the first two terms from Equation (A.8), using $\tan(x) = \frac{\sin(x)}{\cos(x)}$ and the binomial series:

$$\tan(x) \approx \frac{x - \frac{1}{6}x^3}{1 - \frac{1}{2}x^2}.$$

Using Equation (A.7) to get rid of the denominator, we obtain:

$$\tan(x) \approx x \left(1 - \frac{1}{6}x^2\right)\left(1 + \frac{1}{2}x^2\right) \tag{A.9}$$

$$= x + \frac{1}{3}x^3. \tag{A.10}$$

Note that in multiplying out Equation (A.9), we *must* drop the term in x^5, since terms of that order were omitted from the original expression. It is vital to be consistent, and if any terms of a particular order are dropped, *all* terms of that order must be; otherwise, incorrect and sometimes nonsensical results will be obtained. Failure to observe this consistency rule is a major cause of error in the use of series, not only by students, but also by professional scientists.

Another common error is to overlook terms that appear to be of high order, but will be multiplied later by a large number, or whose order is reduced by a subsequent subtraction. This is best illustrated by an example. Consider the function $f(x) = \frac{1}{1-e^{-x}}$ for small x. If we use the linear approximation to the exponential, we find $f(x) \approx \frac{1}{x}$, and we might think that the error is of order x and goes to zero as $x \to 0$. However, this is not correct. If we take the next term in the expansion of e^{-x}, we obtain:

$$\frac{1}{1 - e^{-x}} \approx \left(x - \frac{1}{2}x^2\right)^{-1} = x^{-1}\left(1 - \frac{1}{2}x\right)^{-1} \approx x^{-1}\left(1 + \frac{1}{2}x\right),$$

where in the last step, we have used Equation (A.7). Multiplying out, we find:

$$\frac{1}{1 - e^{-x}} \approx \frac{1}{x} + \frac{1}{2}, \tag{A.11}$$

so that there is a constant term in the expansion, which we would miss if we made the linear approximation to e^{-x}. Chapter 11 discusses a physical application where it is essential to include the constant term.

A.3 Expansion Around an Extremum

Series expansion is often very useful the vicinity of an extremum. At an extremum $\frac{df}{dx} = 0$, so that if we take the extremum to be at $x = 0$, and expand $f(x)$ by Equation (A.1), we obtain:

$$f(x) = f(0) + a_2 x^2 + \cdots, \tag{A.12}$$

where $a_2 = \frac{1}{2}\frac{d^2f}{dx^2}$. Thus if, as is usually the case, $a_2 \neq 0$, $f(x)$ is parabolic in the vicinity of the extremum.

As an example, consider the force between two atoms in a covalently bound molecule as a function of the interatomic separation r. This is derived from a potential $\phi(r)$, which can often be approximated by the Morse curve:

$$\phi(r) = A\left(e^{-2\alpha x} - 2e^{-\alpha x}\right). \tag{A.13}$$

Here, A and α are constants, and $x \equiv r - r_0$, r_0 being the separation at which ϕ has a minimum; that is, the equilibrium separation between the atoms. Expanding about $x = 0$, we find:

$$\phi(r) = -A + A\alpha^2 x^2 + \cdots . \tag{A.14}$$

To lowest order, the potential is quadratic, so that the restoring force $-\frac{d\phi}{dx}$ is proportional to x, and for small displacements from equilibrium, the molecule can be treated as a simple harmonic oscillator.

A.4 A More Complicated Example

As a final example of the use of series expansion to simplify a complicated expression, suppose that we want to find a simple expression for the temperature dependence of the heat capacity of a collection of harmonic oscillators at high temperature. Two expressions for this are given in Equation (6.16). One is $C_V = R_0 x^2 \mathrm{cosech}^2 x$, where $x = \frac{h\nu}{2k_B T}$, and ν is the vibrational frequency. By high temperature, we mean $T \gg h\nu/k_B$; that is, $x \ll 1$.

We simplify this expression by expanding in powers of x. While $\mathrm{cosech}(x)$ is not analytic at $x = 0$, $\sinh(x)$ is. Hence,

$$\mathrm{cosech}(x) \equiv \frac{1}{\sinh(x)} \approx \frac{1}{x + \frac{1}{6}x^3} = \frac{1}{x\left(1 + \frac{1}{6}x^2\right)} \approx \frac{1}{x}\left(1 - \frac{1}{6}x^2\right) = \frac{1}{x} - \frac{x}{6}. \tag{A.15}$$

Squaring this and dropping terms of order higher than x^2, we find:

$$C_V = R_0 x^2 \mathrm{cosech}^2 x = R_0\left(1 - \frac{1}{3}x^2\right)$$

$$= R_0\left[1 - \frac{1}{12}\left(\frac{h\nu}{k_B T}\right)^2\right]. \tag{A.16}$$

Note that if we are to be accurate to order x^2 in the final result, we must retain the cubic term in $\sinh(x)$, because extraction of the common factor x lowers the order to quadratic.

The second expression in Equation (6.16) can be written $C_V = R_0 x^2 \frac{e^x}{(e^x-1)^2}$, where now $x = \frac{h\nu}{k_B T}$. The expansion here turns out to be more complicated, but illustrates the importance of dropping terms of too high order when multiplying series together. Expanding the exponentials gives:

$$C_V \approx R_0 x^2 \frac{1 + x + x^2/2}{\left(x + x^2/2 + x^3/6\right)^2} = R_0 \frac{1 + x + x^2/2}{\left(1 + x/2 + x^2/6\right)^2}. \tag{A.17}$$

Note that in the denominator, we have had to retain the cubic term in the expansion because of the cancelling x^2 in the numerator.

Substituting $y = \frac{x}{2} + \frac{x^2}{6}$, $y^2 \approx \frac{x^2}{4}$ in the denominator of Equation (A.17) and using the binomial expansion $(1 + y)^{-2} \approx 1 - 2y + 3y^2$, we obtain:

$$C_V \approx R_0\left(1 + x + \frac{x^2}{2}\right)\left(1 - x - \frac{x^2}{3} + \frac{3x^2}{4}\right) = R_0\left(1 - \frac{x^2}{12}\right)$$

$$= R_0\left[1 - \frac{1}{12}\left(\frac{h\nu}{k_B T}\right)^2\right],$$

as before.

A.5 Functions of More Than One Variable

We consider only the case of two independent variables x and y, and confine our attention to the first few terms in the expansion. The method can be readily extended to a larger number of variables and to higher terms, but the expressions become rather cumbersome. As before, we assume that we are interested in the region around the point $x = 0$, $y = 0$.

Let:

$$f(x, y) = a_0 + a_1 x + b_1 y + a_2 x^2 + b_2 y^2 + c_2 xy + \text{higher terms.}$$

$$\frac{\partial f}{\partial x} = a_1 + 2a_2 x + c_2 y + \cdots, \qquad \frac{\partial f}{\partial y} = b_1 + 2b_2 x + c_2 y + \cdots$$

$$\frac{\partial^2 f}{\partial x^2} = 2a_2 + \cdots, \qquad \frac{\partial^2 f}{\partial y^2} = 2b_2 + \cdots, \qquad \frac{\partial^2 f}{\partial x \partial y} = \frac{\partial^2 f}{\partial y \partial x} = c_2 + \cdots.$$

Putting $x = y = 0$, we obtain the coefficients in terms of the derivatives at this point. We find, to second order in x and y,

$$\boxed{f(x, y) = f(0, 0) + \frac{\partial f}{\partial x} x + \frac{\partial f}{\partial y} y + \frac{1}{2} \frac{\partial^2 f}{\partial x^2} x^2 + \frac{1}{2} \frac{\partial^2 f}{\partial y^2} y^2 + \frac{\partial^2 f}{\partial x \partial y} xy + \cdots.} \tag{A.18}$$

Appendix B: Review
of Quantized States

B.1 Introduction: Wave-Particle Duality and the Dispersion Relation for a Particle

There are many experiments that show that a particle is associated with a wave, whose intensity (that is, the squared amplitude) at any point is proportional to the probability of observing a particle at that point.[1] The most direct is electron diffraction: A beam of free electrons with velocity v is diffracted from a periodic structure (such as a crystal) as if the beam were a plane wave with wavelength $\lambda = \frac{h}{p}$, where $p = mv$ is the momentum, and h is Planck's constant. This is known as the de Broglie relation, and λ is the de Broglie wavelength. It is often convenient to write this relation as $p = \hbar k$, where $k \equiv \frac{2\pi}{\lambda}$, and $\hbar \equiv \frac{h}{2\pi}$. This relation between wavelength and momentum applies to any particle.

For a non-relativistic particle, the kinetic energy is $\frac{1}{2}mv^2 = \frac{p^2}{2m}$, so that in the absence of a potential, the particle has energy:

$$E = \frac{h^2}{2m\lambda^2} = \frac{\hbar^2 k^2}{2m}. \tag{B.1}$$

A relation such as Equation (B.1), which relates the energy E to k, is called a *dispersion relation*. Two typical dispersion relations, one for a free non-relativistic particle such as a molecule moving at an ordinary thermal velocity in a gas, and the other for an extreme relativistic particle such as the photon, are shown in Figure B.1.

Here, we briefly summarize the stationary states (quantum states) of a particle for some simple cases. Their derivation can be found in any introductory quantum mechanics text.

B.2 Particle in a Box

Consider a non-relativistic spinless particle of mass m confined to a one-dimensional[2] box of length L with impenetrable walls. The potential energy is taken as zero within the box. Since the particle cannot penetrate the walls, its wave function ψ vanishes at each wall, and the de Broglie wavelength is limited to values $\lambda_n = \frac{2L}{n}$, where n is any positive integer: $n = 1, 2, \ldots$ (see Figure 7.1). Here n is the simplest example of a quantum number; that is, a number that specifies the state.

---

[1] See any quantum mechanics textbook.
[2] The two- and three-dimensional cases are discussed in Appendix F.

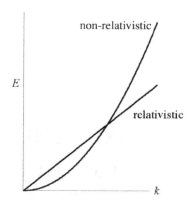

FIGURE B.1   Dispersion relations for free relativistic and non-relativistic particles.

The particle has a series of allowed energy states:

$$E_n = \frac{h^2}{2m\lambda_n^2} = \frac{h^2}{8mL^2}n^2 = \frac{\pi^2\hbar^2}{2mL^2}n^2,$$

(B.2)

and cannot have any other energy if it is inside the box. There is only one state for each energy level; that is, the levels are non-degenerate in the absence of spin.

For a relativistic particle, the dispersion relation is $E = pc = \frac{hc}{\lambda} = \hbar ck$, so that the energy levels are:

$$E_n = \frac{hc}{2L}n.$$

(B.3)

For photons, it is often convenient to use frequency ν, rather than E, where $E = h\nu$. The dispersion relation then becomes $\nu = \frac{c}{\lambda} = \frac{c}{2\pi}k$, and the permitted frequencies are $\nu_n = \frac{c}{2L}n$.

If the walls are truly impenetrable, n can be any positive integer, so that any one particle, relativistic or not, has an infinite number of states, and there is no upper limit to its possible energy.

B.3   Electron Spin in a Magnetic Field

All fermions and many bosons have intrinsic angular momentum (or spin), usually expressed in units of $\hbar$, and this spin is usually associated with a magnetic moment μ. For example, the electron has spin 1/2, and its magnetic moment is approximately equal to the Bohr magneton $\mu_B \equiv \frac{q\hbar}{2m_0}$, where $-q$ is the charge on the electron, and m_0 is its rest mass. The rules of quantum theory require that the component of the spin along any particular axis be quantized, so that the energy of a spin in a magnetic field $\mathbf{B}$ (whose direction defines such an axis) can have only two values, $E_\pm = \pm\mu B$. This splitting into two levels in a field is illustrated in Figure B.2, the arrows indicating the direction of the spin relative to $\mathbf{B}$ in the two states. These states are conventionally called spin up and spin down (often denoted $\uparrow$ and $\downarrow$). Unlike the previous case, there is an upper limit to the energy of any one spin.

Note that if $B = 0$, there are two states with the same energy; the states are then said to be spin degenerate. If the spin is S, the spin degeneracy is $2S + 1$.

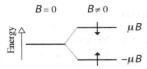

FIGURE B.2 Splitting of a spin 1/2 level in a magnetic field.

B.4 The Harmonic Oscillator

Another important, but simple quantum mechanical system is the harmonic oscillator; for example, a diatomic molecule like HCl or N_2. If one ignores the internal structure of the constituent atoms and treats them as rigid, they oscillate about their mean positions just like masses connected by a spring. The possible energy states of this oscillator are $E_n = (n + 1/2)h\nu$. Here n is any non-negative integer, and $\nu = \frac{1}{2\pi} \left(\frac{K}{m}\right)^{1/2}$ is the frequency of the corresponding classical oscillator, K being the force constant of the spring, and m the reduced mass of the atoms in the molecule. The energy of the $n = 0$ level, $E_0 = h\nu/2$, is called the zero point energy. Note the similarities to and the differences from the case of the particle in a box. Both have zero point energy, which comes from the fact that the particle is restricted to a particular region of space, but in the harmonic oscillator, the excited states are evenly spaced in energy, while the non-relativistic particle in a box has energies proportional to n^2. This equal spacing makes a collection of harmonic oscillators rather simple to treat by statistical mechanics. Real oscillators are not usually exactly harmonic (the force constant varies slightly with displacement), with the result that the levels are not quite evenly spaced.

The quantum states of a three-dimensional isotropic oscillator (that is, one whose force constant is independent of the direction of displacement) need three quantum numbers to specify them and the energy levels are:

$$E(n_1, n_2, n_3) = \left(n_1 + n_2 + n_3 + \frac{3}{2}\right) h\nu, \tag{B.4}$$

where n_1, n_2, n_3 are any three non-negative integers. Since in general many different sets (n_1, n_2, n_3) give the same value of E, the levels (except for the ground state) are degenerate. Such degeneracy is called orbital degeneracy. If the particle has spin, the degeneracy of a level (in zero magnetic field) is the product of the orbital and spin degeneracies. The lower levels of this oscillator, with their orbital degeneracies g and the corresponding quantum numbers, are shown in Table B.1 (the dotted lines indicate permutation of the n_i). Typical values of $h\nu$ for light molecules lie in the range 0.1 to 0.5 eV; for example, the vibrational mode of N_2 has $h\nu \approx 0.3$ eV.

TABLE B.1 Energy Levels and Degeneracies of a Three-Dimensional Harmonic Oscillator

$E/h\nu$	g	n_1, n_2, n_3
11/2	15	013 ..., 022 ..., 112 ..., 400 ...
9/2	10	012 ..., 300 ..., 111
7/2	6	011 ..., 200 ...
5/2	3	100 ...
3/2	1	000

B.5 The Quantum Rotator

Molecules can rotate as well as vibrate. The characteristic energies of rotation are small relative to those of vibration, so that it is usually a good approximation in calculating the rotational energy levels to neglect vibration, treating the molecule as rigid. A rigid diatomic molecule has two equal non-zero moments of inertia $I_{xx} = I_{yy}$, and has energy levels given by:[3]

$$E_j = j(j+1)E_0, \tag{B.5}$$

where j is any non-negative integer, and $E_0 = \frac{\hbar^2}{2I_{xx}}$. The degeneracy of the jth level is $2j + 1$. Typical values of E_0 for moderately light molecules are of order 10^{-4} eV; for example, $E_0 \approx 2.5 \times 10^{-4}$ eV for N_2.

B.6 The Hydrogen-Like Atom

A hydrogen-like atom consists of a nucleus of atomic number Z to which a single electron is bound by the Coulomb interaction. If spin-orbit interaction and other small effects are neglected, the energy levels are given by the Rydberg formula:

$$E_n = -\frac{Z^2 m_0 q^4}{2(4\pi\varepsilon_0)^2\hbar^2}n^2, \tag{B.6}$$

where n is any positive integer, and m_0 is the rest mass of the electron. In this approximation, the orbital degeneracy is n^2, which must be multiplied by the spin degeneracy of 2. For $Z = 1$ (hydrogen), $E_1 \approx -13.6$ eV.

---

[3] See, for example, A. P. French and E. F. Taylor, *Introduction to Quantum Physics*, (Norton, 1978), p. 487–495.

Appendix C: Proof that the Zero'th Law Implies the Existence of Temperature

Here, we put the argument of Section 2.2 into mathematical form.[1] It is convenient to think of the three objects, labeled by $i = 1, 2, 3$, as fixed quantities of a compressible fluid, each of whose states can be defined by measuring the pressure p_i and volume V_i. If two objects are both surrounded by insulating walls, p and V can be varied at will for each. On the other hand, if they are separated by a diathermal wall so that equilibrium between them is established, it is found that p and V can no longer be varied arbitrarily; specification of any three of the values fixes the fourth. We can express this fact mathematically by saying that when objects 1 and 2 are in equilibrium, there exists a functional relation of the form:

$$F_{12}(p_1, V_1, p_2, V_2) = 0, \tag{C.1}$$

where F_{12} is some function which we can determine by measurement.

For example, if object 1 is a mercury thermometer and object 2 is air, F_{12} is found empirically to be:

$$p_2 V_2 - A V_1 + B = 0, \tag{C.2}$$

where A and B are empirical constants which depend on the properties of air and of mercury (in this particular example, p_1 does not enter because mercury is virtually incompressible). It is convenient to use the length of the mercury column as a measure of V_1.

Similarly, if objects 1 and 3 are in equilibrium, there is another relationship:

$$F_{13}(p_1, V_1, p_3, V_3) = 0. \tag{C.3}$$

The zero'th law of thermodynamics states that if objects 2 and 3 are separately in equilibrium with object 1 [that is, if Equations (C.1) and (C.3) both hold], then they are in equilibrium with each other. Thus, there exists a third relation:

$$F_{23}(p_2, V_2, p_3, V_3) = 0, \tag{C.4}$$

and the zero'th law requires that F_{12} and F_{13} must be such that Equations (C.1) and (C.3) together imply Equation (C.4).

We now show that if Equation (C.4) is to follow from Equations (C.1) and (C.3), there must be a property common to all three objects, which we call *temperature*; that is, when two objects are in equilibrium

---

[1] The proof here follows Pippard, pp. 9–10.

with each other, they have the same temperature. In more precise terms, we show that a necessary and sufficient condition[2] for the zero'th law to be true is that there exist function f_i of the state variables p_i and V_i of each object separately, such that in equilibrium:

$$f_1(p_1, V_1) = f_2(p_2, V_2) = f_3(p_3, V_3) \equiv \theta, \tag{C.5}$$

where f_1, f_2, and f_3 may be different functions of their arguments, but in equilibrium are all equal to a quantity which we define as the *empirical temperature* and is denoted θ.

That Equation (C.5) is a sufficient condition can be seen by defining $F_{12} \equiv f_2(p_2, V_2) - f_1(p_1, V_1)$, $F_{13} \equiv f_3(p_3, V_3) - f_1(p_1, V_1)$ in Equations (C.1) and (C.3), where the f satisfy Equation (C.5); then there exists a function $F_{23} = f_2(p_2, V_2) - f_3(p_3, V_3) = 0$ in equilibrium, and Equation (C.4) is satisfied.

That Equation (C.5) is also a necessary condition can be shown as follows. We solve Equation (C.1) for V_1, obtaining $V_1 = G_{12}(p_1, p_2, V_2)$, where G_{12} is a function which can be determined from F_{12}. We do the same for Equation (C.3), obtaining $V_1 = G_{13}(p_1, p_3, V_3)$. Hence, Equations (C.1) and (C.3) imply that:

$$G_{12}(p_1, p_2, V_2) = G_{13}(p_1, p_3, V_3). \tag{C.6}$$

For Equation (C.6) to imply Equation (C.4), which does not contain p_1, all terms in G_{12} and G_{13} that contain p_1 must cancel out, so that the most general possible forms of G_{12} and G_{13} are:

$$G_{12} = g(p_1)f_2(p_2, V_2) + h(p_1), \quad G_{13} = g(p_1)f_3(p_3, V_3) + h(p_1), \tag{C.7}$$

where:

$$f_3(p_3, V_3) = f_2(p_2, V_2), \tag{C.8}$$

and g and h are the same functions of p_1 in both equations. Similarly, we can solve Equations (C.3) and (C.4) for V_3, and after eliminating p_3, we obtain:

$$f_1(p_1, V_1) = f_2(p_2, V_2), \tag{C.9}$$

so that Equation (C.5) is proved.

---

[2] If A is a "sufficient" condition for B, B holds *whenever* A is true. If A is a "necessary" condition, B does *not* hold *unless* A is true. For example, to be sober is a necessary, but not a sufficient condition for one to drive a car legally, while (in most states) driving while drunk is a sufficient, but not a necessary reason for the loss of one's driving license.

Appendix D: Some Results in Probability Theory

D.1 Derivation of the Binomial Coefficients c(N,r)

The *binomial coefficient* $c(N, r)$ is the coefficient of $x^r y^{N-r}$ in the expansion of $(x+y)^N$ and can be obtained as follows. Think of the product of N brackets $(x + y)(x + y) \ldots$. When you multiply the brackets out, you get a term in $x^r y^{N-r}$ by taking x from r of the brackets and y from the rest. How many of these terms are there? Think of each bracket as an object which is assigned either to the x pile (containing r objects) or the y pile (containing $N - r$). We do not care in what order the objects are assigned. The number of terms in $x^r y^{N-r}$ is simply the number of different ways the assignment can be made; call this number $c(N, r)$. Now suppose that the objects are labeled, and we distinguish between different orderings of them. The r objects in the x pile can be permuted to give $r!$ orderings,[1] and those in the y pile $(N - r)!$ orderings. The product $c(N, r)r!(N - r)!$ is the total number of orderings of N objects, which is $N!$. Hence,

$$c(N, r)r!(N - r)! = N!$$

$$c(N, r) = \frac{N!}{(N - r)!r!}. \tag{D.1}$$

Putting $x = y = 1$ in $(x + y)^N = \sum_{r=0}^{N} c(N, r)x^r y^{N-r}$ gives:

$$\sum_{r=0}^{N} c(N, r) = (x + y)^N = 2^N. \tag{D.2}$$

This derivation of $c(N, r)$ can be generalized to m piles, the i'th pile containing r_i objects, where $\sum r_i = N$. In the expansion of $(x + y + z + \cdots)^N$, the coefficient of $x^{r_1} y^{r_2} z^{r_3} \ldots$ is the *multinomial coefficient*:

$$c(N, r_1, r_2, \ldots) = \frac{N!}{\prod r_i!}. \tag{D.3}$$

Equation (D.3) is the traditional starting point for the statistical mechanics of the ideal gas,[2] but we do not use it in this book.

---

[1] This well-known result can be derived as follows. If we draw the objects one at a time from the pile, the first one can be chosen in r ways. For each choice of the first, the second can be chosen in $(r - 1)$ ways, and so on; hence, the total number of permutations (that is, the number of possible orderings), of the r objects, is $r(r - 1)(r - 2) \ldots 1 \equiv r!$.

[2] See, for example, H. Margenau and G. M. Murphy, *Mathematics of Physics and Chemistry*, 2nd ed. (van Nostrand, New York, 1956), Chapter 12.

The above proof of the binomial formula is not logically flawless, and the following alternative proof[3] is preferred by mathematicians.

Let $c(N, r)$ be the number of ways of choosing r objects from a set of N; in the language of set theory, it is the number of r-membered subsets of a set with N members.

Consider a particular object U. It is either in a particular subset or not. The number of subsets not containing U is the number of ways of choosing r objects from $N - 1$; that is, $c(N - 1, r)$. The number containing U is the number of ways of choosing $r - 1$ objects (those objects in the subset which are not U) from $N - 1$; that is, $c(N - 1, r - 1)$. Hence,

$$c(N, r) = c(N - 1, r) + c(N - 1, r - 1).$$

This is known as Pascal's rule.

We now prove Equation (D.1) by induction.

Assume that Equation (D.1) is true for N. Then, by Pascal's rule,

$$
\begin{aligned}
c(N + 1, r) &= c(N, r) + c(N, r - 1) \\
&= \frac{N!}{(N - r)!r!} + \frac{N!}{(N - r + 1)!(r - 1)!} \quad \text{by hypothesis} \\
&= [(N - r + 1) + r]\frac{N!}{(N - r + 1)!r!} \\
&= \frac{(N + 1)!}{(N - r + 1)!r!}.
\end{aligned}
$$

Thus, if the formula is true for N, it is true for $N + 1$.

It is true for $N = 1$, since $c(1, 1) = 1$. Hence, it is true for all N.

D.2 Stirling's Approximation to the Factorial

Stirling's formula for the factorial of a large number is:[4]

$$N! \approx (2\pi N)^{1/2} N^N e^{-N}, \tag{D.4}$$

where the fractional error is $\sim 1/12N$.

We are often only interested in $\ln(N!)$, and Stirling's formula can be written:

$$\ln(N!) = N(\ln N - 1) + \frac{1}{2}\ln(2\pi N). \tag{D.5}$$

If N is large, the second term is much smaller than the first and can often be dropped, so that we obtain the crude version of Stirling's approximation:

$$\ln(N!) \approx N(\ln N - 1). \tag{D.6}$$

---

[3] C. M. Grinstead and J. L. Snell, *Introduction to Probability* (American Mathematical Society, Washington, D.C., 1997), Section 3.2.

[4] Abramowitz and Stegun, p. 257.

Equation (D.6) can be derived directly by approximating the sum over logarithms as an integral and integrating by parts:[5]

$$\ln(N!) = \sum_{x=1}^{N} \ln(x) \approx \int_{1}^{N} \ln(x)\,dx = |x\ln(x) - x|_{1}^{N} \approx N(\ln N - 1),$$

where we have neglected 1 relative to N.

To prove the more accurate expression Equation (D.4), we start from the well-known integral form for the factorial:

$$N! = \int_{0}^{\infty} x^N e^{-x}\,dx = \int_{0}^{\infty} e^{(N\ln x - x)}\,dx.$$

We substitute $x = N + N^{1/2}y$, obtaining:

$$N! = \int_{-\sqrt{N}}^{\infty} \exp\left[N\ln N + N\ln\left(1 + N^{-1/2}y\right) - N - N^{1/2}y\right] N^{1/2}\,dy$$

$$= N^{1/2}N^N e^{-N} \int_{-\sqrt{N}}^{\infty} \exp\left[N\ln\left(1 + N^{-1/2}y\right) - N^{1/2}y\right]\,dy.$$

Since N is large, we can expand the logarithm in the exponent in powers of $N^{-1/2}$:

$$N(\ln\left(1 + N^{-1/2}y\right) \approx N\left(N^{-1/2}y - \frac{1}{2}N^{-1}y^2\right) = N^{1/2}y - \frac{1}{2}y^2,$$

so that the integrand is approximately $e^{-y^2/2}$.

Since N is large, we can take the lower limit of the integral as $-\infty$. Equation (D.4) then follows, since:

$$\int_{-\infty}^{\infty} e^{-y^2/2}\,dy = (2\pi)^{1/2}.$$

Stirling's formula is much more accurate, even for quite small numbers, than the above derivation would suggest; try it for $N = 3$.

D.3 Mean Values (Averages) over a Distribution

To obtain the *mean* (or *average*) grade of a class, we take the number of students who have obtained a particular grade G, multiply it by the grade, sum over all grades, and divide by the total number N of students. If the number of students who get grade G is S_G, then $N = \sum S_G$, and this procedure can be expressed mathematically by:

$$\langle G \rangle = N^{-1} \sum GS_G = \sum GP_G, \tag{D.7}$$

where the mean grade is written $\langle G \rangle$, and P_G is the probability that a randomly chosen student obtained a grade G; this is defined by $P_G \equiv S_G/N$, so that $\sum P_G = 1$.

Equation (D.7) is a special case of the general definition of the mean value of any discrete variable x:

$$\langle x \rangle \equiv \sum xP_x, \tag{D.8}$$

---

[5] For a more rigorous proof, see D. Stirkazer, *Elementary Probability* (Cambridge University Press, Cambridge, UK, 1994), p. 261.

where P_x is the probability that the variable has the value x, $\sum P_x = 1$, and the sums are over all possible values of x.

If x is a continuous variable, P_x is replaced by $P(x)\,dx$, the probability that x lies between x and $x + dx$. $P(x)$ is called a *probability density* (sometimes, less precisely, a *distribution*). When there is no danger of ambiguity, P_x is often written $P(x)$, but it must be remembered that, while P_x is a probability and is dimensionless, $P(x)$ is a probability density and has the dimensions of x^{-1}. The sums in Equations (D.7) and (D.8) are replaced by integrals over the domain of x:

$$\int P(x)\,dx = 1,$$

$$\langle x \rangle \equiv \int xP(x)\,dx. \tag{D.9}$$

One is often interested in the spread in a quantity (that is, the width of its probability distribution) as well as its mean. A useful measure of the spread is the mean square deviation or *variance* $\langle \Delta x^2 \rangle$, where $\Delta x \equiv x - \langle x \rangle$.

$$\langle \Delta x^2 \rangle \equiv \sum \Delta x^2 P_x = \sum (x - \langle x \rangle)^2 P_x$$

$$= \sum x^2 P_x - 2\langle x \rangle \sum xP_x + \langle x^2 \rangle^2 \sum P_x$$

$$= \langle x^2 \rangle - \langle x \rangle^2, \tag{D.10}$$

since $\sum P_x = 1$ and $\sum xP_x = \langle x \rangle$. For continuous distributions, the sums are replaced by integrals and P_x by $P(x)\,dx$.

The *standard deviation* is the square root of the variance, $\langle \Delta x^2 \rangle^{1/2}$. It is the most commonly used measure of the width of a distribution, since it has the same dimensions as x.

D.4 Binomial Distribution

Suppose that we toss a coin many times. If the coin is "fair," so that there is no physical distinction between head and tails, the probability[6] that heads come up r times in N tosses is, from section (a):

$$P(N, r) = \frac{c(N, r)}{\sum_{r=0}^{N} c(N, r)},$$

$$= 2^{-N} c(N, r) \tag{D.11}$$

where, from Equation (D.1),

$$c(N, r) = \frac{N!}{(N - r)!\,r!}.$$

Equation (D.11) is called the *binomial distribution*.

If the coin is biased, so that the probability of getting heads is P_0 and of tails $1 - P_0$, Equation (D.11) can be generalized to give the asymmetric binomial distribution:

$$P(N, r) = c(N, r)P_0^r(1 - P_0)^{(N-r)}. \tag{D.12}$$

---

[6] Note that N and r are discrete variables, so that there is no ambiguity in writing $P(N, r)$ for the probability $P_{N,r}$.

D.5 Gaussian Distribution

When N is large, it is convenient to symmetrize Equation (D.11) by substituting $r = \frac{1}{2}(N - n)$, so that $N - r = \frac{1}{2}(N + n)$, and use the fact, which we prove below, that $P(N, n)$ is only significantly different from zero when $|n| \ll N$.

Equation (D.11) then becomes:

$$P(N, n) = 2^{-N} \frac{N!}{[(N + n)/2]! \, [(N - n)/2]}. \tag{D.13}$$

This distribution is symmetric about its maximum at $n = 0$, so that $\langle n \rangle = 0$. If N is large, $P(N, n)$ is only significantly different from zero if $n \ll N$, and we would like to find an analytically more tractable form of Equation (D.13) which is valid for small n/N. The logarithm of $P(N, n)$ is easier to work with than P itself. Equation (D.13) can be written:

$$\ln P(N, n) = - N \ln 2 + \ln(N!)$$
$$- \ln \left\{ \left[\frac{N}{2} \left(1 + \frac{n}{N} \right) \right]! \right\} - \ln \left\{ \left[\frac{N}{2} \left(1 - \frac{n}{N} \right) \right]! \right\}. \tag{D.14}$$

Substituting Equation (D.5) in Equation (D.14) gives:

$$\ln P(N, n) \approx - \frac{1}{2} \ln \left(\frac{\pi N}{2} \right) - N \ln 2 + N \ln(N)$$
$$- \frac{1}{2} N \left\{ \left(1 + \frac{n}{N} \right) \left[\ln \left(\frac{N}{2} \right) + \ln \left(1 + \frac{n}{N} \right) \right] \right.$$
$$\left. + \left(1 - \frac{n}{N} \right) \left[\ln \left(\frac{N}{2} \right) + \ln \left(1 - \frac{n}{N} \right) \right] \right\}, \tag{D.15}$$

where the neglect of n^2 relative to N^2 in the small first term introduces no significant error.

In the region where $n \ll N$, we can expand the logarithms in powers of n/N, using Equation (A.5):

$$\ln \left(1 \pm \frac{n}{N} \right) \approx \pm \frac{n}{N} - \frac{1}{2} \left(\frac{n}{N} \right)^2.$$

After some cancellation, Equation (D.15) then becomes:

$$\ln P(N, n) \approx - \frac{1}{2} \ln \left(\frac{\pi N}{2} \right) - \frac{n^2}{2N}, \tag{D.16}$$

where terms of order (n^3/N^2) have been neglected.

Hence, the probability distribution of the discrete variable n is:

$$P(N, n) \approx \left(\frac{2}{\pi N} \right)^{1/2} \exp \left(- \frac{n^2}{2N} \right). \tag{D.17}$$

Since we are interested in large numbers, it is convenient to treat n as a continuous variable. In doing this, we have to be careful because in Equation (D.17), n has the same parity as N; it is always odd or always even. Hence, half the integers are missing, and the probability density of n, regarded as a continuous variable, is:

$$P(N, n) \approx P(N, 0) \exp \left(- \frac{n^2}{2N} \right), \tag{D.18}$$

where $P(N, 0) = (2\pi N)^{-1/2}$. Whereas in Equation (D.17), $P(N, n)$ is a discrete probability, only defined for even (or odd) integer values of n, in Equation (D.18) it is a probability *density* and is defined for all values of the *continuous* variable n. The normalization of the probability density is exact for large N:

$$\int_{-\infty}^{\infty} P(N, n)\, dn = (2\pi N)^{-1/2} \int_{-\infty}^{\infty} \exp\left(\frac{-n^2}{2N}\right) dn = 1. \tag{D.19}$$

The probability distribution given by Equation (D.18) is called the Gaussian distribution (or simply the Gaussian) in physics and the normal distribution in statistics. It is shown for $N = 10$ by the continuous curve in Figure D.2 at the end of this appendix. Note that even for so small a value of N, $P(n)$ becomes vanishingly small as $n \to \pm N$. N in a macrosystem is, of course, enormously larger, and a graph on the horizontal scale of Figure D.2 would look like a vertical line at the origin.

The variance of a distribution is defined as $\langle \Delta n^2 \rangle = \langle n^2 \rangle - \langle n \rangle^2$, where $\langle n \rangle$ is the mean value of n (see section D.3). In the present case $\langle n \rangle = 0$, so that:

$$\begin{aligned}
\langle \Delta n^2 \rangle = \langle n^2 \rangle &\equiv \int_{-\infty}^{\infty} n^2 P(N, n)\, dn \\
&= (2\pi N)^{-1/2} \int_{-\infty}^{\infty} n^2 \exp\left(-\frac{n^2}{2N}\right) dn \\
&= N.
\end{aligned} \tag{D.20}$$

The standard deviation is thus $N^{1/2}$.

The negative exponential in Equation (D.18) ensures that if n is much greater than $N^{1/2}$, $P(N, n)$ is extremely small relative to $P(N, 0)$.

D.6 Poisson Distribution

Suppose we have some event that occurs at random (for example, the decay of a radioactive nucleus), with a mean rate of λ events per unit time, and we want to calculate the probability $P(r)$ of r such events occurring in an arbitrarily chosen time interval, of length t. This can be derived from the asymmetric binomial distribution [Equation (D.12)].

We divide up the interval t into N intervals, where t/N is so small that the chance of two events within this time is negligible (we will ultimately let $N \to \infty$). The probability of there being an event in this short interval is $\lambda t/N$, so that the probability of there being r events in N such intervals is, from Equation (D.12),

$$P(N, r) = c(N, r) \left(\frac{\lambda t}{N}\right)^r \left(1 - \frac{\lambda t}{N}\right)^{N-r}, \tag{D.21}$$

where:

$$c(N, r) = \frac{N(N - 1)(N - 2)\ldots(N - r + 1)}{r!} \approx \frac{N^r}{r!}, \tag{D.22}$$

since $N \gg r$.

Furthermore, since $N \gg r$, $\left(1 - \frac{\lambda t}{N}\right)^{N-r} \approx \left(1 - \frac{\lambda t}{N}\right)^{N} \to e^{-\lambda t}$ as $N \to \infty$.

Hence, in the limit $N \to \infty$,

$$\begin{aligned}
P(r) &= \frac{N^r}{r!} \left(\frac{\lambda t}{N}\right)^r e^{-\lambda t} \\
&= e^{-\langle r \rangle} \frac{\langle r \rangle^r}{r!},
\end{aligned} \tag{D.23}$$

where $\langle r \rangle = \lambda t$, the mean number of events in interval t. Equation (D.22) is known as the Poisson distribution. It is easily shown [see Problem 10.2 (c), (d)] that $\sum P(r) = 1$, that $\sum r P(r) = \langle r \rangle$, that the variance $\langle \Delta r^2 \rangle = \langle r \rangle$, and that for large $\langle r \rangle$, Equation (D.22) becomes a Gaussian in Δr.

D.7 Derivation of the Multiplicity Function for a System of Identical Oscillators

We consider a collection of N identical harmonic oscillators of frequency ν. We want to calculate the number of different microstates (that is, the multiplicity) of the i'th macrostate of this system, $\Omega(N, i)$. This is the number of different ways that the system can have a total energy $ih\nu$ (the zero point energy is not included).

The method is adapted from Bose's original derivation. We suppose that we have i identical packets of energy and ask how they can be distributed over N oscillators. Imagine the packets to be laid out along a horizontal line, with a vertical bar separating the packet assigned to one oscillator from the packet assigned to the next. Two such orderings of packets and bars, each with $N = 9$, $i = 12$, are shown in Figure D.1.

There are $N - 1$ bars. The total number of packets plus bars is $N + i - 1$, and they can be permuted in $(N + i - 1)!$ different ways. However, the packets are indistinguishable from each other, so that for any particular ordering of packets and bars, the packets can be permuted in $i!$ ways without changing the microstate. Similarly, the bars are indistinguishable, so they can be permuted in $(N - 1)!$ ways. Hence, the number of microstates is:

$$\Omega(N, i) = \frac{(N + i - 1)!}{i!(N - 1)!}. \tag{D.24}$$

Bose derived the Bose-Einstein distribution for photons from Equation (D.23) by maximizing Ω, but the derivation from the grand partition function is simpler and more general (see Chapter 9).

D.8 Random Walk Program

This program illustrates graphically how the random walk (or spin $1/2$) problem leads to a binomial or Gaussian distribution. It sums N (≤ 2000) values of a number which can be either $+1$ or -1 at random, the sum being n, where n can run from $-N$ to N. This sum is repeated m times, and a histogram showing the number of times the different values of n have been obtained is plotted continuously during the calculation. The results are compared with the properly scaled binomial coefficients [Equation (D.11)] and the Gaussian approximation [Equation (D.17)]. For large values of N (>90), the factorials in the binomial coefficients become too large and only the Gaussian is calculated. One can also plot the Gaussian logarithmically in order to show the behavior in the tails of the distribution, where the probabilities are too small to be visible on a linear plot. The larger the number of trials, the closer the results approach the theoretical distribution. By successively increasing the value of N, say from 10 to 2000, one can see how the distribution sharpens up.

$$\cdots | \cdot | \cdot \cdot | | \cdot | \cdot \cdot \cdot | | | \cdot \cdot \qquad \cdot | \cdot \cdot \cdot | | \cdot \cdot | \cdot | \cdot \cdot | \cdot \cdot | | \cdot$$

FIGURE D.1 Two possible distributions of energy $12h\nu$ over 9 oscillators. Each dot represents a packet of energy $h\nu$, and the bars separate the packets assigned to different oscillators. Note that if a particular oscillator is in its ground state and so has no energy, the corresponding bars are adjacent to each other.

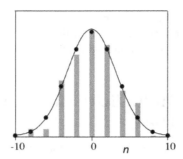

FIGURE D.2 Output of the random walk program. The vertical bars are a histogram of the results of 10000 random walks of 10 steps each, the circles the binomial distribution, and the curve the Gaussian distribution.

Typical output of this program, for $N = 10$ and $m = 10000$, is shown in Figure D.2, in which the log plot has been omitted for clarity. The vertical bars are the histogram. The binomial coefficients are indicated by circles, and the Gaussian by the continuous curve. Even for this small value of N, the binomial is almost indistinguishable from the Gaussian.

Appendix E: Differentials, Partial Derivatives, and the Maxwell Relations

E.1 Differentials

One of the peculiarities of the literature of thermodynamics is its widespread use of differentials such as dx. Differentials are infinitesimal quantities and are regarded by most mathematicians, and some physicists,[1] as very dubious entities, since it is difficult to define them a logically precise way. If, for example, one defines an infinitesimal as a quantity smaller than any non-zero number, it has to be zero, and a ratio of two differentials, such as $\frac{dy}{dx}$, becomes meaningless. However, in this book we define dx as *a change in x so small that quantities of the order of $(dx)^2$ can be neglected*. Such a quantity can, with caution, be treated as an ordinary variable so long as one does not change its order of magnitude; for example, $\sqrt{dx}$ is meaningless, since the square root of a very small number is much larger than that number. One way to avoid the use of differentials is to follow Bohren and Albrecht[2] in recognizing that any change must take some time, so that instead of writing dx, one writes $\frac{dx}{dt}$, which can be defined rigorously in terms of a limit (see any textbook of analysis). This is logical, but it complicates the notation and introduces time as a variable where it is not relevant; we are not usually interested in the rate of change, but only in the fact that there is a change. However, anyone who is bothered by the presence of differentials in an equation in this book is welcome to divide the entire equation through by dt.

That said, consider an analytic function of two variables:[3] $u = u(x, y)$. From the definition of a partial derivative, the differential of u is:

$$du = \left(\frac{\partial u}{\partial x}\right)_y dx + \left(\frac{\partial u}{\partial y}\right)_x dy. \tag{E.1}$$

What this means is that if we make very small changes dx in x and dy in y, the resulting small change in u is du and is given by this equation. Note that no meaning can be attached to ∂x; this is *not* a differential.

---

[1] See, for example, the section entitled "Those accursed differentials" in Chapter 3 of Bohren and Albrecht.

[2] Ibid.

[3] Here we commit another sin against mathematical rigor, using the same symbol (here u) for a variable and for the function which defines that variable. This is standard practice in thermodynamics, since it causes less confusion than would a proliferation of different symbols. In fact, no confusion need arise if one is always clear which independent variables are being considered. For example, we can write the internal energy U of a compressible fluid as a function of temperature and volume, $U(T, V)$ or as a function of pressure and volume, $U(p, V)$. These are two *different* functions and have different partial derivatives with respect to V. Throughout this book, unless there is no possibility of confusion, we specify which variable is being kept constant in the differentiation; for example writing $\left(\frac{\partial U}{\partial V}\right)_T$ for $\frac{\partial U(T,V)}{\partial V}$ and $\left(\frac{\partial U}{\partial V}\right)_p$ for $\frac{\partial U(p,V)}{\partial V}$.

E.2 Triple Product of Partial Derivatives

Figure E.1 shows part of a contour $u(x, y) = $ constant in the (x, y) plane, indicated by the heavy line. As one moves along the contour from the point (x, y) to $(x + dx, y + dy)$, u is constant so that $du = 0$. However, we could have made the same move in two steps: horizontally by dx, and then vertically by dy. From Equation (E.1), the change in u in the first step is $\left(\frac{\partial u}{\partial x}\right)_y dx$, while in the second step it is $\left(\frac{\partial u}{\partial y}\right)_x dy$. The sum of these is du, which is zero, so that:

$$\left(\frac{\partial u}{\partial x}\right)_y dx + \left(\frac{\partial u}{\partial y}\right)_x dy = 0.$$

The slope of the contour in the (x, y) plane is $\left(\frac{\partial y}{\partial x}\right)_u$, so that $dy = \left(\frac{\partial y}{\partial x}\right)_u dx$.
  Hence,

$$\left(\frac{\partial u}{\partial x}\right)_y + \left(\frac{\partial u}{\partial y}\right)_x \left(\frac{\partial y}{\partial x}\right)_u = 0,$$

or:

$$\boxed{\left(\frac{\partial x}{\partial u}\right)_y \left(\frac{\partial u}{\partial y}\right)_x \left(\frac{\partial y}{\partial x}\right)_u = -1.}$$ (E.2)

This *triple product rule* is an important result that we use a lot.

E.3 Relation Between Cross Derivatives

Here we consider the second derivatives of u with respect to x and y, making it necessary to consider small, but finite changes δx and δy, and to retain terms of second order in δx and δy.
  As seen in Appendix A, if $u(x, y)$ is analytic it can be Taylor expanded about the point (x, y):

$$u(x + \delta x, y + \delta y) = u(x, y) + a_1\delta x + b_1\delta y + a_2\delta x^2 + b_2\delta y^2 + c_2\delta x\delta y + \ldots..$$

Differentiating with respect to x and then with respect to y, and putting $\delta x = \delta y = 0$, gives:

$$\frac{\partial^2 u}{\partial y\partial x} = c_2,$$

while differentiating with respect to y and then with respect to x gives:

$$\frac{\partial^2 u}{\partial x\partial y} = c_2.$$

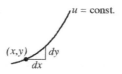

FIGURE E.1   Contour $u(x, y) = $ constant in the (x, y) plane.

Hence,

$$\boxed{\frac{\partial^2 u}{\partial x \partial y} = \frac{\partial^2 u}{\partial y \partial x}.}$$ (E.3)

The second derivatives in Equation (E.3) are called *cross derivatives*, and they must be equal if u is analytic. This fact is used in the next two sections.

E.4 Exact and Inexact Differentials

Ordinary first order differential equations often take the form:

$$du = f(x)\, dx, \quad \text{with the solution} \quad u(x) = \int f(x)\, dx.$$

Similarly, first order partial differential equations in two independent variables may have the form:

$$du = f(x, y)\, dx + g(x, y)\, dy,$$ (E.4)

where $f(x, y)$ and $g(x, y)$ are specified functions.

If the function $u(x, y)$ exists, comparing Equations (E.4) and (E.1) shows that:

$$f = \left(\frac{\partial u}{\partial x}\right)_y, \qquad g = \left(\frac{\partial u}{\partial y}\right)_x,$$

so that, by Equation (E.3),

$$\left(\frac{\partial f}{\partial y}\right)_x = \frac{\partial^2 u}{\partial x \partial y} = \frac{\partial^2 u}{\partial y \partial x} = \left(\frac{\partial g}{\partial x}\right)_y.$$

Thus, u exists only if:

$$\left(\frac{\partial f}{\partial y}\right)_x = \left(\frac{\partial g}{\partial x}\right)_y.$$ (E.5)

In other words, if the function $u(x, y)$ exists, f and g in Equation (E.4) cannot be chosen arbitrarily, but must obey Equation (E.5). If f and g do obey Equation (E.5), du is called an *exact* (or *perfect*) differential. Only exact differentials can be integrated to give a well-defined function, commonly called a *state function*.

In thermodynamics, we come across quantities like "heat" and "work" which are not state functions, since they change by different amounts according to the path taken from one state of the system to another (see Section 2.3). It is convenient, nevertheless, to write equations in which differentials of these quantities appear, and we distinguish these *inexact* differentials from the exact differentials defined in the previous paragraph by substituting đ for d. For instance, we write đQ for a small quantity of heat, meaning that if the rate of supply of heat is q per unit time, $đQ = q\, dt$. See Reif, Section 2.11, for a more extended discussion of exact and inexact differentials, including a demonstration of the fact that the integral of an inexact differential depends on the path taken.[4]

---

[4] A careful discussion of the use of exact and inexact differentials in thermodynamics is given by K. Stowe, *Introduction to Statistical Mechanics and Thermodynamics* (John Wiley & Sons, New York, 1984), pp. 94–97.

It is often the case that an inexact differential can be turned into an exact one by multiplying it by an integrating factor. In the case of $đQ$, this integrating factor is T^{-1}, and the resulting state function is the entropy S. It requires quite sophisticated mathematics to prove that the existence of such an integrating factor follows rigorously from the existence of irreversible processes.[5]

E.5 Maxwell Relations

A *Maxwell relation* is any relation between thermodynamic variables which follows by Equation (E.5) from the existence of a state function.

For example, the free energy $F(V, T)$, defined in Chapter 5, is a state function, and, for a given number of particles, Equation (5.20) gives:

$$dF = -S\,dT - p\,dV.$$

Thus,

$$S = -\left(\frac{\partial F}{\partial T}\right)_V, \qquad p = -\left(\frac{\partial F}{\partial V}\right)_T,$$

so that:

$$\left(\frac{\partial S}{\partial V}\right)_T = -\frac{\partial^2 F}{\partial V \partial T},$$

and:

$$\left(\frac{\partial p}{\partial T}\right)_V = -\frac{\partial^2 F}{\partial T \partial V}.$$

From Equation (E.5), it follows that:

$$\left(\frac{\partial S}{\partial V}\right)_T = \left(\frac{\partial p}{\partial T}\right)_V. \tag{E.6}$$

This Maxwell relation is very useful since it gives a quantity that is difficult to measure in terms of one that is relatively easy.

We can combine Equation (E.6) with the triple product rule [Equation (E.2)] in the form:

$$\left(\frac{\partial p}{\partial T}\right)_V \left(\frac{\partial T}{\partial V}\right)_p \left(\frac{\partial V}{\partial p}\right)_T = -1,$$

to get:

$$\left(\frac{\partial S}{\partial V}\right)_T = -\left(\frac{\partial V}{\partial T}\right)_p \left(\frac{\partial p}{\partial V}\right)_T. \tag{E.7}$$

The right-hand side of Equation (E.7) is the product of the bulk thermal expansion coefficient α_b and the isothermal bulk modulus B_T, so that by measuring these, one can deduce the entropy as a function of volume without having to do a calorimetric measurement.

---

[5] See Max Born, *Natural Philosophy of Cause and Chance* (Claredon Press, Oxford, UK, 1949), pp. 39–42, for the proof and for a critique of the way that the relation $dS = đQ/T$ is usually derived in textbooks of thermodynamics. The existence of the integrating factor $1/T$ was rigorously derived, at Born's instigation, by the Greek-German mathematician Constantin Carathéodory (1873–1950).

Equation (E.6) can also be derived directly from the fact that in a closed cycle, energy conservation requires that $\oint p\, dV = \oint T\, dS$ (see Chapter 3), but the derivation requires a knowledge of Jacobians.[6]

In systems in which the number of particles varies, a variety of Maxwell relations can be obtained from any state function whose differential is known; for example, $U(S, V, N)$, $H(S, p, N)$, or $G(T, p, N)$. For example, from Equation (7.13) and the definition of free enthalpy G [Equation (5.26)], the differential of G in a system containing only one type of particle is:

$$dG = -S\, dT + V\, dp + \zeta\, dN, \tag{E.8}$$

where N is the number of particles, and ζ the chemical potential. Three Maxwell relations can be obtained from the fact that G is a state function. Here we derive one of them:

$$S = -\left(\frac{\partial G}{\partial T}\right)_{p,N}, \qquad \zeta = \left(\frac{\partial G}{\partial N}\right)_{p,T},$$

so that:

$$\left(\frac{\partial S}{\partial N}\right)_{p,T} = -\left(\frac{\partial \zeta}{\partial T}\right)_{p,N}. \tag{E.9}$$

This shows that if, as is usually the case, the addition of a particle to a system increases the entropy, the derivative of ζ with respect to temperature is negative.

One can also obtain corresponding results for derivatives with respect to magnetic field. For example, in an incompressible magnetic material:

$$dF = -S\, dT - M\, dB,$$

whence:

$$\left(\frac{\partial S}{\partial B}\right)_{T} = \left(\frac{\partial M}{\partial T}\right)_{B} = B\frac{d\chi}{dT}, \tag{E.10}$$

if the isothermal susceptibility $\chi = \left(\frac{\partial M}{\partial B}\right)_{T}$ is independent of B. Note that S usually decreases with B, since the field reduces the number of states available (see Chapter 4), so that $\frac{d\chi}{dT}$ is negative.

---

[6] D. J. Ritchie, "A simple method for deriving Maxwell's relations," *Am. J. Phys.* **36**, 760–761 (1968). See also *Am. J. Phys.* **70**, 104–105 (2002) and references therein.

Appendix F: Standing Wave Solutions of the Wave Equation

In the text we consider two types of wave: an electromagnetic wave, associated with photons, and particle waves. In this appendix, we further assume that there are no media or potentials and look only for stationary solutions of the wave equation; that is, for standing rather than for propagating waves. In free space, the electromagnetic field $E(x, y, z, t)$ obeys the equation of wave motion:

$$\nabla^2 E = c^{-2} \frac{\partial E}{\partial t^2}, \tag{F.1}$$

where the Laplacian operator $\nabla^2 \equiv \frac{\partial^2}{\partial x^2} + \frac{\partial^2}{\partial y^2} + \frac{\partial^2}{\partial x^2}$.

If we assume a time dependence $e^{\pm i\omega t}$, we obtain the Helmholtz equation: for $E(x, y, z)$:

$$\left(\nabla^2 + \kappa^2\right) E = 0, \tag{F.2}$$

where $\kappa \equiv \frac{\omega}{c}$.

The particle wave function $\psi(x, y, z)$ obeys Schrödinger's time-independent equation, which in free space reduces to Equation (F.2) with E replaced by ψ and κ by $\frac{(2mE)^{1/2}}{\hbar}$, where E is the energy of the particle.

In this treatment, we ignore the vector nature of the electromagnetic field, substitute ψ for E in Equation (F.2), and look for solutions that satisfy the condition that $\psi = 0$ at the boundaries.

We first consider the one-dimensional problem; for example, waves on a string of length L fixed at both ends or particles in a box. Then Equation (F.2) becomes:

$$\frac{d^2\psi}{dx^2} + \kappa^2 \psi = 0, \tag{F.3}$$

with the general solution:

$$\psi(x) = A \sin \kappa x + B \cos \kappa x, \tag{F.4}$$

where A and B are constants.

We now apply the boundary conditions $\psi(0) = 0$, $\psi(L) = 0$. Since $\cos(0) = 1$, the first condition requires that $B = 0$. The second condition requires that $\sin \kappa L = 0$, and thus restricts the possible values of κ to:

$$\kappa_n = \frac{n\pi}{L}, \tag{F.5}$$

where n is an integer.

Thus, the normal modes of the string are given by:

$$\psi(x) = A \sin(\kappa_n x), \tag{F.6}$$

where the possible values of κ_n are given by Equation (F.5).

We now jump to the three-dimensional problem, since the two-dimensional case can easily be obtained from it. We look for solutions to Equation (F.2) of the form:

$$\psi(x, y, z) = X(x)Y(y)Z(z). \tag{F.7}$$

Substituting in Equation (F.2), we obtain:

$$YZ\frac{d^2X}{dx^2} + ZX\frac{d^2Y}{dy^2} + XY\frac{d^2Z}{dz^2} + \kappa^2 XYZ = 0. \tag{F.8}$$

Dividing through by XYZ gives:

$$X^{-1}\frac{d^2X}{dx^2} + Y^{-1}\frac{d^2Y}{dy^2} + Z^{-1}\frac{d^2Z}{dz^2} + \kappa^2 = 0. \tag{F.9}$$

Each of the first three terms in Equation (F.9) is a function of a different independent variable, so that each of these terms must be a constant if the equation is to be satisfied for all possible (x, y, z). We call the constants[1] $-\kappa_x^2$, $-\kappa_y^2$, $-\kappa_z^2$, where in order to satisfy Equation (F.9), we must have:

$$\kappa_x^2 + \kappa_y^2 + \kappa_z^2 = \kappa^2. \tag{F.10}$$

Thus, Equation (F.9) splits up into three ordinary differential equations:

$$\frac{d^2X}{dx^2} + \kappa_x^2 X = 0; \quad \frac{d^2Y}{dy^2} + \kappa_y^2 Y = 0; \quad \frac{d^2Z}{dz^2} + \kappa_z^2 Z = 0. \tag{F.11}$$

The boundary conditions require that $\psi(x, y, z) = 0$ on the six planes defined by:

$$x = 0, \quad x = L_x, \quad y = 0, \quad y = L_y, \quad z = 0, \quad z = L_z.$$

Applying the boundary conditions, we obtain solutions of each of these three equations exactly as in Equation (F.6) (we temporarily drop the arbitrary constant A):

$$X = \sin(\kappa_x x), \quad Y = \sin(\kappa_y y), \quad Z = \sin(\kappa_z z),$$

where $(\kappa_x, \kappa_y, \kappa_z)$ are restricted to the values:

$$\kappa_x = \frac{n_x \pi}{L_x}, \quad \kappa_y = \frac{n_y \pi}{L_y}, \quad \kappa_z = \frac{n_z \pi}{L_z}, \tag{F.12}$$

(n_x, n_y, n_z) being any three positive non-zero integers. Thus, the solution to Equation (F.2) is:

$$\psi(x, y, z) = A \sin(\kappa_x x) \sin(\kappa_y y) \sin(\kappa_z z), \tag{F.13}$$

where κ_x, κ_y, and κ_z are given by Equation (F.12), and A is any constant.

---

[1] The limitation to negative definite constants ensures that we will get harmonic solutions. Positive values give exponential solutions, which cannot satisfy the boundary conditions.

Note that $\psi(x, y, z)$ is the *product* of the individual sine functions. This product form is necessary to ensure that the boundary conditions are satisfied. For example, the boundary condition $\psi = 0$ for $x = 0$ requires that ψ vanish over the entire $x = 0$ plane, whatever the values of y and z; only a product has this property.

By the same reasoning that led to Equation (F.13), the solution to the two-dimensional Helmholtz equation in a rectangular box is:

$$\psi(x, y) = A \sin(\kappa_x x) \sin(\kappa_y y), \tag{F.14}$$

where $\kappa_x^2 + \kappa_y^2 = \kappa^2$.

Appendix G: The Equipartition Theorem

Classical statistical mechanics starts with the concept of *phase space*. This is a multidimensional space whose axes represent the *canonical coordinates* (or *degrees of freedom* in the statistical mechanical sense; see Footnote 9 of Chapter 7). For our purposes, these are the Cartesian coordinates specifying the position and velocity of every particle in the system.[1]

A point in this space, the *phase point*, represents one possible set of values of these degrees of freedom. A very simple example is the one-dimensional harmonic oscillator, which has two degrees of freedom, and whose phase space is a plane with axes x (displacement) and $v = \frac{dx}{dt}$ (velocity). If the oscillator has a constant energy, the phase point describes an ellipse in this plane, as shown in Figure G.1.

However, if the oscillator is in contact with a heat bath at temperature T, its energy is constantly fluctuating, and the phase point moves irregularly in the plane. We then need to know the probability of the phase point being found within a given "cell" of phase space. Such a cell (illustrated in the top right-hand corner of Figure G.1) is a square of sides dx and dv, with its corner at the point (x, v), so that when the phase point is within the cell, the displacement lies between x and $x + dx$ while the velocity lies between v and $v + dv$. By essentially the same argument that we derived the Boltzmann factor in Section 6.3, the probability that the phase point is within the cell is:

$$P(x, v) \, dx \, dv = Z^{-1} e^{-\beta E(x,v)} \, dx \, dv, \tag{G.1}$$

where $E(x, v)$ is given by Equation (6.50), $\beta = 1/k_B T$, and:

$$Z = \int \int e^{-\beta E(x,v)} \, dx \, dv, \tag{G.2}$$

where the double integral is over the entire range of possible x and v.

Equations (G.1) and (G.2) can be generalized to any number of degrees of freedom. We write the energy as a function of the canonical coordinates q_i:

$$E = E(q_1, q_2, \ldots q_i, \ldots). \tag{G.3}$$

According to the canonical distribution (see Chapter 6), the probability that the phase point is within the cell whose corner is at the point $(q_1, q_2, \ldots, q_i, \ldots)$ and whose volume is $dq_1 \, dq_2 \, \ldots \, dq_i \, \ldots$ is:

$$P(q_1, q_2, \ldots) \, dq_1 \, dq_2 \, \ldots \, dq_i \, \cdots = Z^{-1} e^{-\beta E} dq_1 \, dq_2 \, \ldots \, dq_i \, \ldots, \tag{G.4}$$

---

[1] In a system of canonical coordinates $\{q_i\}$, the volume of a cell in phase space is the product of the dq_i. This is the case for Cartesian coordinates, but it is not in general the case for other coordinate systems. For instance, in spherical coordinates (r, θ, ϕ), the cell volume is $r^2 \sin \theta \, dr \, d\theta \, d\phi$, and (r, θ, ϕ) are *not* canonical coordinates.

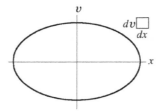

FIGURE G.1 Phase space for the one-dimensional harmonic oscillator.

where:

$$Z = \int \int \ldots \int e^{-\beta E} \, dq_1 \, dq_2 \, \ldots \, dq_i \, \ldots \ldots \tag{G.5}$$

The range of integration is $-\infty$ to ∞ throughout. In fact, this is an unnecessary restriction; all we need is that at the limits of each q_i, $q_i e^{-\beta E} = 0$.

Consider the quantity:

$$\left\langle q_i \frac{\partial E}{\partial q_i} \right\rangle \equiv \int \int \ldots \int q_i \frac{\partial E}{\partial q_i} P(q_1, q_2, \ldots) \, dq_1 \, dq_2 \, \ldots$$

$$= Z^{-1} \int \int \ldots \int q_i \frac{\partial E}{\partial q_i} e^{-\beta E} \, dq_1 \, dq_2 \, \ldots \tag{G.6}$$

$$= -(\beta Z)^{-1} \int \int \ldots \int q_i \frac{\partial}{\partial q_i} \left(e^{-\beta E} \right) \, dq_1 \, dq_2 \, \ldots \ldots \tag{G.7}$$

Now integrate with respect to q_i by parts:

$$\int q_i \frac{\partial}{\partial q_i} \left(e^{-\beta E} \right) \, dq_i = - \int e^{-\beta E} \, dq_i \tag{G.8}$$

since $q_i e^{-\beta E} = 0$ at the limits of integration.

Hence,

$$\left\langle q_i \frac{\partial E}{\partial q_i} \right\rangle = -(\beta Z)^{-1} \int \int \ldots \int e^{-\beta E} \, dq_1 \, dq_2 \, \ldots$$

$$\boxed{\left\langle q_i \frac{\partial E}{\partial q_i} \right\rangle = \beta^{-1} = k_B T.} \tag{G.9}$$

Equation (G.9) is the *generalized equipartition theorem*.[2]

---

[2] Readers familiar with classical mechanics will notice a formal resemblance between the equipartition theorem and the virial theorem (see, e.g., J. B. Marion, *Classical Dynamics of Particles and Systems*, Academic Press, New York 2nd ed., 1970, p. 234). However, the two must not be confused. The virial theorem relates average values of the potential and kinetic energies over mechanically determined orbits, so that the concept of temperature does not enter. The equipartition theorem is concerned with an average over thermal motion and relates it to temperature. Furthermore, the equipartition theorem holds *only* in the classical limit of continuous energy levels, while the virial theorem, *mutatis mutandis*, applies to a quantized system as well as a classical one [see, e.g., D. Park, *Introduction to the Quantum Theory*, 3rd ed. (McGraw-Hill, New York, 1992), p. 85].

In many cases, a system of coordinates exists in which E can be written as a sum of quadratic functions of the coordinates:

$$E = \sum_i a_i q_i^2, \tag{G.10}$$

so that:

$$\frac{\partial E}{\partial q_i} = 2a_i q_i.$$

The average contribution of the term in q_i to the energy is then given by:

$$\boxed{\langle a_i q_i^2 \rangle = \frac{1}{2} \left\langle q_i \frac{\partial E}{\partial q_i} \right\rangle = \frac{1}{2} k_B T.} \tag{G.11}$$

Hence, each degree of freedom q_i that appears quadratically in the energy function contributes an amount $\frac{1}{2} k_B T$ to the mean energy of the system. If there are n_f degrees of freedom for each particle and there are N particles, the total energy is:

$$U = \frac{1}{2} N n_f k_B T. \tag{G.12}$$

The kmolar heat capacity at constant volume is:

$$C_V = \left(\frac{\partial U}{\partial T} \right)_V = \frac{1}{2} n_f R_0, \tag{G.13}$$

where $R_0 = N_A k_B$.

The fact that the contribution to the internal energy of the i'th degree of freedom is independent of the magnitude of the coefficient a_i has a curious and (in classical terms) paradoxical consequence. Suppose that a_i tends to zero. However small a_i may be, so long as it is non-zero, the equipartition theorem says that the contribution to U is $Nk_B T/2$. On the other hand, if a_i actually *is* zero, that degree of freedom does not appear in the energy expression and cannot contribute to U.

This non-analytic behavior is a consequence (like Gibbs paradox; see Section 9.4) of the internal inconsistency of classical statistical mechanics. Quantum theory resolves this paradox, since as a_i becomes smaller, the corresponding quantum levels become further and further apart, so that a higher and higher temperature is needed for the assumption of continuous energy levels to be valid and for equipartition to hold. As a_i tends to zero, the required temperature tends to infinity. Take, for example, the rotational degrees of freedom of a diatomic molecule, discussed in Section 9.5. It is customary to treat the atoms in the molecule as mass points, so that the moment of inertia about the molecular axis is strictly zero, and there is no term in the energy expression corresponding to rotation about this axis. Hence, the diatomic molecule has only two rotational degrees of freedom, and Equation (G.13) gives a rotational contribution to the molar heat capacity equal to R_0, as observed. However, real atoms are not mass points, and, however small the moment of inertia about the molecular axis may be, it is not zero. Equipartition requires that rotation about this axis contribute to the heat capacity. In fact, of course, as pointed out in Section 9.5, the moment of inertia is so small that quantum theory predicts energy levels which are usually separated by many eV, enormously greater than $k_B T$ at any terrestrial temperature, so that the equipartition theorem does not hold for this degree of freedom.

Appendix H: Some Results in the Kinetic Theory of Gases

H.1 Introduction

While the equilibrium properties of perfect gases can be obtained as special cases of the general statistical approach of the previous chapters, the transport processes considered in Chapter 16 are more easily discussed in terms of kinetic theory. The basic assumptions of the theory are: (1) that the molecules in a gas are small hard particles which do not interact except through collisions, (2) the interval between collisions is long relative to the time taken by a collision, (3) their collisions with each other and with the walls of the container are elastic, and (4) that the molecules move at random, their motion being describable entirely in terms of averages over different powers of their velocities. The results are independent of the actual distribution of velocities, except insofar as this distribution determines the averages and are applicable to a degenerate gas such as the electron gas in a metal as well as to the ideal gas.

H.2 Particle Flux in a Perfect Gas

The *flux* of anything (e.g., of particles or of energy) is defined as the quantity of that thing crossing unit area per unit time. Suppose we have a gas of particles in a container with planar walls. The particle density is n. We define the *particle flux* J_n as the number of particles striking unit area of the wall per unit time. We take the z-axis perpendicular to the wall (see Figure H.1) and first ask how many particles strike the wall at angle θ to the normal. The particles are moving in random directions, so the number per unit volume whose velocities lie between v and $v + dv$, and whose velocity vectors lie within a solid angle $d\Omega$ is:

$$n(v, \theta)\, d\Omega\, dv = (4\pi)^{-1} nP(v)\, d\Omega\, dv, \tag{H.1}$$

where $P(v)\, dv$ is probability that the velocity lies between v and $v + dv$, and n is the particle density. The factor $(4\pi)^{-1}$ comes from the fact that the velocity can be in any direction, while Equation (H.1) only includes those velocity vectors within the solid angle $d\Omega$.

From Figure H.1, it can be seen that $d\Omega = 2\pi \sin\theta\, d\theta$. In unit time, the particles within a distance $v \cos\theta$ of the wall reach it, so that the number of particles with velocity between v and $v + dv$ that strike the wall per unit time and area is:

$$n(v)\, dv = \frac{1}{4\pi} nP(v)v\, dv \int_0^{\pi/2} 2\pi \sin\theta \cos\theta\, d\theta \tag{H.2}$$

$$= \frac{1}{4} nP(v)v\, dv. \tag{H.3}$$

353

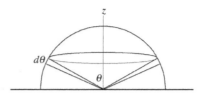

FIGURE H.1   Geometry for calculation of particle flux.

Integrating Equation (H.3) over v gives the total number of particles striking unit area of the wall per unit time; that is, the particle flux:

$$J_n = \frac{1}{4}n \int_0^\infty P(v)v\, dv = \frac{1}{4}n\langle v \rangle, \qquad (H.4)$$

where $\langle v \rangle$ is the mean velocity. Note that the probability distribution $P(v)$ does not enter the final result, so that Equation (H.4) is valid for any perfect gas, not necessarily ideal (e.g., the photon gas, for which $\langle v \rangle = c$, the velocity of light).

If each particle carries an energy E, which is in general a function of v, the energy flux is:

$$J_u = \frac{1}{4}n \int_0^\infty vEP(v)\, dv = \frac{1}{4}n\langle vE \rangle. \qquad (H.5)$$

H.3 Pressure Exerted by a Gas

A particle with mass m and velocity v, whether relativistic or not, has momentum perpendicular to the wall $mv\cos\theta$. If we assume for simplicity that when a particle strikes a wall, it is specularly reflected (that is, it rebounds with momentum perpendicular to the wall $-mv\cos\theta$), the total momentum transferred to the wall by each particle striking it is $2mv\cos\theta$. This momentum change is directed perpendicular to the wall.

The number of particles for given v and θ is given by Equation (H.1), so the average momentum transferred to unit area of the wall per unit time, which, by Newton's Second Law, is equal to the pressure p, is:

$$p = \frac{1}{4\pi}2nm \int_0^\infty v^2 P(v)\, dv \int_0^{\pi/2} 2\pi \sin\theta \cos^2\theta\, d\theta$$

$$= \frac{1}{3}nm\langle v^2 \rangle. \qquad (H.6)$$

To go further, we need the relation between velocity and energy. The kinetic energy of a non-relativistic free particle is $K_e = \frac{1}{2}mv^2$, so that:

$$p = \frac{2}{3}n\langle K_e \rangle = \frac{2}{3}u, \qquad (H.7)$$

where u is the kinetic energy density. If we give K_e its equipartition value of $\frac{3}{2}k_B T$ (see Appendix G), Equation (H.7) becomes the equation of state for an ideal gas [Equation (9.11)].[1]

For photons, $\langle v^2 \rangle = c^2$ and $E = mc^2$, so that:

$$p = \frac{1}{3}n\langle E \rangle = \frac{1}{3}u. \tag{H.8}$$

Equation (H.8) can also be derived by using classical electromagnetic theory to calculate the pressure exerted on a perfectly reflecting surface by electromagnetic waves falling on it from random directions.[2]

---

[1] This kinetic derivation of the ideal gas law was obtained in 1845 by John James Waterston (1811–1883), a Scottish civil engineer who instructed British naval cadets in India. His paper "On the physics of media that are composed of free and perfectly elastic molecules in a state of motion," was rejected by the Royal Society of London, one of the referees describing it as "nothing but nonsense." The manuscript was never returned to its author, and many years later it was discovered in the Royal Society archives. It was eventually published, after Waterston's death, in the *Phil. Trans. Roy. Soc.* **193A**, 5–79 (1893). The distinguished historian of science Stephen Brush says that, "One might claim that the history of statistical mechanics really began when Waterston completed (this) paper" [S. G. Brush, *Statistical Physics and the Atomic Theory of Matter* (Princeton University Press, Princeton, NJ, 1983), p. 265].

[2] See, for example, B. I. Bleaney and B. Bleaney, *Electricity and Magnetism* (Oxford University Press, Oxford, UK, 1976), p. 248.

Appendix I: The Correspondence Principle

I.1 Introduction

One reason there is likely no one "go to" textbook on thermodynamics and statistical mechanics is there is not an exclusive perspective for looking at systems thermodynamically. Much of what we know historically was begun from a macroscopic point of view and reconciled by quantum mechanics and statistical mechanics with the microscopic point of view. Yet, regardless of one's point of view, the microscopic and macroscopic manifestations of thermodynamics must hold simultaneously. Here, we wish to connect the two manifestations as an example of the *correspondence principle* in order to demonstrate something profound upon which one should return to think about frequently, regardless of whether the particular problem or issue you are dealing with is on the macroscopic or microscopic scale.

On quantum mechanical scales, we can only speak of the energy of the system in terms of its expectation, or average, value:

$$\bar{U} = \sum_i p_i E_i. \tag{I.1}$$

Here, we use a sum over discrete states for convenience. If we were to think of changes in this expectation value $\bar{U}$, then we could write:

$$d\bar{U} = \sum_i (dp_i E_i + p_i dE_i) = \sum_i dp_i E_i + \sum_i p_i dE_i. \tag{I.2}$$

The implication is that any changes in the expectation value of the energy is due to a change in the probability of the states being occupied or a change in the energy states themselves while the occupancy remains fixed.

From Appendix B, let us recall that the energy states for a single particle in a one-dimensional box are given by:

$$E_n = \frac{\pi^2 \hbar^2}{2mL^2} n^2, \tag{I.3}$$

(or for a relativistic particle):

$$E_n = \frac{hc}{2L} n. \tag{I.4}$$

In either case, the n energy states depend only upon L, the volume of the box in one dimension. From this observation, we can proceed in one of two ways. Either $dE = (\partial E/\partial V)dV$ or more directly, fixing the volume of the box fixes the energy levels so that $dE = 0$ if $dV = 0$.

Now we return a version of the first law of thermodynamics:

$$dU = \dj Q + \dj W = \dj Q - pdV, \tag{I.5}$$

where we substitute in $-pdV$ for the mechanical work done near equilibrium for a gas. Again, if we hold the volume of the gas constant, so that $dV = 0$, only one term for dU remains.

Microscopically, $d\bar{U} = \sum_i dp_i E_i$, and macroscopically, $dU = \dj Q$. One could then also reasonably conclude that the macroscopic term $-pdV$ plays the same role that $\sum_i p_i dE_i$ contributes to the change in the energy at microscopic scales. So, let us imagine two instances that we can simultaneously reflect upon at both the microscopic and macroscopic scales.

If we imagine small changes near equilibrium to a gas system, then the addition of heat (at fixed volume) to our system macroscopically corresponds to changing the occupancy numbers of the microscopic states of the system. Alternately, if we imagine adiabatically making small changes to the macroscopic volume of our gas system near equilibrium, this corresponds to the occupancy numbers of the states remaining fixed and the energy levels themselves undergoing small shifts. Doing work to shrink the volume of the gas is then the energy added to the system to raise the energy levels to their new values as the volume (of the microscopic box) is decreased.

Finally, if one examines $dE = (\partial E/\partial V)dV$, then for the non-relativistic particle in the box where $L = V^{1/3}$ in the isotropic 3D case:

$$dE_n = (\partial E/\partial L)\,(\partial L/\partial V)\,dV$$

$$= (-2E_n/L)\left(1/3V^{-2/3}\right)dV$$

$$= (-2/3E_n)\,(dV/V).$$

From this $-pdV = \sum_i p_i dE_i = -2/3\bar{E}/VdV$ and by this $PV = 2/3\bar{U}$. As $PV = Nk_BT$, then $\bar{U} = 3/2Nk_BT$. The three degrees of freedom each contributing $1/2Nk_BT$ to the gas being the same equipartition result discussed in Appendix G.

Appendix J: The $\mathcal{H}$ Function

J.1 Introduction

After more than a century of success in equilibrium thermodynamics and statistical mechanics, it is quite natural to ask if a more fundamental relationship than Boltzmann's assertion has been found on the microscopic lengthscales to explain the relationship between entropy and the microstates of a thermodynamic system. In more than one textbook on thermodynamics, it is often the practice to assume the form of the ideal gas law and eventually work to the point in quantum mechanics where one can derive the ideal gas law from the single particle in the 3D box. So the question is, having dealt with so much success from Boltzmann's assertion that $S = k_B \ln \Omega$, have we been able to work to a point where we can actually derive the relationship from something more fundamental? Is there a more fundamental definition of entropy from the quantum mechanical scale?

What follows is an overview for a much more involved set of calculations that are usually handled in a graduate course on statistical mechanics.[1] For the purposes of this text, we wish to present a more qualitative introductory treatment that demonstrates that Boltzmann's assertion is aptly named an *assertion* and not something stronger deserving the moniker of "law" or "axiom." Regardless of the name, it should not be lost on the reader that more than 100 years of equilibrium thermostatistics rests on Boltzmann's groundbreaking work.

In a gaseous system, we consider the non-ideal gas comprised of two types of molecules in the gas. One type of molecules are described by f with velocities v and the other are described by F with velocities V. We define two functions, H_B and H such that:

$$H_B \equiv \int f \, \ln(f) \, d^3v + \int F \, \ln F \, d^3V, \tag{J.1}$$

and:

$$H \equiv \int f \, \ln(f) \, d^3v + \int F \, \ln F \, d^3V = H_B - n_m - n_M, \tag{J.2}$$

where:

$$\int f \, d^3v = n_m, \tag{J.3}$$

---

[1] See the treatment in a variety of references ranging from older [W. Pauli, *Pauli Lectures on Physics Volume 4: Statistical Mechanics* (Dover, New York, 1973)] to newer [J. R. Dorfman, *An Introduction to Chaos in Nonequilibrium Statistical Mechanics* (Cambridge University Press, Cambridge, UK, 1999)] references.

and:

$$\int F \, d^3 V = n_M, \tag{J.4}$$

so that n_m and n_M are the number of particles in each type in the gas.[2] We then proceed by noting that:

$$\frac{dH_B}{dt} = \frac{dH}{dt}, \tag{J.5}$$

and:

$$\frac{d(f \ln(f))}{dt} = \ln(f)\frac{df}{dt} + \frac{f}{f}\frac{df}{dt}, \tag{J.6}$$

by the chain rule (and with an identical relationship for the time rate of change of $F \ln(F)$.)

Hence,

$$\frac{dH}{dt} = \int \ln(f)\frac{df}{dt} \, d^3 v + \int \ln(F)\frac{dF}{dt} \, d^3 V. \tag{J.7}$$

In order to evaluate the integrals, several arguments based upon symmetry and the nature of the collisions are then made. One of the most important, and still debated, assumptions by Boltzmann, is referred to as the "Stosszahlansant," which assumes that the density of particles (and by that, the density of collisions) in the volume of question within the gas is the same as that of the remaining gas. The assumption is further strengthened by assuming the gas is dilute and therefore the collision "cylinders" of the particles in question do not overlap within the gas. By this, the collisions between more than two particles are not included and the idea of density fluctuations (which would be present in the real gas) are suppressed.

Our point here, however, is only to set up the treatment so as to observe that equilibrium within the gas necessitates the condition that:

$$\frac{dH}{dt} = 0, \tag{J.8}$$

thus implying that both the time rate of change of both df/dt and dF/dt must likewise be zero. This can then be used to examine the details between the three types of collisions within the gas (neglecting the collisions with the containing walls) that determine the energy and momentum balance for collisions between like particles within f and F, that is collisions between two particles within either f or F, and the non-homogeneous collisions between a particle in f and a particle in F.

For our purposes, we will instead examine the result for then integrating spatially over the entire gas, which is defined as:

$$\mathcal{H} \equiv \int d^3 r \int d^3 v f \ln f - f = \mathcal{H}_B - N, \tag{J.9}$$

and it can be shown that:

$$S = -k_B \mathcal{H} + constant. \tag{J.10}$$

Here, we simply use the result to show that the $\mathcal{H}$ function that would seem to buttress a more fundamental foundation of entropy is something we have already seen in Chapter 6 in our treatment of

---

[2] The odd formulation here in terms of the differential elements proceeds from Pauli's treatment and can be thought of as integrating over the velocity-space of the gas.

Gibbs-Helmholtz probabilities. That is, $\mathcal{H}$ is a quantity that is minimized as entropy is maximized in the system. Recalling from Chapter 6 that:

$$S = -k_B \sum_i P_i \ln P_i, \tag{6.31}$$

we see that a version of the $\mathcal{H}$ function is simply $\sum_i P_i \ln P_i$, where the constant is zero. (The reader is reminded that such overall constants do not contribute thermodynamically.) In Chapter 6, there was a discussion about the relationship between S and $\mathcal{M}$ from informational theory. Here, $\mathcal{H}$ differs by a minus sign so that it is a minimized quantity when the entropy is maximized. A discussion of the further relationship between entropy and the Lyapunov exponent of chaotic maps is available elsewhere.[3]

---

[3] See, for example, G. L. Baker and J. P. Gollub *Chaotic Dynamics: An Introduction* (Cambridge University Press, Cambridge, UK, 1990).

Afterword

This is the original Preface from the first edition of M.D. Sturge's text presented here in its entirety[1]:

This book began as a set of notes for a course for physics majors entitled *Statistical Physics* which I have taught for a number of years at Dartmouth. The course is designed to provide an introduction to the principles and applications of statistical mechanics and thermodynamics in one term (twenty-eight 65 minute lectures). The "Dartmouth plan," under which every student is off campus for at least one quarter in the sophomore or junior year, limits the prerequisites to courses that can be taken in the freshman year. The only prerequisites to the course on which this book is based are: an introductory course in mechanics, in electricity and magnetism, in modern physics, and in multi-variable calculus. In fact, the only essential prerequisite to most of the material in this book is an understanding of what is meant by a "quantum state," of partial derivatives, and of series expansion in a small parameter. Appendices give a brief review of what the student needs to know on these topics.

The mathematics have been kept as simple as possible, and I have made no attempt to achieve a mathematical rigor which, as Max Born has said, is anyway illusory in physics. Nor have I delved into the philosophical implications or logical underpinnings of the basic assumptions of statistical mechanics. As in other branches of physics, for example quantum theory, a student should first achieve facility in handling the theory, and use it to derive concrete results, before inquiring too deeply into its foundations.

Students are often confused by the terminology of thermodynamics and statistical mechanics; all the more since in the literature the same term is used with different meanings, and different terms are used for the same quantity, sometimes even in the same book. I have tried to use a consistent terminology and notation and to draw careful distinctions where necessary. For example, I distinguish between a "perfect" (often called "non-interacting") gas and an "ideal" gas, which is a perfect gas in the dilute (that is, classical) limit. In order to guide the student through the terminological and notational maze, the book ends with a comprehensive glossary and list of symbols. I have used SI units consistently throughout, even when such consistency is not customary; for example, in this book N_A stands for "Avogadro's kmolar number," that is, the number of molecules in one *kilo*mole (kmole).

It has become customary in recent years to end each chapter of a textbook with a summary of the main points made. I believe that this practice encourages laziness on the part of the student, and instead I have boxed the most important equations in the text. The chapters end with an "Envoi," which is a signpost rather than a summary, showing the reader the point that the argument has reached and indicating where it is going.

---

[1] Rather than change the chapter references in his original preface, note that his original Chapter 15 is now Chapter 3, meaning that his Chapters 3 through 14 are now Chapters 4 through 15, and Chapters 17 and 18 are new to the text.

In the first eight chapters, I introduce the general concepts and methods of thermodynamics and statistical mechanics, while the remaining chapters apply these ideas to various classes of systems. However, this division is not rigid, and simple applications which illustrate the principles are freely introduced wherever appropriate.

Chapter 1 is a brief introduction in which the objectives of thermodynamics and statistical mechanics are outlined and compared. Chapter 2 reviews the basic thermodynamic concepts of thermal equilibrium, empirical temperature, heat and work, internal energy and enthalpy, and introduces the zero'th and first laws of thermodynamics. Chapter 3 introduces the probabilistic concepts of macrostate, microstate, and multiplicity (the number of microstates in a given macrostate), and illustrates them by considering the simplest possible quantum system, the spin 1/2 magnet. Chapter 4 introduces Boltzmann's statistical definition of entropy and shows how the entropy can in general be found without actually calculating the multiplicity. All the thermodynamic quantities of a system with a fixed number of particles, such as absolute temperature and free energy, are then defined in terms of energy and entropy. The second and third laws of thermodynamics are introduced as statements about entropy.

Chapter 5 introduces the important concept of "heat bath" and derives the canonical or "Boltzmann" distribution, which gives the probability of occupation of any particular state of a system with a fixed number of particles in thermal equilibrium with a heat bath. The partition function is introduced, and we derive the thermal properties of some simple systems with discrete energy levels, such as the harmonic oscillator and the paramagnet. Chapter 6 treats systems with continuous energy levels, introduces the concept of density of states, and discusses the conditions under which the classical equipartition theorem holds. The derivation of this theorem in its most general form, which requires a familiarity with the concept of phase space, is given in Appendix G.

The next two chapters extend the analysis to systems in which the number of particles can change. Chapter 7 introduces the concepts of diffusive equilibrium and chemical potential. The partition function is generalized to the grand partition function, from which we derive the thermal properties of some simple systems in which the number of particles can vary. In Chapter 8, the grand partition function is applied to perfect gases, and the Fermi-Dirac, Bose-Einstein, and Maxwell-Boltzmann distributions are derived.

Chapter 9 uses the Maxwell-Boltzmann distribution to derive the thermal properties of ideal gases, both monatomic and polyatomic. We discuss entropy of mixing and show that the so-called "Gibbs paradox" is a consequence of the internal inconsistency of classical statistical mechanics.

The next two chapters deal with perfect boson gases. Chapter 10 introduces the thermodynamics of radiation and derives the Stefan-Boltzmann law. It then treats the photon gas as a Bose gas in which the number of particles is not conserved and derives Planck's radiation law. Chapter 11 treats the perfect Bose gas in which particle number is conserved and briefly discusses Bose-Einstein condensation, the only phase transition which a perfect gas can undergo.

The next two chapters deal with perfect Fermi gases. Chapter 12 applies the theory to electrons in metals and white dwarf stars. Chapter 13 extends the treatment to semiconductors, discusses the conditions under which the electrons and holes in a semiconductor behave as ideal gases, and gives a brief account of the physics of the p–n junction.

Chapter 14 goes beyond the perfect gas approximation to include interparticle interactions, which lead to first and second order phase transitions. Mean field theory and Landau's generalization of it are developed and applied to ferromagnets, order-disorder transitions, and van der Waals' model of an imperfect gas. The limitations of the theory and the importance of critical fluctuations are briefly discussed.

Chapter 15 applies thermodynamic principles to heat engines. The Carnot cycle and some of the principal cycles used in practice, such as the Rankine, Otto, Diesel, and vapor-compression cycles, are described. Chapter 16 uses kinetic theory to derive simple approximations to the transport properties (electrical and thermal conductivity, diffusivity, and viscosity) of quasi-perfect gases.

The appendices include background and ancillary material, most of which is essential to understanding the main text. Since Appendices A and B are relevant to the entire book they should be read first and referred back to when necessary. Appendix C gives the formal proof that the zero'th law implies the existence of temperature. Appendices D through F review the mathematical background to Chapter 3 and subsequent chapters. Appendix G uses the concept of phase space (not needed for the rest of the book) to justify and extend the equipartition theorem, which is only superficially treated in Chapter 6, but is used extensively thereafter. Appendix H gives some results of kinetic theory, essential to Chapter 16.

In order to provide a choice of illustrative material, this book contains more on the applications of the theory than can be covered in a one-semester course, and it could in fact be used, with some supplementary reading in the references provided, in a two-semester course. Chapter 11 and Chapters 13 through 16 are self-contained and any of these can be omitted, according to the taste of the user, without affecting the understanding of later material.[2] Sections of the other chapters which can also be omitted at first reading are marked with a border. The problems at the end of each chapter are designed to extend and consolidate the student's understanding of the subject and to apply it to real problems. As a result, many of them are quite difficult, but in my experience students get much more satisfaction, and learn more, from (for example) using a simple model to derive the Chandrasekhar limit for the stability of a white dwarf, than from solving an artificial problem designed only to illustrate the course material.

As in any textbook, references to original work are partial and incomplete. However, I do not share the common aversion to footnotes.[3] References in footnotes are intended to be what Peter Scott calls "rabbit holes," through which readers can find more detail about, or alternative approaches to, topics which interest them. Names in the text without bibliographic data, for example, "Pippard, Chapter 7," refer to books listed in the bibliography.

As will be obvious to any reader who knows Kittel and Kroemer's *Thermal Physics*, I owe that book a great debt. I have followed its general approach to statistical mechanics, but have tried to clarify and extend the treatment in the light of my students' difficulties with that book. The thermodynamics sections of this book owe much to Pippard's *Classical Thermodynamics*, which is based on a course which I was privileged to take as an undergraduate at Cambridge. The whole book has benefited greatly from the incisive criticism of many of the Dartmouth students who have taken this course over the years, in particular Sing-Foong Cheah, Jeffrey Hannisian, Kara Relyea, Thomas Vikoren, Theodore Yuo, Niteen Bhatia, Imran Ansari, Thomas Levy, Ajibayo Ogunshola, Jacob Waldbauer, and especially Karen Glocer. Joan Thompson helped with the graphics, Jaclyn Lewin and Caroline Kauffmann with the index, while Kangbin Chuah patiently converted the entire manuscript to TeX format and made many valuable suggestions. I owe a special debt to the detailed criticisms of Harvey Gould (Clark University), and of Carmen Gagne, who was teaching assistant for the course in 1997. I have benefited from the advice and encouragement of Robert Knox (University of Rochester), Peter Scott (UCSD) and Arnold Dahm (Case Western Reserve University), and of many of my faculty colleagues at Dartmouth, in particular Jane Lipson, Walter Stockmeyer, Dean Wilcox, David Lemal, and Robert Cantor of the Chemistry Department, Christopher Levey and Horst Richter of the Thayer School of Engineering, and Joseph Harris, W.E. ("Jay") Lawrence, David Montgomery, Geoffrey Nunes, John Thorstensen, Gary Wegner, and the late John Walsh of the Physics Department. I owe much to the late John Kidder, who was department chair when I came to Dartmouth in 1986, for asking me to teach this course, thus making me take a new and more careful look at the fundamentals of statistical physics. Finally, I must thank my wife, Mary, for her unfailing support and, in particular, for taking time off from her own writing to help me with mine.

---

[2] Chapter 15 (heat engines) uses some of the results obtained in the first two sections of Chapter 14.
[3] "Reading a footnote is like going downstairs to answer the doorbell while making love." Noel Coward.

Glossary

Italicized words or phrases are common usages that are not used in this book.

Word or phrase	Symbol	Meaning
absolute activity	α	$e^{\beta\zeta}$, where $\beta = 1/k_B T$ and ζ is the chemical potential.
absorptivity	α_r	The fraction of incident radiation that is absorbed by a surface.
acceptor		An impurity in a semiconductor that takes an electron from the valence band, creating a hole.
activity	α	In this book, short for absolute activity (q.v.).[1]
adiabatic process		A process in which the occupancies of quantum states do not change.
adiabatic		Adiathermal.
adiathermal		Permitting no heat flow, thermally isolating or isolated.
analytic function		A function all of whose derivatives exist.
angular frequency	ω	$2\pi\nu$. Rate of oscillation in radians per second.
anisotropic		With properties that depend on direction.
atomic mass unit	amu	Unit of mass in which the proton mass $M_H = 1$.
Avogadro's kmolar number	N_A	Number of particles in one kmole of a substance.[2]
Avogadro's number		Number of particles in one mole of a substance.
band		A continuum of allowed electron states in a solid.
band edge		The lowest point in the conduction band or the highest in the valence band.
bandgap		Energy gap (q.v.).

---

[1] "q.v." stands for *quod vide*, meaning "which see".
[2] We use kmoles, not moles; see Footnote 1, below the list of physical constants.

barrier height		Potential barrier opposing diffusion of carriers across p–n junction.
binomial expansion		The series $(1 + x)^a = 1 + ax + a(a - 1)x^2/2! + \cdots$, where x is assumed to be small.
black body		A body with unit absorptivity and emissivity at all relevant frequencies.
black body radiation		Electromagnetic radiation emitted by a hot body.
Boltzmann's constant	k_B	Constant converting the "natural" unit of entropy, $\ln \Omega$, to the practical unit (J/K).
Boltzmann distribution		Maxwell-Boltzmann distribution (q.v.).
Boltzmann factor		$e^{-\beta E}$, where E is the energy of a state and $\beta = 1/k_B T$.
Boltzmann law		Canonical distribution (q.v.).
Bose-Einstein condensation		Condensation of a perfect Bose gas into the lowest single particle state.
Bose-Einstein distribution		Distribution function of a Bose gas.
Bose gas		A gas of bosons.
boson		A particle which does not obey Pauli's exclusion principle, so that any number of particles can occupy a single particle state. Bosons have zero or integer spin.
Boyle's law		$p \propto n$ at constant T, for an ideal gas.
brightness		Radiation emitted from a surface per unit time, area, and solid angle.
brightness theorem		The brightness of an image in an optical system cannot exceed that of the object.
built-in voltage		Barrier height (q.v.).
bulk expansion coefficient	α_b	$V^{-1}(\partial V/\partial T)_p = -n^{-1}(\partial n/\partial T)_p$.
bulk modulus, isentropic	B_S	$-V(\partial p/\partial V)_S = n(\partial p/\partial n)_S$.
bulk modulus, isothermal	B_T	$-V(\partial p/\partial V)_T = n(\partial p/\partial n)_T$.
canonical coordinates	q_i	A system of coordinates in which the volume of a cell in phase space is the product of the dq_i.
canonical distribution		The probability that a state of energy E be occupied is $Z^{-1}e^{-\beta E}$.

Carnot cycle		Cycle undergone by the working substance in a Carnot engine.
Carnot efficiency	η_c	Efficiency of a Carnot engine.
Carnot engine		An ideal reversible heat engine working between two fixed temperatures.
carrier		Electron or hole.
cavity		An enclosure in an opaque body, filled with radiation.
chemical potential	ζ	$(\partial F/\partial N)_{V,T}$. What two systems at the same temperature have in common when in diffusive equilibrium.
coefficient of performance	h	For a heat pump: ratio of heat delivered to work done. For a refrigerator: ratio of heat extracted to work done.
coexistence curve or line		Locus of points in the (p, T) plane where two phases can coexist.
coexistence line		See coexistence curve.
coexistence curve		Limits of coexistence region.
coexistence region		Range of density, pressure, and temperature where two phases can coexist.
collision cross-section	σ_c	"Target" area of a particle: if the center of another particle passes within this area, a collision occurs.
compressibility		Inverse of the bulk modulus (q.v.).
compression ratio		Ratio between the maximum and minimum volume of the working substance in a heat engine.
conduction band		Lowest unoccupied band in a semiconductor or insulator.
conduction band edge (CBM)	E_c	Energy of the lowest state in the conduction band.
conductivity	σ_e	Electrical conductivity. Ratio of current density to electric field.
conservative force		A forice that can be derived from a potential.
critical exponent		Parameter specifying the rate of change of a physical quantity near a critical point.
critical point		Density, pressure, and temperature at which a second order transition occurs.
critical temperature	T_c	Temperature at which a second order transition occurs.
cross-section		Collision cross-section (q.v.).
Debye frequency	ν_D	Frequency of the highest vibrational mode in the Debye model.

Debye model		A model of a solid in which vibrational normal modes are approximated by sound waves of constant velocity.
Debye temperature	θ_D	$h\nu_D/k_B$: Debye frequency expressed as a temperature.
degeneracy	g	The number of quantum states in an energy level (q.v.). See also multiplicity.
degenerate gas		A gas whose particle density is so high that the difference between bosons and fermions is significant.
degree of freedom		Any canonical coordinate that appears in the expression for the energy of a system.
density		Number or quantity per unit volume, short for particle density n.
density of states	$g(E)$	The density with respect to energy of single particle states, per unit volume.[3]
	$g(\nu)$	The density, with respect to frequency, of normal modes of a set of coupled oscillators, per unit volume.[3]
diamagnet		A substance with negative magnetic susceptibility.
diathermal		Permitting unlimited heat flow, thermally conducting.
diatomic gas		A gas of molecules which consist of two atoms (e.g., N_2).
differential	dx	A change in x so small that quantities of the order of $(dx)^2$ can be neglected.
diffusion		Flow of particles under a density gradient.
diffusion constant		Diffusivity (*q.v.*).
diffusive equilibrium		The situation where the average number of particles in each of two or more systems that can exchange particles remains constant.
diffusivity	D	Ratio of particle flux to density gradient.
dimensions		The units in which a quantity is expressed. Besides the fundamental units of mass, length, and time, it is often convenient to include other units, such as charge and temperature.
dispersion relation		Dependence of the frequency of a wave on wave vector. Dependence of the energy of a quantum state on wave vector or momentum.
distinguishable microsystem		A microsystem that can in principle be distinguished from other microsystems (e.g., by its location).
distribution function	$f(E)$	Mean occupancy of single particle states.

---

[3] Per unit area in a two-dimensional system, per unit length in a one-dimensional system.

donor		An impurity in a semiconductor that gives up an electron to the conduction band.
doping		The deliberate addition of impurities to a semiconductor to control its electrical properties.
effective density of states		Quantum concentration (q.v.).
effective mass	m^*	$\hbar^2\|d^2E/dk^2\|^{-1}$ at a band edge in a semiconductor.
efficiency	η	Ratio of work out to heat in (in a heat engine).
eigenmode		Normal mode (q.v.).
eigenstate		Quantum state (q.v.).
Einstein model		A model of a solid in which all the atoms are assumed to vibrate at the same frequency.
Einstein temperature	θ_e	$h\nu/k_B$, where ν is the vibrational frequency in the Einstein model.
electric dipole		Pair of equal and opposite charges, separated by a small distance.
electron volt	eV	Energy gained by an electron falling through a potential difference of 1 V.
emissivity	e_r	Ratio of the radiation emitted by a surface, per unit area and time at a given frequency, to that emitted by a black body at the same temperature.
energy gap	E_g	Energy difference between the highest occupied and lowest unoccupied electron states in a semiconductor or insulator: $E_g = E_c - E_v$.
energy level		Level (q.v.).
enthalpy	H	$U + pV$.
entropy	S	$k_B \ln \Omega$.
equation of state		Relation between p, V (or n), and T for a substance (e.g., the ideal gas law).
equilibrium (thermal)		State of a system in which all spontaneous directed change has ceased.
equipartition		A theorem of classical statistical mechanics, which holds only if the energy levels are continuous. In its restricted form, it states that every degree of freedom that enters quadratically into the expression for the energy of a system adds $k_B T/2$ to the average thermal energy.
exact differential		Differential which can be integrated to give a state function.

exchange interaction		Quantum mechanical effect that produces an effective interaction between spins.
exciton		Electron bound to a hole in a solid.
extensive quantity		Quantity whose value is proportional to the size of the system; for example, S, U, V.
extrinsic conduction		Conduction due to the presence of impurities in a semiconductor.
Fermi-Dirac distribution		Distribution function of a Fermi gas.
Fermi energy	E_F	Fermi level at $T = 0$; that is, the energy of the highest occupied state in a perfect Fermi gas at $T = 0$.
Fermi gas		Gas of fermions.
Fermi level		Chemical potential of a Fermi gas.
fermion		Particle obeying Pauli's exclusion principle. Fermions have half-integer spin.
ferromagnet		Substance with a magnetic moment even in the absence of an applied magnetic field.
first law of thermodynamics		Energy is conserved if heat is taken into account.
first order transition		Transition between phases which is accompanied by latent heat.
fluctuation		Spontaneous deviation of an observable property of a macrosystem from its equilibrium value.
fluid		Liquid or gas.
flux	J	Rate of transfer of some quantity (particles, energy, charge, momentum) across unit area.
free energy	F	$U - TS$.
free enthalpy	G	$H - TS$.
frequency	ν	Rate of oscillation in cycles per second.
fugacity		Activity relative to that at standard pressure and temperature, multiplied by the standard pressure. Common in the chemical literature, but not used in this book.
Gibbs factor		$e^{\beta(n\zeta - E)}$, where n is the occupancy.
Gibbs free energy, Gibbs function		Free enthalpy (q.v.).

Gibbs sum		Grand partition function (q.v.).
grand partition function	Ξ	$\sum e^{\beta(n\zeta - E)}$, where the sum is over all states and all possible occupancies.
harmonic oscillator		Oscillator in which the restoring force is proportional to displacement.
heat	Q	Energy that flows between two objects in thermal contact, independent of any directed motion of material or charge.
heat bath		System of effectively infinite heat capacity in thermal contact with the system(s) under consideration.
heat capacity	C, C_V, C_p	Quantity of energy needed to raise the temperature of a body by one unit.
heat engine		Device for converting heat into work.
heat pump		Device to deliver heat at a high temperature, extracting it from a source at a lower temperature.
Helmholtz free energy		Free energy (q.v.).
hole		Missing electron in an otherwise filled band of states. It behaves like a positively charged particle.
hysteresis		Dependence of some property (e.g., magnetization) on previous history.
hysteretic		Showing hysteresis.
ideal gas		Perfect gas (q.v.) whose density is so low that the difference between bosons and fermions is of no importance.
ideal gas law		$p = nk_B T$.
imperfect gas		Gas in which the interactions between molecules play a significant role in the thermodynamic properties.
indistinguishable particles		Particles which cannot be distinguished from each other, even in principle. Their many-particle wave function is either symmetric (bosons) or antisymmetric (fermions) in exchange of the particles.
indicator diagram		$p(V)$ graph for a heat engine.
inexact differential		A differential expression which cannot be integrated to give a state function.
insulated, insulating		Not thermally conducting, thermally isolated.
insulator		A solid in which the electrons are not free to move, so that it does not conduct electricity.

intensive quantity		A quantity independent of the size of the system considered; for example, T, p.
internal energy	U	The energy stored in a macrosystem; that is, the sum of the kinetic and potential energies of the microsystems comprising it, excluding any energy that is macroscopically observable by mechanical or electrical means.
intrinsic conductivity		Conductivity of a pure semiconductor due to thermal excitation of carriers.
inversion temperature		Temperature where the Joule-Thomson coefficient changes sign.
irreversible		Describes a process which can only go in one direction.
isentrope		Line of constant entropy.
isentropic		At constant entropy.
isobar		Line of constant pressure.
isotherm		Line of constant temperature.
isothermal		At constant temperature.
isotropic		With properties that are independent of direction.
Joule-Thomson coefficient		$(\partial T/\partial p)_H$.
kilocalorie, kcal		4.18×10^3 J. The energy required to raise 1 kg of liquid water by 1°C. The unit of energy commonly used in chemistry and also in dietary recommendations (where is it miscalled the calorie).
kmolar		Pertaining to 1 kmole of a substance.
kmole		M_w kg of a substance, where M_w is the molecular weight (a kmole contains N_A molecules).
latent heat	L	In this book, the change in enthalpy when 1 kmole of material transforms from one phase to another at constant pressure.
lattice		The atoms (or the arrangement of atoms) in a solid.
level		Set of one or more quantum states with the same energy.
macrostate		Set of microstates which are indistinguishable from each other by any macroscopic measurement.
macrosystem		System with so many degrees of freedom that its individual quantum states cannot be calculated, and must be treated statistically.

Maxwell- Boltzmann distribution		Distribution function of an ideal gas.
Maxwell relation		Relation between partial derivatives of state functions which follows from the existence of another state function.
mean		Average.
mean field theory		Model in which interactions between microsystems are approximated by an average interaction proportional to the order parameter; for example, the Weiss model (q.v.).
mean free path	l	Average distance traveled between collisions.
metal		Solid in which electrons are free to move, so that it conducts electricity.
metastability		The continued existence of a phase under conditions where another phase is more stable; for example, superheating of a liquid above its boiling point or supercooling below its freezing point.
microstate		Quantum state of a macrosystem.
microsystem		System such as a particle or oscillator that is sufficiently small for its quantum states to be calculable.
mobility		Ratio of drift velocity to driving force or field.
mode		Normal mode (q.v.).
molar		Pertaining to 1 mole of a substance.
mole		M_w gram of a substance, where M_w is the molecular weight, $10^{-3}N_A$ particles.
molecular weight	M_w	Mass of a molecule in atomic mass units.
monatomic gas		A gas of free atoms.
multiplicity	Ω	Number of microstates in a macrostate. Note that "multiplicity" and "degeneracy" are usually considered to be synonymous, but in this book we use multiplicity when referring to macrosystems and degeneracy when referring to microsystems.
normal mode		Stationary state of a vibrating system. Often called a mode or eigenmode.
n-type semiconductor		Doped with donors.
nucleation		The formation of a small volume of a stable phase in a matrix of a metastable phase.

occupancy	n	Number of particles in a particular single particle state.
orbital		Single particle quantum state, neglecting spin.
order parameter	ψ	A parameter specifying the ordering that occurs at a second order phase transition; for example, the magnetization of a ferromagnet.
Otto cycle		Ideal thermal cycle of the spark-ignition internal combustion engine.
paramagnet		Substance with positive magnetic susceptibility.
partition function	Z	$\sum e^{-\beta E}$, where the sum is over all states for a fixed number of particles.
perfect differential		Exact differential.
perfect gas		A gas in which all interparticle interactions can be neglected.
phase		A particular form of a substance, which under the right conditions can change into other forms of the same substance with different macroscopic properties.
phase space		A multidimensional Cartesian space whose axes are the canonical coordinates of a system.
photovoltaic effect		Generation of a voltage by light.
Planck's radiation law		The distribution of energy in black body radiation as a function of frequency and temperature: $u_s(\nu, T)$.
polyatomic gas		A gas made up of molecules that consist of two or more atoms.
Poisson's equation		$\nabla^2 V = -\rho_e/\epsilon_0$, where ρ_e is the charge density.
pressure	p	Force normal to a surface, per unit area of surface.
probability density	$P(x)$	$P(x)\,dx$ is the fraction of a population with property x between x and $x + dx$.
p-type semiconductor		Doped with acceptors.
quantum concentration	n_q	$(2S + 1)\left(\frac{mk_B T}{2\pi\hbar^2}\right)^{3/2}$: density of a gas at which quantum effects become important.
quantum number		One of the numbers specifying a quantum state.
quantum state		A solution of Schrödinger's equation. Often called an eigenstate.
quantum volume	V_q	$1/n_q$. The smallest volume in which a particle with energy equal to the thermal energy can be confined.

quasi-Fermi level	ζ_e, ζ_h	Chemical potential of electrons or holes considered separately, in a semiconductor out of thermal equilibrium.
quasi-ideal gas		A gas that is ideal in its thermodynamic properties, but in which collisions between particles occur so that its transport coefficients are finite.
Rankine cycle		Ideal thermal cycle of the steam engine.
Rayleigh-Jeans law		The classical expression for the distribution of energy in black body radiation $u_s(\nu, T)$, derived from the equipartition theorem.
reservoir		A heat bath that is in diffusive as well as thermal contact with the system(s) under consideration, so that it can exchange particles as well as energy.
reversible process		A process that can be carried out in either direction.
scattering time	τ_s	Average time between collisions.
second law of thermodynamics		The entropy of an isolated system cannot decrease.
second order transition		A transition between phases unaccompanied by latent heat.
semiconductor		An insulator with a small energy gap. It conducts electricity when impure or when its temperature is high.
shear stress		Force parallel to a surface, per unit area of surface.
SI units		Système Internationale. The system of units based on the meter, kg, second, and Coulomb, which is used consistently in this book.
singularity		A point in the domain of a function where it ceases to be analytic.
specific heat		Heat capacity, usually per mole.
spectral energy density	u_s	Energy per unit volume per unit frequency interval.
spin		The intrinsic angular momentum of an elementary particle. The term is also used for the associated magnetic moment.
standard deviation	$\langle \Delta x^2 \rangle^{1/2}$	The square root of the variance of a probability distribution.
standard temperature and pressure	STP	273.16 K, 1 bar.

state function		Single-valued function of the observable properties of a macrosystem.
stationary state		Quantum state of well-defined energy, a solution of Schrödinger's time-independent equation.
Stefan-Boltzmann law		Radiation from a hot surface varies as T^4. Often called Stefan's law.
sticking coefficient		Fraction of incident gas molecules that remain on a surface.
supercooling		See "metastability."
superfluidity		Transport of a fluid with zero viscosity and zero entropy transport.
superheating		In a steam engine: heating steam, at a given pressure above atmospheric, to a temperature significantly above its boiling point at that pressure. In a phase transition: see "metastability."
surface energy	σ	Energy per unit area of a free surface.
surface tension		Surface energy.
susceptibility	χ	Ratio of magnetization (or dielectric polarization) to magnetic (electric) field.
Taylor expansion		$\phi(x) = \phi(x_0) + (x - x_0)\phi'(x_0) + \cdots$, where $x - x_0$ is assumed to be small.
temperature (empirical)	θ	What two bodies in thermal equilibrium have in common.
temperature (absolute)	T	$(\partial U/\partial S)_V$.
thermal conductivity	K_{th}	Ratio of heat flux to temperature gradient.
thermal equilibrium		See equilibrium.
thermodynamic limit		Limit $k_B \to 0$ and $N \to \infty$, keeping Nk_B finite. In this limit, the atomic nature of matter becomes irrelevant and thermal fluctuations disappear.
third law of thermodynamics		It is impossible to reach the absolute zero of temperature.
total heat		Enthalpy (q.v.).
transition temperature	T_t	Temperature at which a first order transition occurs.
valence band		Highest occupied band in a semiconductor or insulator.
valence band edge (VBM)	E_v	Energy of the highest state in the valence band.

van der Waals model		A mean field model of an imperfect gas.
variance	$\langle \Delta x^2 \rangle$	Mean square deviation of the probability distribution of quantity x.
virial coefficient	B_r	Coefficient of n^r ($r \geq 2$) in the equation of state of an imperfect gas (there are other, logically equivalent, definitions in the literature).
viscosity	η	Ratio of shear stress to velocity gradient (in a fluid).
Weiss model		Mean field model of a ferromagnet.
Wien's law		An empirical approximation to Planck's law, valid at high frequencies.
work	W	Mechanical or electrical energy.
work function	ϕ_w	Energy needed to remove an electron from a solid.
working substance		Substance used, but not consumed in a heat engine; for example, the water/steam in a steam engine.
zero'th law of thermodynamics		Two bodies that are in thermal equilibrium with a third are in equilibrium with each other.

List of Symbols

See glossary for definitions.

Symbol	Meaning	Unit
A	Area	m^2
A	Ampere	
amu	Atomic mass unit $\approx M_H$	
a, b	Parameters in van der Waals' equation of state	
B	Magnetic flux density	T
B_r	The r'th virial coefficient	
B	Bulk modulus	$N\,m^{-2}$
B_S	Isentropic bulk modulus $-V(\partial p/\partial V)_S$	$N\,m^{-2}$
B_T	Isothermal bulk modulus $-V(\partial p/\partial V)_T$	$N\,m^{-2}$
bar	10^5 Pa: approximately 1 atmosphere	
C	Coulomb	
C	Heat capacity per unit volume or per kmole	$J\,K^{-1}\,m^{-3}$ $J\,K^{-1}\,kmole^{-1}$
C_p	Kmolar heat capacity at constant pressure	$J\,K^{-1}\,kmole^{-1}$
C_V	Kmolar heat capacity at constant volume	$J\,K^{-1}\,kmole^{-1}$
C_e	Electronic heat capacity per unit volume	$J\,K^{-1}\,m^{-3}$
C_m	Magnetic heat capacity per unit volume or per kmole	$J\,K^{-1}\,m^{-3}$ $J\,K^{-1}\,kmole^{-1}$
c	Velocity of light	m/s
c_s	Velocity of sound	m/s
$c(N, r)$	Binomial coefficient $N!/r!(N-r)!$	
$cosech(x)$	Hyperbolic cosecant $2/(e^x - e^{-x})$	
$\cosh(x)$	Hyperbolic cosine $(e^x + e^{-x})/2$	
$\coth(x)$	Hyperbolic cotangent $(e^x + e^{-x})/(e^x - e^{-x})$	
D	Diffusivity	$m^2\,s^{-1}$
d	Dimensionality	
$đ$	Inexact differential	
E	Electric field	V/m

E	Energy of a microsystem	J
E_i	Energy of the i'th quantum state of a microsystem	J
E_F	Fermi energy	J
E_g	Energy gap (bandgap)	J
E_v	Energy of valence band maximum	J
E_c	Energy of conduction band minimum	J
e_r	Emissivity	
eV	Electron volt	
F	Free energy $U - TS$	J
f	Free energy density F/V	$J\,m^{-3}$
f	Fractional concentration	
$f, f(E)$	Distribution function	
$f'(x)$	df/dx where $f(x)$ is any function	
$f''(x)$	d^2f/dx^2	
f	Force	N
f_e	Distribution function of electrons	
f_h	Distribution function of holes	
G	Free enthalpy $H - TS$	J
G	Gravitational constant	$N\,m^2\,kg^{-2}$
$G(E)$	Number of states per unit volume/area/length with energy less than E	m^{-d}
$G(\nu)$	Number of modes per unit volume/area/length with frequency less than ν	m^{-d}
$G(k)$	Number of modes/states per unit volume/area/length with wave vector less than k	m^{-d}
g	Acceleration due to gravity	$m\,s^{-2}$
g	Degeneracy of an energy level	
g_j	Degeneracy of the j'th energy level of a microsystem	
g_0	Degeneracy of the lowest energy level of a microsystem	
$g(E)$	Density of states with respect to energy	$J^{-1}\,m^{-d}$
$g(\nu)$	Density of normal modes with respect to frequency	$s\,m^{-d}$
$\mathcal{H}$	the (Boltzmann) H function	
H	Enthalpy	J
H	Magnetic field	Ampere-turn/m
h	Coefficient of performance of heat pump or refrigerator	
h	Planck's constant	J s
$\hbar$	$h/2\pi$, where h is Planck's constant	J s
h_c	Coefficient of performance of Carnot heat pump	
I	Electric current	A
I	Ionization potential	J

I_{ij}	Component of moment of inertia	kg m^2
idf	Internal degrees of freedom (of a molecule)	
$\text{int}(x)$	Integer part of x	
J	Joule	
J_u	Energy flux	W m^{-3}
J_n	Particle flux	$\text{m}^{-2}\,\text{s}^{-1}$
J_e	Charge flux (current density)	A m^{-2}
J_p	Momentum flux (shear stress in viscous flow)	N m^{-2}
j	Rotational quantum number	
j	Energy level label	
K	Kelvin (unit of absolute temperature)	
K	Torque	N m
K	Force constant of a spring	N m^{-1}
K_e	Kinetic energy	J
K_{th}	Thermal conductivity	$\text{W m}^{-1}\,\text{K}^{-1}$
k	Wave vector	m^{-1}
kcal	Energy needed to raise temperature of 1 kg of water from 4°C to 5°C (4.18 kJ)	
kmole	M_w kg of a substance (contains N_A molecules or atoms)	
k_B	Boltzmann's constant	J/K
L	Length	m
L	Latent heat (molar)	J/kmole
L_0	Lorentz number	$\text{V}^2\,\text{K}^{-2}$
$\mathcal{L}(x)$	Langevin function $\left[\coth(x) - x^{-1}\right]$	
l	Length	m
l	Mean free path	m
ln	Natural logarithm	
M	Magnetic moment of a macroscopic sample	J T^{-1}
M_H	Mass of hydrogen atom (1 amu)	kg
M_w	Molecular weight (mass of a molecule in amu)	
m	Mass of a particle	kg
m	Meter	
m	Magnetization (magnetic moment per unit volume)	$\text{J T}^{-1}\,\text{m}^{-3}$
m_0	Rest mass of free electron	kg
m^*	Effective mass (in a solid)	kg
m_e^*	Effective mass of electron	kg
m_h^*	Effective mass of hole	kg
N	Number of particles, or other independent microsystems, in a macrosystem	
N	Newton (SI unit of force)	
$N(k)$	Number of modes or states with wave vector less than k	

N_A	Avogadro's kmolar number (number of molecules in 1 kmole)[1]	kmole^{-1}
n	Occupancy	
n	Net spin (in Chapters 3 and 4 only)	
n	A number, with various meanings, defined in the section where used	
n_f	Number of degrees of freedom of a microsystem	
n_r	Refractive index	
n	Particle density N/V	m^{-3}
n_A	Density of acceptor impurities	m^{-3}
n_c	Critical particle density (in Chapter 14)	m^{-3}
n_c	Quantum concentration in conduction band (in Chapter 13)	m^{-3}
n_{cr}	Critical density for Bose-Einstein condensation	m^{-3}
n_D	Density of donor impurities	m^{-3}
n_e	Electron density	m^{-3}
n_h	Hole density	m^{-3}
n_i	Intrinsic electron or hole density	m^{-3}
n_i	Particle density of the i'th component of a system	m^{-3}
n_I	Density of impurities	m^{-3}
n_q	Quantum concentration	m^{-3}
n_v	Quantum concentration in valence band	m^{-3}
P	Dielectric polarization of a macroscopic sample	C m
P	Probability (discrete)	
$P(x)$	Probability density	
Pa	Pascal (SI unit of pressure: N m^{-2})	
p	Pressure (usually of a gas)	Pa or bar
p	Electric dipole moment of a molecule	C m
p_c	Critical pressure	Pa or bar
p_m	Maximum pressure	Pa or bar
Q	Energy in the form of heat	J
Q	Thermoelectric power	V/K
q	Charge (absolute) on the electron	C
R	Electrical resistance	
R_0	Gas constant (kmolar)[2] $k_B N_A$	J K^{-1} kmole^{-1}
R_c	Rate of a chemical reaction	kmole s^{-1}
R_c	Critical radius below which a liquid drop evaporates	m
R_H	Relative humidity	
r	$p/p_c - 1$	

---

[1] See Footnote 1, below the list of physical constants.
[2] See Footnote 1, below the list of physical constants.

r	Compression ratio	
S	Entropy	$J K^{-1}$
S_0	Entropy of a heat bath	$J K^{-1}$
s	Spin of a quantum state	
STP	Standard temperature and pressure (273.16 K, 1 bar)	
s	Second	
s	$n/n_c - 1$	
sech(x)	Hyperbolic secant $2/(e^x + e^{-x})$	
sinh(x)	Hyperbolic sine $(e^x - e^{-x})/2$	
T	Absolute temperature	K
T	Tesla	
T_{inv}	Inversion temperature	K
T_t	Transition temperature	K
T_c	Critical temperature	K
T_{BE}	Bose-Einstein condensation temperature	K
t	Time	s
t	$T/T_c - 1$	
tanh(x)	Hyperbolic tangent $(e^x - e^{-x})/(e^x + e^{-x})$	
U	Internal energy	J
U_i	Interaction energy	J
U_0	Internal energy of a heat bath	J
U_0	Internal energy of a Fermi gas at 0 K	J
U_0	Zero point energy of a collection of oscillators	J
U_{int}	Contribution to U of the internal degrees of freedom of molecules	J
u	Internal energy density, U/V	$J m^{-3}$
u_s	Spectral energy density	$J m^{-3} s$
V	Electrostatic potential	V
V	Volt	
V	Volume	m^3
V_q	Quantum volume	m^3
V_m	Kmolar volume[3] (volume of one kmole of gas at given T and p) For an ideal gas at STP $V_m = 22.7 \ m^3$	m^3
v	Velocity of a particle	$m \ s^{-1}$
v_d	Drift velocity	$m \ s^{-1}$
v_{th}	Mean thermal velocity $\langle v^2 \rangle^{-1/2}$	$m \ s^{-1}$
W	Work	J
W	Watt $(1 \ J \ s^{-1})$	

---

[3] See Footnote 1, below the list of physical constants.

Z	Partition function	
Z_1	Partition function of a single particle	
Z_{int}	Partition function for the internal degrees of freedom of a single molecule	
Z_N	Many-particle partition function of a macrosystem containing N particles	
α	Activity (absolute), $e^{\beta\zeta}$	
$\boldsymbol{\alpha}, \boldsymbol{\alpha'}$	Critical exponents	
α_b	Bulk thermal expansion coefficient	K^{-1}
α_r	Absorptivity of a surface	
β	$(k_B T)^{-1}$	J^{-1}
$\boldsymbol{\beta}$	Critical exponent	
γ	Specific heat ratio C_p/C_V	
γ	Coefficient of T in the electronic heat capacity per unit volume	
$\boldsymbol{\gamma}$	Critical exponent	
Δx	Change in x	
δx	Small change in x, but large enough that δx^2 cannot be neglected	
ϵ	Characteristic energy of a microsystem (*e.g.,* $h\nu$ in an oscillator)	J
ϵ	Energy measured from some standard energy, such as the Fermi energy	J
ϵ_0	Dielectric permittivity of free space	$\text{C}^2\,\text{m}^{-1}\,\text{J}^{-1}$
ϵ_e	Energy of electron measured from E_{c}	J
ϵ_h	Energy of hole measured from E_{v}	J
ζ	Chemical potential	J
ζ_i	Chemical potential of i'th species	J
η	Efficiency (in Chapters 2 and 15)	
η	Viscosity (elsewhere)	$\text{kg}\,\text{m}^{-1}\,\text{s}^{-1}$
η_c	Carnot efficiency	
θ	Angle in radians	
θ	Empirical temperature	Arbitrary
θ_D	Debye temperature	K
θ_e	Einstein temperature	K
κ	Wave vector	m^{-1}
λ	Wavelength, $2\pi/\kappa$	m
λ_w	Weiss molecular field parameter	m^{-3}
μ	Elementary magnetic moment	$\text{J}\,\text{T}^{-1}$
μ_B	Bohr magneton, $q\hbar/2m_0$	$\text{J}\,\text{T}^{-1}$
μ_e	Electrical mobility	$\text{m}^2\,\text{V}^{-1}\,\text{s}^{-1}$

μ_n	Nuclear magneton, $q\hbar/2M_H$	$J\,T^{-1}$
μ_u	"Universal" mobility	$s\,kg^{-1}$
μ_0	Magnetic permeability of free space	$kg\,m\,C^{-2}$
μ_z	Elementary magnetic moment projected on the field direction	$A\,m^2$
ν	Frequency	Hz
ν	N/N_A: the number of kmoles in a system	
ν_D	Debye frequency	Hz
ν_m	Frequency of the maximum of the Planck radiation curve	Hz
Ξ	Grand partition function	
ξ	Correlation length	m
ρ	Mass density	$kg\,m^{-3}$
$\sum$	Sum	
$\sum$	Label for a system	
σ	Surface energy	$J\,m^{-2}$
σ	Areal particle density in a two-dimensional system	m^{-2}
σ_B	Stefan's constant	$W\,m^{-2}\,K^{-4}$
σ_c	Collision cross-section	m^2
σ_e	Electrical conductivity	$W\,m^{-1}$
τ_s	Scattering time	s
ϕ	Potential energy or electrostatic potential	J or V
ϕ_w	Work function	J or eV
ϕ_b	Barrier potential	J or eV
χ	Magnetic susceptibility	
χ_e	Dielectric susceptibility	
ψ	Wave amplitude	
ψ	Order parameter	
Ω	Multiplicity (number of microstates in a macrostate)	
Ω	Solid angle	
Ω	Ohm	
ω	Angular frequency $2\pi\nu$	rad/s
$\langle x \rangle$	Mean (average) of x over a distribution	
$\langle \Delta x^2 \rangle$	Mean square deviation of x from its mean (variance)	
∇	Gradient operator $\partial/\partial x,\ \partial/\partial y,\ \partial/\partial z$	
∇^2	Laplacian operator $\partial^2/\partial x^2 + \partial^2/\partial y^2 + \partial^2/\partial z^2$	

Physical Constants

These constants are given to a precision which is sufficient for the purposes of this book. For more precise values, see the American Institute of Physics Handbook.

Acceleration due to gravity at latitude 45°, g	9.8	m/s^2
Avogadro's kmolar number,[1] N_A	6.0×10^{26}	kmole^{-1}
Bohr magneton, μ_B	9.3×10^{-24}	J T^{-1}
Boltzmann's constant, k_B	1.38×10^{-23}	J/K
Temperature at which $k_B T = 1$ eV, q/k_B	11600	K/eV
Dielectric permittivity of free space, ϵ_0	8.8×10^{-12}	C^2 m^{-1} J^{-1}
Electron volt, eV	1.60×10^{-19}	J
Electronic charge, q	1.60×10^{-19}	C
Gas constant, $R_0 \equiv k_B N_A$	8.3×10^3	J K^{-1} kmole^{-1}
Gravitational constant, G	6.67×10^{-11}	N m^2 kg^{-2}
Kilocalorie (kcal)	4.18×10^3	J
Magnetic permeability of free space, μ_0	$4\pi \times 10^{-7}$	kg m C^{-2}
Mass (rest mass) of free electron, m_0	9.1×10^{-31}	kg
Mass of hydrogen atom (1 amu), M_H	1.67×10^{-27}	kg
Kmolar volume at STP,[1] V_m	22.7	m^3
Nuclear magneton, μ_n	5.0×10^{-27}	J T^{-1}
Planck's constant, h	6.6×10^{-34}	J s
"Dirac h" $(h/2\pi)$, $\hbar$	1.05×10^{-34}	J s
Stefan's constant, σ_B	5.7×10^{-8}	W m^{-2} K^{-4}
Velocity of light, c	3.00×10^8	m/s
Velocity of sound in air at 300 K, c_s	350	m/s

---

[1] Note that for consistency with the SI system of units we use kmoles, not moles. One kmole of a substance has mass M_w kg, where M_w is the molecular weight. Hence N_A, V_m, and R_0 are a factor 10^3 greater than the conventional values.

Bibliography

Abramowitz, M., and I. A. Stegun, *Handbook of Mathematical Functions* (Dover, New York, 1965).

Bailyn, M., *A Survey of Thermodynamics* (American Institute of Physics, New York, 1994).

Berg, J. M., J. L. Tymoczko and L. Stryer, *Biochemistry*, 6th Ed. (W. H. Freeman Co., New York, 2007).

Bohren, C. F., and and B. A. Albrecht, *Atmospheric Thermodynamics* (Oxford University Press, Oxford, UK, 1998).

Born, M., *Atomic Physics*, 8th ed., trans. by J. Dougall, revised by R. J. Blin-Stoyle and J. M. Radcliffe (Dover, New York, 1989).

Cooper, N. G., and G. B. West, eds., *Particle Physics, a Los Alamos Primer* (Cambridge University Press, Cambridge, UK, 1988).

Edsall, J. T., and H. Gutfreund, *Biothermodynamics: The Study of Biochemical Processes at Equilibrium* (John Wiley & Sons, New York, 1983).

Flowers, B. H., and E. Mendoza, *Properties of Matter* (John Wiley & Sons, New York, 1970).

Goodstein, D. L., *States of Matter* (Dover, New York, 1985).

Haywood, R. W., *Analysis of Engineering Cycles*, 4th ed. (Pergamon Press, New York, 1991).

Kittel, C., and H. Kroemer, *Thermal Physics* (W. H. Freeman, San Francisco, CA, 1980).

Mandl, F., *Statistical Physics*, 2nd ed. (Wiley, New York, 1988).

McGlashan, M. L., *Chemical Thermodynamics* (Academic Press, 1979).

Park, D., *The Image of Eternity: Roots of Time in the Physical World* (University of Massachusetts Press, Amherst, MA, 1980).

Pippard, A. B., *The Elements of Classical Thermodynamics* (Cambridge University Press, Cambridge, UK, 1957).

Reif, F., *Fundamentals of Statistical and Thermal Physics* (McGraw-Hill, New York, 1965).

Index

Note: Page numbers in italic and bold refer to figures and tables, respectively. Page numbers followed by n refer to footnote.